GLENCOE TECH SERIES

ELECTRONIC DEVICES

Joseph J. Carr

With Contributions By
Abraham I. Pallas, Ph.D.
Columbia State Community College
Columbia, Tennessee

GLENCOE

Macmillan/McGraw-Hill

New York, New York Columbus, Ohio Mission Hills, California Peoria, Illinois

ELECTRONIC DEVICES
International Editions 1993

1 2 3 4 5 6 7 8 9 0 BJE UPE 9 8 7 6 5 4 3

Library of Congress Cataloging-in-Publication Data

Carr, Joseph J.
Electronic devices / Joseph J. Carr; with contributions by Abraham I. Pallas.
p. cm. - (Glencoe tech series)
Rev. ed. of: Mastering IC electronics. 1st ed. c1991.
Includes index.
ISBN 0-02-801302-6 (softcover)
1. Integrated circuits. 2. Electronics. I. Pallas, Abraham. II. Carr, Joseph J. Mastering IC electronics. III. Title. IV. Series.
TK7874.C378 1993
621.3815-dc20 92-34442
CIP

When ordering this title, use ISBN 0-07-113245-7

Printed in Singapore

CONTENTS

EDITOR'S FOREWORD

The Glencoe Tech Series is an economically packaged sequence of books tailored for use in an educational setting. The series is designed to provide the student with graduated learning experiences, ranging from dc and ac fundamentals through advanced electronics devices and integrated circuits, often with a minimum of mathematics. Each book contains chapter objectives, summaries, questions, and problems, all designed to enhance the student's learning. An instructor's manual with answers to questions and solutions to even-numbered problems is also available.

This text introduces the student to semiconductor electronics, starting from basic concepts and continuing to advanced circuits. Both discrete and integrated devices are discussed, with linear ICs being emphasized. Laboratory experiments and projects are included within the text. The material is suitable for a one- or two-semester course for nonmajors, with the only mathematics requirement being elementary algebra.

Abraham I. Pallas
Project Editor

GLENCOE TECH SERIES

Books in this series:

Basic Electricity and Electronics by Delton T. Horn
DC/AC Electricity by Victor F. Veley
Electronic Devices by Joseph J. Carr
Digital Electronics by Ray Ryan and Lisa A. Doyle
Electronic Power Control by Irving M. Gottlieb

INTRODUCTION

Electronics is an exciting field that accommodates interests from the newcomer to the seasoned veteran, from the hobbyist to the graduate engineer, from children to the elderly, and from both genders. Electronics is open to nearly everyone, and it can be practiced either with cheap or with very expensive instruments and components.

The purpose of this book is to provide you with information and training to better understand the subject matter. The approach of this book is different from some others in that it combines theoretical and practical information. Some books are overly theoretical, others are overly practical, and still others are written at such a level as to be useless (except as a picture book). In this book, however, I seek to combine the best elements of each approach. For example, you will find some arithmetic in this book, for electronics cannot be done without it. In every case, however, the arithmetic is either well within the ability of a high-school graduate, or is taken step-by-painful-step in order to show the calculation.

Experiments in this book are selected in order to illustrate the principles discussed in the accompanying text. The experiments will require some electronic instrumentation, but it is kept simple wherever possible. For example, a voltmeter is easy to obtain at low cost, so nearly all experiments will use a voltmeter. Oscilloscopes, on the other hand, are a bit more costly. Although oscilloscope experiments are used, several alternative experiments are also included.

Also, several construction projects are presented in this book. They are selected to further illustrate the principles being discussed, but at the same time provide both a working experience in practical electronics and a useful project for later on.

If you are an electronic hobbyist, or a student preparing for a career in electronics technology, you will find this book a unique resource. If you are an engineer or engineering student, then you will find this book a great augmentation to your theoretical training.

Joseph J. Carr

PROJECTS AND EXPERIMENTS

1 INTRODUCTION TO ELECTRONICS

The electronics industry is based on radio and the primitive electrical industry that existed at the turn of the century. The first radio receivers were not terribly sensitive because they used simple circuits and devices, such as Branly coherers and galena crystals to detect signals. A critical need at that time was for a better detector and, above all, an amplification device. The device that provided both of these functions was the vacuum tube, the first true electronic component.[1]

For many years, the electronics industry was wedded to radio, but that situation began to change as World War II approached. It turned out that many of the components, circuits, and techniques of radio were equally useful for other purposes as well. For example, the amplifier could be used for audio, motion pictures, electronic instruments, recording devices, and other entertainment products. The first scientific and medical instruments appeared during World War I, and by World War II, the industry was firmly established. Indeed, these devices were separate from radio.

World War II presented the electronics industry with an unprecedented opportunity for advancement. Many new military applications of electronics were invented, and many of these translated into civilian technology after the war: communications, radar, control systems, etc. During the war,

---

[1]Readers interested in old-time radios or vacuum tube theory might wish to see *Old Time Radios! Restoration and Repair* (TAB Books, No. 3342).

the first solid-state signal diodes were created for the VHF through microwave regions of the electromagnetic spectrum. These advances led directly to the invention of the transistor in 1948.

The transistor was invented by a team of physicists at Bell Laboratories, a research arm of American Telephone and Telegraph's (AT&T) Bell System. The device was actually postulated in scientific circles prior to the war, but the mighty research effort of the war years was required to produce pure semiconductor materials for these new devices.

Perhaps one of the most momentous decisions of the electronics era was the decision by AT&T to license the transistor to other companies. Although it was arguably in their own best interest, it spurred on the new technology at an unprecedented pace. This decision opened the solid-state era at a much earlier date than would have been otherwise possible. Within ten years, vacuum tubes disappeared from large areas of electronics technology.

The first transistor radios, mostly portable AM radios, appeared in 1954. The low power consumption and low voltage requirements of transistors made them ideal for portable radios. Motorola produced a universal car radio with transistor technology, but it was not until 1962 that the first transistor car radio was produced for a major automobile manufacturer. That year, the Delco Radio division of General Motors produced all transistor radios for GM cars. From 1957 to 1962 (or 1963 for other brands) car radios used low-voltage vacuum tubes for all stages except the detector and audio stages, where diodes and transistors were used. By 1963, however, all car radios were transistorized, including AM/FM models.

When I first became interested in electronics as a high school student in the late 1950s, one teacher who I fondly remember, Mr. Bruno Paras, predicted that semiconductor manufacturers were working on technology that would permit putting an entire audio amplifier chain, or even an entire radio, on a single crystal chip not larger than a thumbnail. Although that prediction seemed a little on the wild side in 1959, by 1969 it was a fact in home, portable, and car radio receivers.

The prediction that my electronics teacher mentioned was the invention of the *integrated circuit*, or "chip." These devices integrate all of the transistors, diodes, and resistors needed to make a circuit onto a single monolithic crystal of semiconductor material. Some modern circuits also include inductances and capacitances, but only at very high frequencies. These components, built onto a small crystal, are very dense, so the size of electronic devices can be reduced considerably.

Digital electronics also progressed at a tremendous pace. The modern computer would not be practical for most users without the semiconductor industry. Digital circuits obey the rules of the binary ("base-2") arithmetic system. In these devices, the numerals "0" and "1" are presented by two different voltage levels; all other voltage levels are prohibited. The first digital devices in large-scale production were the *resistor-transistor logic (RTL)* and *diode-transistor logic (DTL)*. In the 1960s, the now ubiquitous *transistor-transitor logic (TTL)* devices were produced. The TTL devices were followed soon thereafter with the *complementary metal-oxide semiconductor (CMOS)* devices. CMOS chips require such low power that many previously impossible portable devices became easy to manufacture.

One of the startling aspects of integrated electronics that has been repeated endless times is the well-known "learning-curve" phenomenon. The price of IC devices typically starts out very high when the technology is new, and then drops rapidly. For example, the μA-709 operational amplifier cost over $120 in the early 1960s (when dollars were bigger!), but now cost less than 50¢. The μA-703 RF/IF amplifier cost $15 to replace in the middle to late 1960s when I was servicing electronic equipment, but now cost two-bits when you can find them. Similarly, I paid $10 for a 741 operational amplifier in the mid-1960s, but now they cost 10 for $2 in commercial plastic packages. The low cost of these devices makes it reasonable for hobbyists, amateurs, and students to experiment with ICs very practical projects.

Today, electronics cover a bewildering variety of functions in radios, many types of communications, control systems, instruments, and thousands of other applications. Even your automobile is highly dependent on electronics. Although older cars had only a radio as an electronics device, modern cars are run on central computers and have dozens of other electronics devices on board. Even the ignition system is now an electronics circuit.

The focus of this book is on *analog electronics*, more specifically *linear analog electronics*. You will first be introduced to electronics components and electronic construction practices. You will also be introduced to the electronic instruments that are needed to work the experiments, build the projects, or enjoy electronics in any deeper sense. Indeed, the newcomer to electronics should build or buy a number of electronic instruments for his or her own use.

A large portion of this book is devoted to the *operational amplifier* because the "op amp" is very popular. It is used in many different applications, and perhaps is the easiest flexible and sufficiently well-behaved IC device for the newcomer to use. Also, the device really defines the field of linear integrated circuits. Even many supposed "non-op-amps" are actually special-purpose op amps in disguise (e.g., internal connections or components that make the device work in a special way).

However, op amps do not define the entire universe of integrated circuit devices, so we will also take a look at certain nonoperational amplifier devices, such as the *current difference amplifier* (CDA) (also called the *Norton amplifier*) and *operational transconductance amplifier* (OTA).

This exploration of integrated-circuit electronics will end with a look at IC timers, such as the 555, XR-2240, and LM-320 devices. These are not linear devices, but are so ubiquitous, and flexible, and easy to use, that any IC discussion would be incomplete without them.

In several chapters, certain applications of IC devices will be discussed theoretically. After all, electronic theory that is divorced from practical considerations is little more than a mind exercise. So, you will learn to design and build amplifiers, multivibrators, oscillators, and other practical, very useful electronic circuits.

This book is intended to be open-ended, with hopes that you will pursue electronics in ever deeper circles. This publisher has an outstanding collection of electronics textbooks—from the most-elementary levels through the intermediate and most-advanced levels. Once you master a level of difficulty on any given topic in electronics, then you might want to either go deeper into the same topic or explore another topic. Practical books will help you to gain this deeper understanding.

If I ever have the pleasure of visiting your lab or workshop, then I hope to find a copy of this book on the workbench, bent, dirtied, and well-marked as you pursue the material presented herein. It is no dishonor to a "how-to" author to find that readers have been using the book "where the rubber meets the road . . . on the bench."

Now that we've laid some foundation, let's start the journey! The first topic is electronic components. The discussion in chapter 2 is not exhaustive, but it deals with the principal types of components that are needed for IC electronics.

2 ELECTRONIC COMPONENTS

AFTER COMPLETING THIS chapter, you will be able to:

- ☐ Understand the construction and ratings of capacitors
- ☐ Describe resistors and their ratings
- ☐ Discuss integrated-circuit (IC) package styles

Capacitors

Capacitors (once upon a time called *condensers*) are used in a wide variety of electronic circuit applications including ac bypass, decoupling between circuits that share a common dc power supply, dc blocking, tuning, timing, and more. Chapter 16 includes information on the workings of resistor-capacitor networks.

The capacitor is an energy-storage device. Although inductors store electrical energy in a magnetic field, capacitors store energy in an *electrical* (or *electrostatic*) field; electrical charge (Q) is stored in the capacitor.

The basic capacitor consists of a pair of metallic plates that face each other, and are separated by an insulating material, called a *dielectric*. The dielectric can be any insulating material, including air (it can also be a vacuum). This arrangement is shown schematically in Fig. 2-1A and in a more physical sense in Fig. 2-1B.

The fixed capacitor (Fig. 2-1B) consists of a pair of square metal plates that are separated by a dielectric. Although this type of capacitor is not terribly practical, it was once frequently used in transmitters. Spark-gap radio transmitters of the 1920s often used a glass and tin-foil capacitor that was fashioned very much like Fig. 2-1B. Layers of glass and foil are sandwiched together to form a high-voltage capacitor.

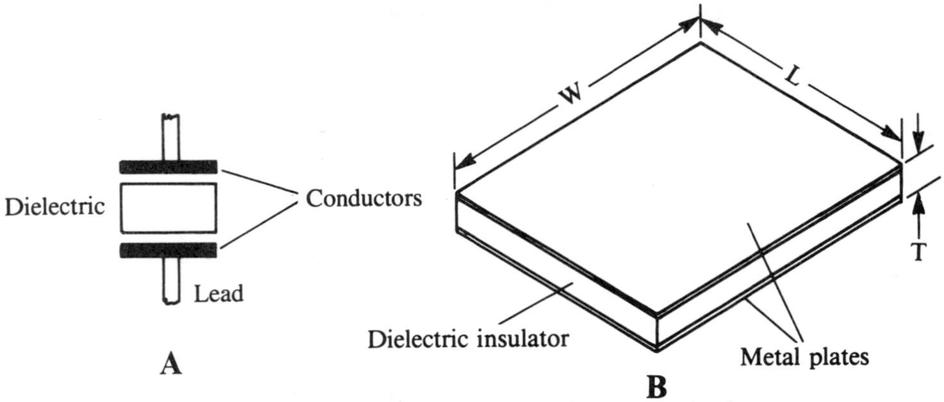

Figure 2-1. Structure of a capacitor: A) schematic form, B) physical form.

Units of capacitance

The *capacitance* (C) of the capacitor is a measure of its ability to store an electrical charge. The principal unit of capacitance is the *farad* (named after physicist Michael Faraday). One farad is the capacitance that will store one coulomb of electrical charge (6.28×10^{18} electrons) at an electrical potential of one-volt. Or, in math form:

$$C_{farads} = \frac{Q\ \text{(Coulombs)}}{V\ \text{(volts)}} \tag{2-1}$$

The farad is far too large for practical electronics work, so subunits are used. The *microfarad* (μF or mF) is 0.000001 F or 10^{-6} farads. The *picofarad* (pF) is 0.000001 μF, or 10^{-12} farads. In older radio texts and schematics, the picofarad was called the *micromicrofarad* (μμF or mmF), but never fear: 1 μμF = 1 pF.

The capacitance of the capacitor is directly proportional to the area of the plates (in terms of Fig. 2-1B, $L \times W$), inversely proportional to the *thickness* (T) of the dielectric (or the spacing between the plates, if you prefer), and directly proportional to the *dielectric constant* (K) of the dielectric.

The *dielectric constant* is a property of the insulator material used for the dielectric. The dielectric constant is a measure of the material's ability to support electric flux, and it is thus similar to the permeability of a magnetic material. The reference standard for the dielectric constant is a perfect vacuum, which has a value of 1.00000. Other materials are compared to this vacuum. The values of K for some common materials are:

Vacuum	1.0000
Dry air	1.0006
Paraffin (wax) paper	3.5
Glass	5 to 10
Mica	3 to 6
Rubber	2.5 to 35
Dry wood	2.5 to 8
Pure (distilled) water	81

The value of capacitance in any given capacitor is found from:

$$C = \frac{0.0885KA(N - 1)}{T} \tag{2-2}$$

Where:

C is the capacitance (in *picofarads*)
K is the dielectric constant
A is the area of one of the plates ($L \times W$), assuming that the two plates are identical
N is the number of identical plates
T is the thickness of the dielectric

Breakdown voltage

The capacitor works by supporting an electrical field between two metal plates. This potential, however, can become too large. When the electrical potential (i.e., the voltage) becomes too large, free electrons in the dielectric material (a few, but not many, exist in any insulator) might flow. If a stream of electronics starts, then the dielectric might break down and allow a current to pass between the plates. The capacitor is then shorted. The maximum breakdown voltage of the capacitor must not be exceeded. However for practical purposes, a smaller voltage, called the *dc working voltage* (wvdc) rating, defines the maximum safe voltage that can be applied to the capacitor. Typical values for common electronic circuits range from 8 wvdc to 1,000 wvdc.

Circuit symbols for capacitors

The circuit symbols that are used to designate fixed-value capacitors are shown in Fig. 2-2A, and the symbols for variable capacitors are shown in Fig. 2-2B. Both types of symbol are common. In certain types of capacitors, the curved plate shown on the left in Fig. 2-2A is usually the outer plate, i.e., the one closest to the outside package of the capacitor. This end of the capacitor is often indicated with a color band next to the lead that is attached to that plate.

The symbols for the variable capacitor are shown in Fig. 2-2B. This symbol is the fixed-value symbol, except with an arrow through the plates. Small trimmer and padder capacitors are often denoted by the symbol of Fig. 2-2C. The variable set of plates is designated by the arrow.

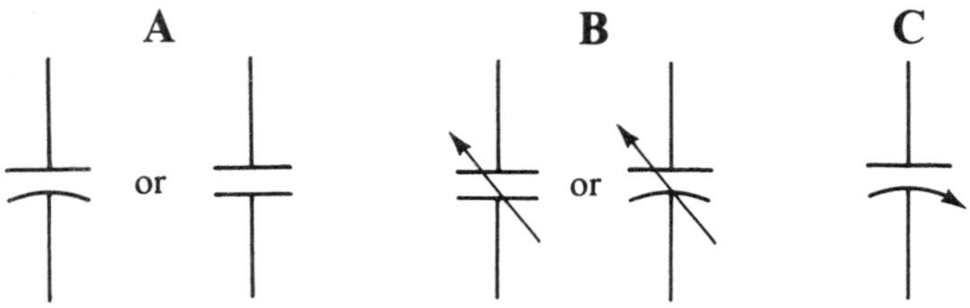

Figure 2-2. Capacitor symbols: A) fixed, B) variable, C) trimmer variable.

Fixed capacitors

Several types of fixed capacitors are found in typical electronic circuits; these are classified by the dielectric type: paper, mylar, ceramic, mica, polyester, and others.

The construction of old-fashioned *paper capacitors* is shown in Fig. 2-3. Two strips of metal foil are sandwiched on both sides of a strip of paraffin wax paper. The strip sandwich is then rolled into a tight cylinder. This cylinder is then packaged in either a hard plastic, bakelite, or paper-and-wax case. If the case is cracked or the wax end plugs are loose, replace the capacitor even if it tests good . . . it won't be for long. Paper capacitors range in value from about 300 pF to about 4 μF. The breakdown voltages are from 100 to 600 wvdc.

The paper capacitor is used for a number of different applications in older circuits, such as bypassing, coupling, and dc blocking. Unfortunately, no component is perfect. The long rolls of foil that were used in the paper capacitor exhibit a significant amount of stray inductance. As a result, the paper capacitor was not used for high frequencies. Although they are found in some shortwave receiver circuits, they were rarely or never used at VHF. Today, the paper capacitor is all but obsolete, and is rarely (perhaps never) used.

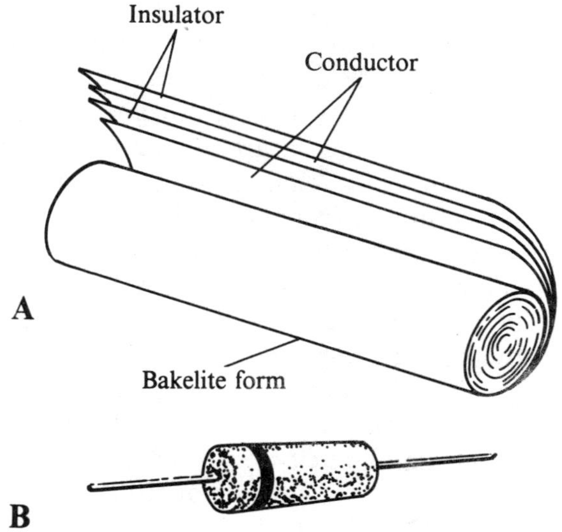

Figure 2-3. Physical structure of a paper/mylar capacitor: A) internal, B) external.

In modern applications, or when servicing older equipment that used paper capacitors, use a *mylar dielectric capacitor* in place of the paper capacitor. Select a unit with exactly the same capacitance rating and a wvdc rating that is equal to or greater than the original wvdc rating.

Several different forms of *ceramic capacitors* are shown in Fig. 2-4. These capacitors range in value from a few picofarads up to 0.5 μF. The working voltages range from 400 to more than 30,000 wvdc. The common "garden-variety" disk ceramic capacitors are usually rated at either 600 or 1,000 wvdc. Tubular ceramic capacitors are typically much smaller in value than disk or flat capacitors, and are used extensively in VHF and UHF circuits for blocking, decoupling, bypassing, coupling, and tuning.

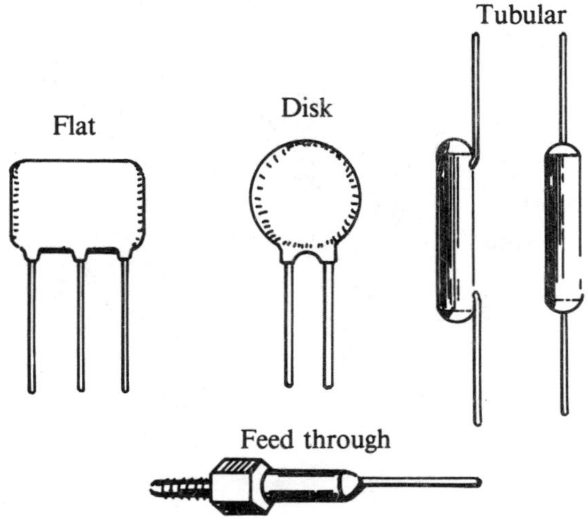

Figure 2-4. Forms of ceramic capacitors.

The feedthrough type of ceramic capacitor is used to pass dc and low-frequency ac lines through a shielded panel. These capacitors are often used to filter or decouple lines that run between circuits that are separated by the shield to reduce electromagnetic interference (EMI).

Ceramic capacitors are often rated by the *temperature coefficient*. This specification is the change of capacitance per change of temperature in degrees Celsius. A *P* prefix indicates a positive temperature coefficient, an *N* indicates a negative temperature coefficient, and the letters *NPO* indicate a zero temperature coefficient (NPO stands for *negative positive zero*. Do not ad-lib these ratings when servicing a piece of electronic equipment. Use exactly the same temperature coefficient as the original manufacturer used. Nonzero temperature coefficients are often used in oscillator circuits to temperature compensate the oscillator's frequency drift.

Several different types of mica capacitors are shown in Fig. 2-5. The fixed mica capacitor consists of either metal plates on both sides of a mica sheet or a sheet of mica that is silvered with a metal deposit on both sides. Mica capacitors range in value from 50 pF to 0.02 μF at voltages from 400 to 1,000 wvdc. The mica capacitor in Fig. 2-5C is a *silvered mica capacitor*. These capacitors have a low temperature coefficient. However, for most applications, an NPO disk ceramic will service better than all but the best silvered-mica units. Mica capacitors are typically used for tuning and other higher frequency applications.

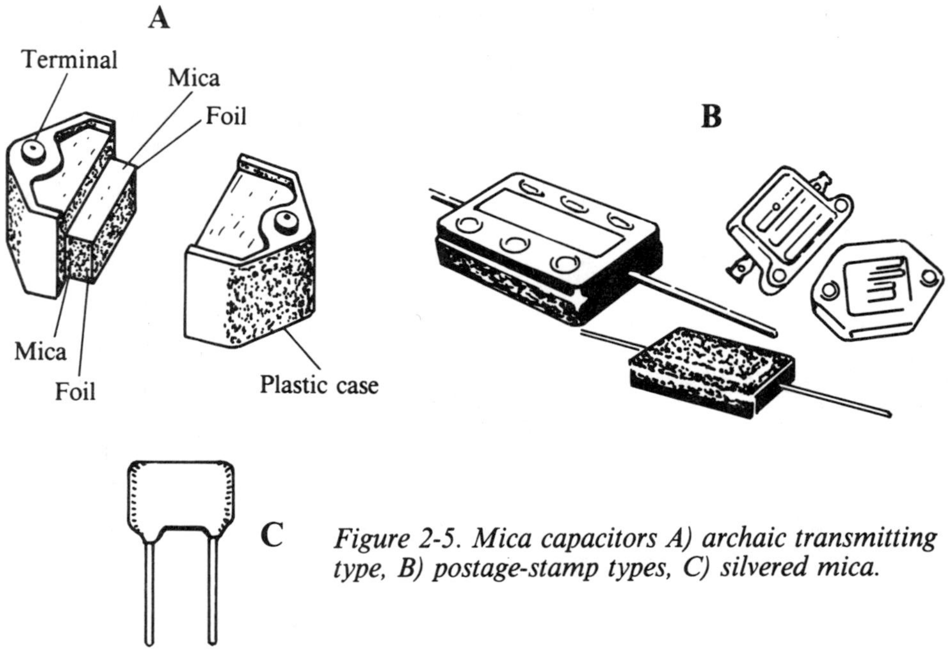

Figure 2-5. Mica capacitors A) archaic transmitting type, B) postage-stamp types, C) silvered mica.

Electrolytic capacitors

It is very difficult to achieve large values of capacitance in small packages using any of the previously discussed standard dielectrics. In the *electrolytic capacitor*, an electrolyte is used to achieve a high dielectric constant. Theoretically, either liquid (wet electrolytic capacitors) or paste (dry electrolytic capacitors) can be used, although the dry electrolyte has been used exclusively for the past several decades. In the common aluminum electrolytic, two pieces of aluminum foil are sandwiched between the aluminum-oxide film electrolyte (Fig. 2-6). The assembly is then rolled and inserted into an aluminum cylinder ("can"), which is also the negative terminal of the capacitor.

Figure 2-6. Internal structure of an electrolytic capacitor.

Electrolytic capacitors are polarity sensitive. Not only will they fail to work if connected into the circuit backwards, it is likely that the *capacitor will explode* if connected reverse polarity. This situation is dangerous, so be careful. Also, aluminum electrolytics are not very effective at frequencies above 100 kHz. It is common in radio circuits to find a 0.01 to 0.1 μF paper or mica capacitor shunted across the electrolytic to handle high frequencies.

Figure 2-7 shows several different forms of the electrolytic capacitor. The tubular electrolytic capacitor (Fig. 2-7A) has a pair of axial leads that protrude from the ends of the cylinder. The negative lead is directly attached to the metal can, but the positive lead is connected to a terminal on an insulating plug of cardboard or another material. A single-section electrolytic capacitor is shown in Fig. 2-7B. In this case, the electrodes protrude from the same end of the capacitor. Notice that one electrode is marked "+" to indicate polarity. The heavier single-section capacitor of Fig. 2-7C will have either heavy-duty terminals or screw terminals. These capacitors tend to have very high values (e.g., 2,000 μF and up), but with low wvdc ratings (10 to 100 wvdc).

Multi-section electrolytic capacitors are shown in Figs. 2-7D, 2-7E, and 2-7F. These units have two or more electrolytic capacitors in the same package, which share a common negative lead. The version shown in Fig. 2-7D is a chassis-mounted capacitor, and those in Fig. 2-7E and 2-7F are basically multi-section tubular capacitors. The version shown in Fig. 2-7E was popular in radios well into the solid-state era, but the capacitor in Fig. 2-7F was archaic by World War II.

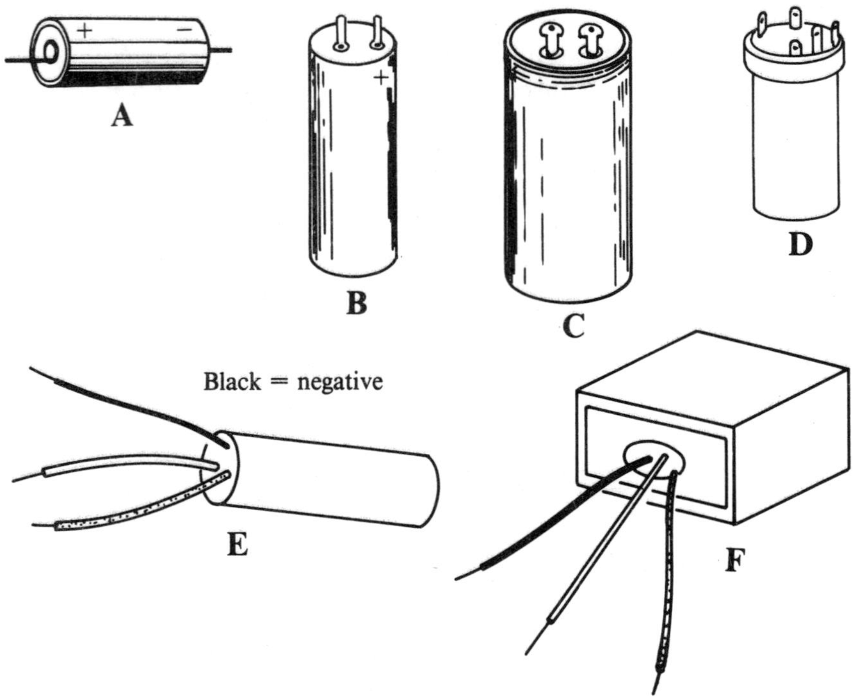

Figure 2-7. Examples of electrolytic capacitors: A) axial-lead tubular, B) radial-lead tubular, C) high-value radial-lead tubular, D) Multi-section chassis mount, E) multi-section tubular, F) obsolete multi-section.

Aluminum electrolytics are used for dc power-supply ripple reduction, bypassing, audio coupling, and stage-to-stage decoupling in audio and low-frequency circuits. The aluminum electrolytic was used almost exclusively for many years, but recently more circuits have been using *tantalum dielectric electrolytics.* These capacitors offer higher frequency operation than aluminum electrolytics and are physically much smaller. One lead is marked with a "+" sign or some other mark to indicate either the lead polarity.

Other capacitors

Today, the equipment designer has a number of different dielectric capacitors available that were not commonly available (or available at all) a few years ago. *Polycarbonate, polyester,* and *polyethelyne capacitors* are used in a wide variety of applications where the previously discussed capacitors once ruled supreme. In digital circuits, tiny 100-wvdc capacitors carry ratings of 0.01 to 0.1 μF. These capacitors are used to decouple the noise on the +5-volt dc power-supply line. In circuits such as timers and op amp Miller integrators, where the leakage resistance across the capacitor becomes terribly important, polyethylene capacitors are useful. Check current catalogues for various old-, and new-style capacitors.

Resistors

Resistance is a basic property of all conductors above the temperature of absolute zero (≈ -272.16 °C) and it refers to the tendency to oppose the flow of electrical current. The basic unit of electrical resistance is the *ohm*. One ohm is the resistance in which a potential difference ("voltage drop") of one volt occurs when exactly one ampere of dc current is flowing through the resistance (recall Ohm's law: $E = IR$). For any given conductor (2-8),

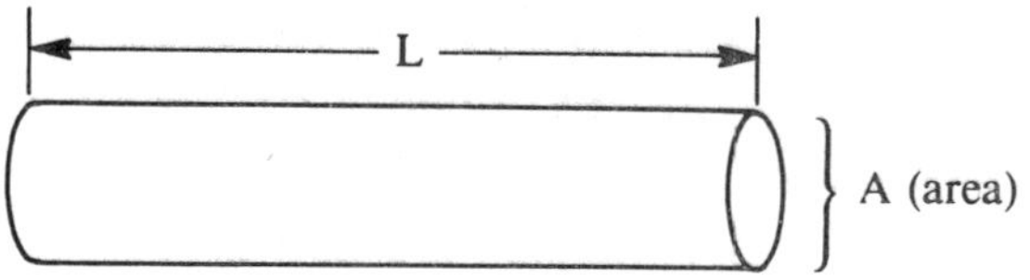

Figure 2-8. Electrical conductor with length (L) and cross sectional area (A).

the electrical resistance is directly proportional to the length of the conductor (L) and inversely proportional to the cross-sectional area (A). The resistance is also directly proportional to a property of the material used for the conductor, called *resistivity* (ρ). Resistivity is expressed in ohm-meters when the length is in meters and the area is in square meters. Put into equation form:

$$R = \rho \frac{L}{A} \tag{2-3}$$

All conductors have resistance and the function of providing resistance in an electronic circuit can (and sometimes is) provided by wires of the correct length and cross-sectional area. But practical considerations require a lumped resistance in the form of a small component. A *resistor* is an electrical component that opposes the flow of both ac and dc electrical currents. The usual circuit symbol for resistors is shown in Fig. 2-9. The normal fixed resistor is shown in Fig. 2-9A, a variable resistor is shown in Fig. 2-9B, and a potentiometer in Fig. 2-9C.

The difference between the potentiometer and the variable resistor is that the potentiometer uses three terminals and is shaft adjustable. The variable resistor might have only two terminals, and can be either shaft adjustable like the potentiometer (in which case it is called a *rheostat*), or adjustable with a slider or tapper device. A *tapped resistor* (Fig. 2-9D) is a fixed resistor on which a series of several resistance taps are available. You can solder to any or all of the taps that are on the body of the resistor.

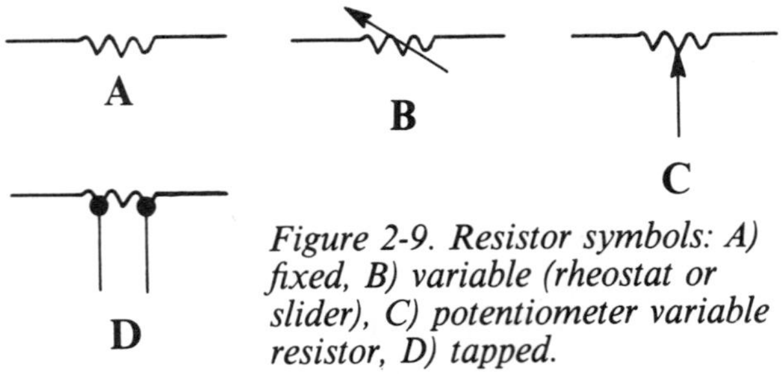

Figure 2-9. Resistor symbols: A) fixed, B) variable (rheostat or slider), C) potentiometer variable resistor, D) tapped.

Multiples of the ohm, commonly encountered in electronics, are the *kilohm* and the *megohm.* The kilohm is 1,000 ohms and the value is abbreviated with the letter *k* or the word *kohm.* In other words, 1 kohm = 1,000 ohms. For example, a 1,200-ohm resistor is a 1.2-kohm (or 1.2 k) resistor and a 470,000-ohm resistor is labelled 470 kohm (or 470 k).

"Megohm" refers to millions of ohms; that is, 1 megohm = 1,000,000 ohms. Thus, a 3.3-megohm resistor has a value of 3,300,000 ohms and a 10-megohm resistor has a value of 10,000,000 ohms. Resistors to about 18 megohms are commonly available and much higher values can be ordered for special purposes.

Resistor color code

Common carbon and metallic-film resistors are usually color coded to indicate the values (Fig. 2-10). The colors and their assigned values are shown in Table 2-1. The value is constructed by taking the first two digits and multiplying these two by the third digit. For example, suppose a resistor is found with the following colors.

First band color: red
Second band color: violet
Third band color: orange
Fourth band color: gold

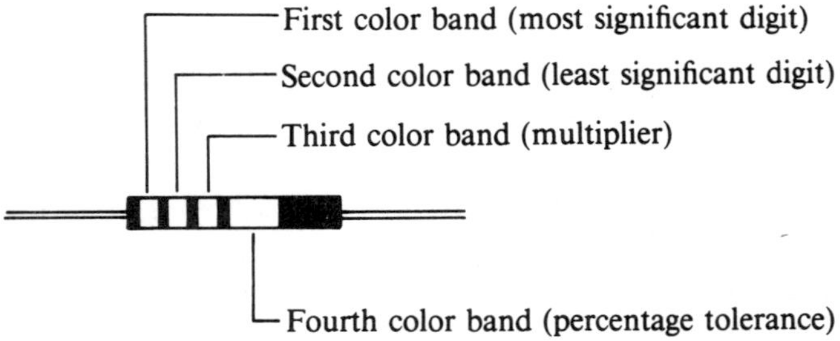

Figure 2-10. Resistor color code.

Table 2-1. Resistor color codes.

Color	1st digit	2nd digit	Multiplier	Tolerance
Black	0	0	1	–
Brown	1	1	10	–
Red	2	2	100	–
Orange	3	3	1,000	–
Yellow	4	4	10,000	–
Green	5	5	100,000	–
Blue	6	6	1,000,000	–
Violet	7	7	–	–
Gray	8	8	–	–
White	9	9	–	–
None	–	–	–	20%
Silver	–	–	–	10%
Gold	–	–	–	5%

The actual value of this resistor is: (red + violet) × orange = 27 × 1000 = 27,000 ohms. The fourth band is gold, so the tolerance is 5 percent. In other words, the resistor has a resistance of 27,000 ±5%, or an actual value between 25,650 and 28,350 ohms. With a little practice, you can recognize the various resistor values almost instinctively.

Power ratings

Resistors are available in power ratings from ⅛ watt to hundreds of watts. However, the most common ratings for electronic circuits considered in this book are ¼ watt, ½ watt, 1 watt, 2 watts, and 5 watts. Of these, all but the latter are carbon-film or metal-film resistors; resistors rated at more than 2 watts are usually wirewound, rather than carbon. For all projects and experiments in this book, the ¼-watt carbon-composition and metallic-film resistors are used, except as otherwise noted.

Common integrated-circuit package styles

The integrated circuit is formed on a tiny "chip" of silicon material by a photolithographic process. The typical chip "die" is approximately 100 mils (0.100 inch); some are larger and others are smaller. The die is typically mounted inside of a package and connected to the package pins by fine wires.

Figure 2-11A shows a die with wire attached and Fig. 2-11B shows a packaged die with a see-through window for illustration purposes. The connecting wires between the package pins and "solder" pads on the die are approximately 10 mils (0.010 inch) in diameter, and are made of either gold or aluminum in most cases. Either an electric current or a thermosonic process is used to melt the end of the wire onto (and bond it with) the connecting pad on the semiconductor die.

A

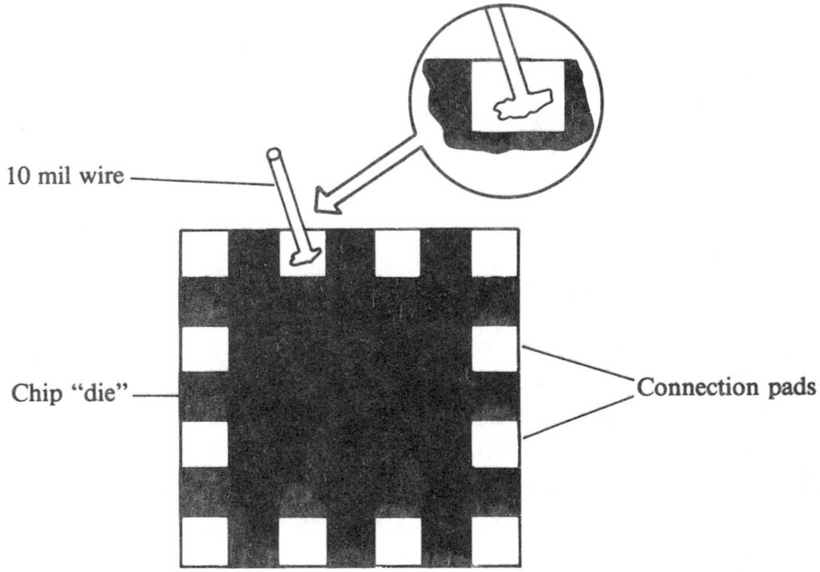

B

Figure 2-11. A) Integrated circuit die, B) integrated circuit.

The particular package style selected for any given IC device depends in part on the intended application and the number of pins required. For many IC devices several different packages are available. The earliest IC packages were the 6-, 8-, 10-, and 12-lead metal-can devices (Fig. 2-12A).

These packages were redesigns of (and similar to) the TO-5 metal-transistor package.

When viewed from the bottom of the package, the keyway marks the highest number lead or pin (e.g., pin 8 in Fig. 2-12A) and pin 1 is the next pin clockwise from the keyway. Be careful when looking at IC base diagrams to know whether a top or bottom view is depicted.

Perhaps the largest number of IC devices on the market today are sold in *dual in-line packages (DIP)*, examples of which are shown in Fig. 2-12B. DIP packs are available in a wide variety of sizes from 4 to more than 48 pins. Although many devices are available in other sizes of DIP packs, most linear devices are contained in 8-, 14-, or 16-pin DIP packs.

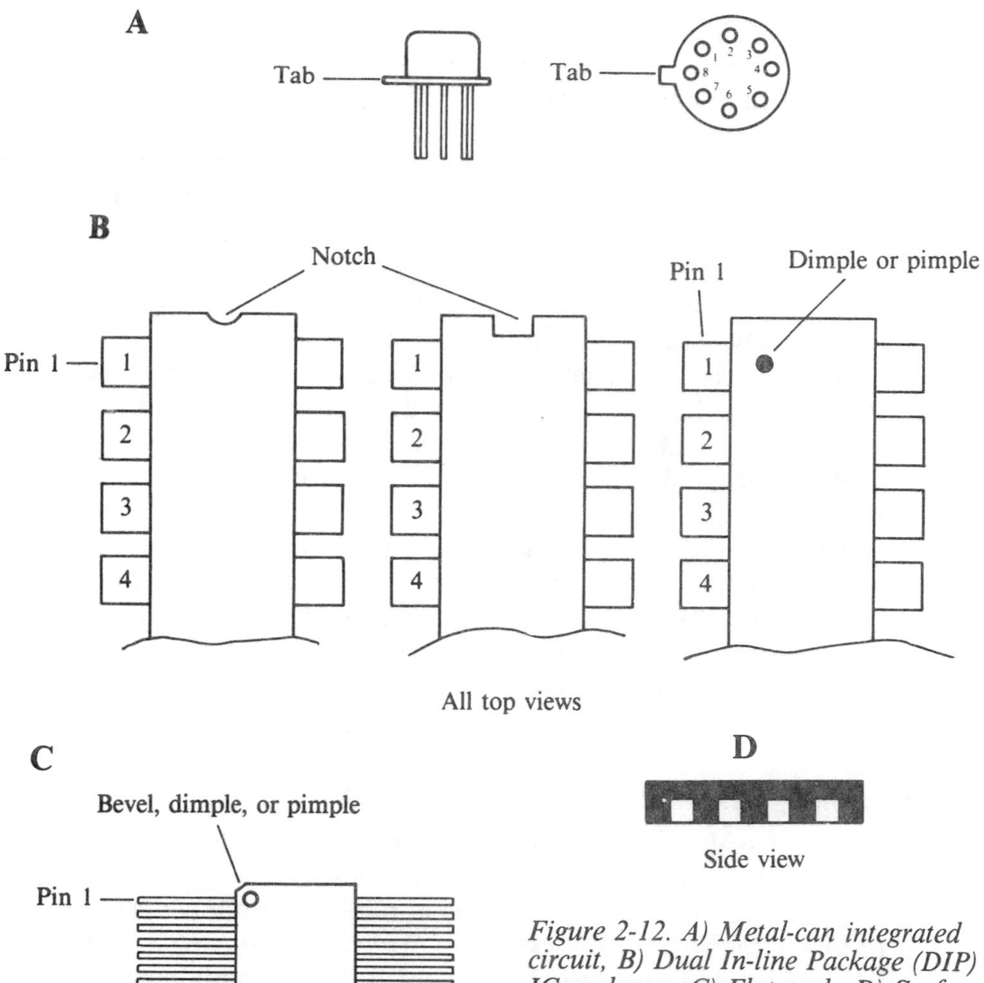

Figure 2-12. A) Metal-can integrated circuit, B) Dual In-line Package (DIP) IC packages, C) Flat pack, D) Surface mount.

The DIP pack is symmetrical with respect to pin count regardless of the package size, so some other means is needed to designate pin 1; Figure 1-12B shows several common methods. In all cases, the IC DIP pack is viewed from the top. In some devices, a paint dot, "pimple," or "dimple" will mark pin 1. In other cases, a semi-circular or square notch marks the end where pin 1 is located. When viewed from the top, with the notch pointed away from you, pin 1 is to the left of the notch and the highest-numbered pin is to the right of the notch.

Both plastic and ceramic materials are used in DIP-pack construction. In general, the plastic packages are used in consumer and noncritical commercial (or industrial) equipment and the ceramic packs are used in military and critical commercial equipment. The principal difference between the plastic and ceramic packages is the intended temperature range. Although exceptions exist, typical temperature-range specifications are 0 to +70 degrees Celsius for commercial plastic devices, and −10 to +80 degrees Celsius for ceramic. Military and some critical commercial ceramic devices are rated from −55 to +125 degrees Celsius. The principal difference between military chips and the highest-grade commercial or industrial chips is the amount of testing, "burn-in," and documentation that accompanies each device.

An example of an IC flat-pack is shown in Fig. 2-12C. This type of package is typically used where very high component density is required. Most flat-pack devices are digital ICs, although a few linear devices are also offered. DIP-package IC devices are mounted either in sockets or by inserting the pins through holes drilled in a printed circuit board (PCB). Flat-packs, on the other hand, are mounted on the surface by directly soldering them to the conductive track of the PCB.

A relatively new style of package is the *surface mounted device (SMD)*, shown in Fig. 2-12D. SMD technology represents a significant improvement in packaging density. SMD components can be mounted closer together than other types of packages and they are amenable to automatic PCB production methods. It is expected that the SMD package will eventually overtake and replace other forms, especially in VLSI applications.

Scales of integration

Several different scales of integration are found amongst IC devices. The ordinary *Small-Scale Integration (SSI)* device consists of single gates, small amplifiers, and other smaller circuits. The number of components on each chip is on the order of 20 or less. *Medium-Scale Integration (MSI)* devices are slightly more complex and can have about 100 or so components on the chip. Devices such as operational amplifiers, shift registers, counters, and so forth are usually classified as MSI devices. *Large-Scale Integration (LSI)* devices are mostly digital ICs and include calculators, microprocessors, and so forth. Typical LSI devices contain from about 100 to 1000 components. Some newer devices, called *Very Large-Scale Integration (VLSI)*, include some of the latest computer chips. The numbers and descriptions that are listed for SSI, MSI, and LSI devices are only approximations, but serve to provide guidelines. Most linear IC devices are either SSI or MSI, with the latter predominating.

Summary

Circuits involving ICs often contain capacitors and resistors as well. Capacitors are devices capable of storing energy in the form of an electric field. They are typically made up of two or more parallel plates, separated by an insulator, or dielectric. The important ratings of a capacitor are its capacitance, measured in farads, and the breakdown voltage. Capacitors may be either fixed or variable. Resistors oppose current in either a dc or an ac circuit. The important ratings for a resistor are its resistance, measured in ohms, and the power rating, measured in watts. ICs contain many individual components on a single chip. The number of components on a chip is known as the *level of integration*, which may range from small scale to very large scale. The most common package for ICs is the dual in-line package (DIP).

Questions

1. What is capacitance? In what units is it measured?
2. What factors determine the value of capacitance for a capacitor?
3. What is a dielectric?
4. What is the difference between breakdown voltage and dc working voltage?
5. Discuss the differences between an electrolytic capacitor and a nonelectrolytic capacitor.
6. What is resistance? In what units is it measured?
7. What factors determine the value of resistance for a resistor?
8. What is an integrated circuit?
9. What is meant by level of integration?
10. What is the most common package for ICs?

Problems

1. A 1.1-μF capacitor has 10 volts across it. What is the charge on each plate?
2. A 2200-pF capacitor has 25 volts across it. What is the charge on each plate?
3. A capacitor consists of two plates with area of 50 cm^2 separated by 4 cm of mica ($K = 6$). What is the value of the capacitance?
4. A capacitor consists of two plates with area of 75 cm^2 separated by 2 mm of paraffin. What is the value of the capacitance?
5. A resistor has colored bands of red, yellow, orange, and gold. Calculate **a.** nominal resistance value **b.** acceptable range of resistance
6. Repeat Problem 5 for a resistor with bands of green, brown, black, and silver.

3 ELECTRONIC CONSTRUCTION

AFTER COMPLETING THIS chapter, you will be able to:

- □ Describe the types of power supplies used for laboratory experiments
- □ List the types of media used for designing circuits
- □ Outline the steps involved in designing a circuit

Power supplies for laboratory experiments

Many readers will perform laboratory experiments—both independently and as part of a formal class. Guidelines for experiments are given at various points in this book. A power supply must be selected (or built) to work these experiments.

Unless otherwise specified, the experiments in this book are designed to use either ±12-volt dc or +5-volts dc regulated power supplies. The selected power supply should offer either a single bipolar power supply or two independent 12-volt dc supplies that are not ground referenced. The "non-grounded" feature allows you to create a bipolar supply by connecting the positive output terminal of one supply to the negative output terminal of the other (see Fig. 3-1).

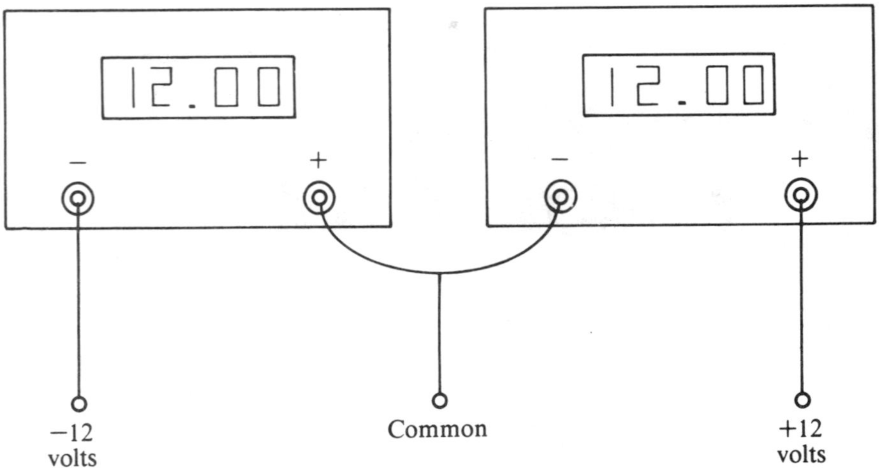

Figure 3-1. Power supplies needed for bipolar linear ICs, such as operational amplifiers.

Desirable features to have on a bench power supply include the following:

- Output voltages fixed at ±12 volts dc (or ±15 volts) or
- Output voltages adjustable from 0 to greater than 12 volts dc
- Current available from each polarity not less than 100 milliamperes
- Metered outputs (*I* and *V*)
- Voltage regulation
- Current limiting for output short-circuit protection and
- Overvoltage protection

Although not all good dc power supplies will have all of these features, those that do are clearly superior for most laboratory applications. All of the experiments in this book can be run on a power supply that substantially meets these requirements.

Electronic construction media

Electronic projects must be constructed on some sort of base or media. Although novices tend to build their first few projects rat's-nest style (in sort of a three-dimensional wire sculpture form), more experienced people use a chassis, printed-circuit board, or other media.

Successfully performing experiments and designing or debugging circuits on a professional level requires certain skills and knowledge of construction practices. This section looks at some of these practical matters.

Four major categories of construction practices exist: partial breadboard, complete breadboard, brassboard, and full-scale development model. The two *breadboard* categories require special construction methods to check circuit validity and make certain preliminary measurements. The breadboard is not usable in actual equipment, but rather is a bench or laboratory device.

The difference between partial and complete breadboards is merely a matter of scale. The partial breadboard is used to check a partial circuit or a small group of circuits. The complete breadboard contains the entire circuit; it is much more extensive than the partial breadboard. Fundamental differences exist in the type of breadboarding hardware that is used for both types, especially in large circuits.

The two commonly used types of hardware are shown in Figs. 3-2 and 3-3. The device shown in Fig. 3-2 is an IC pin-socket breadboard; this model is the Heath/Zenith Educational Systems ET-3300. It, like others of its type, offers three built-in dc power supplies: +12 volts at 100 mA, −12 volts at 100 mA, and +5 volts at 1.5-amperes (for TTL digital-logic IC devices). Located between the large multipin IC sockets are bus sockets in which all pins in a given row are strapped together. These sockets are generally used for power distribution and common ("ground") busses. Interconnections between sockets, power sources, and components are made with #22 insulated hook-up wire. The bared ends of the wires are pressed into the socket holes.

Figure 3-2. Typical IC breadboard with internal power supplies.

The other method of breadboarding is wire wrapping (Fig. 3-3A). Sockets with special square or rectangular wirewrap pins are mounted on a piece of "universal" printed-circuit board. These boards are perforated to accept wirewrap IC sockets, and usually have "printed" power distribution and ground busses. With a special tool (Fig. 3-3B), the wire interconnects can be strapped from point-to-point, as needed by the circuit. The wire is wrapped around each contact tightly enough to break through the insulation and make the electrical connection.

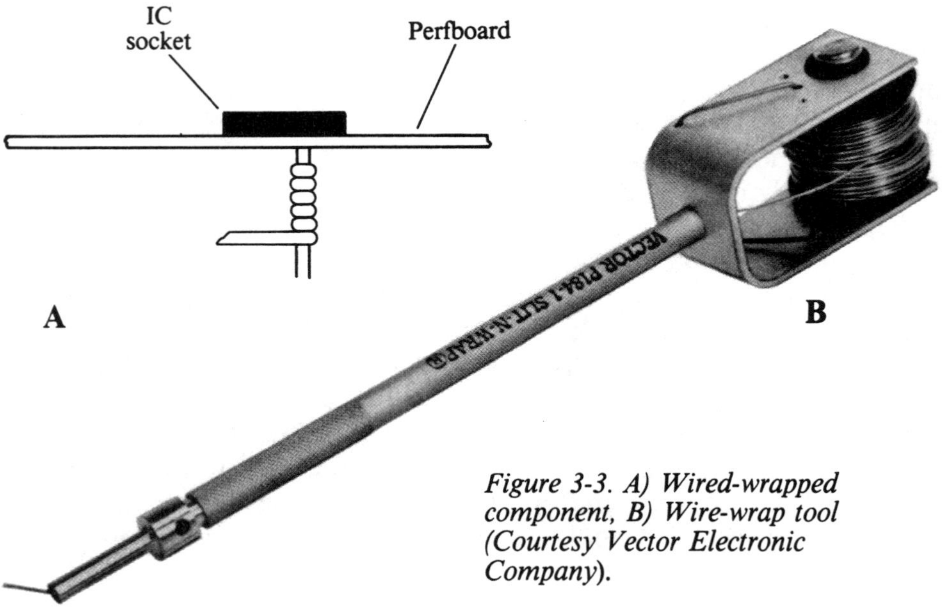

Figure 3-3. A) Wired-wrapped component, B) Wire-wrap tool (Courtesy Vector Electronic Company).

In general, socketed breadboards are used for student applications, for partial breadboarding in professional laboratories, and for small design projects that don't justify the cost of the wirewrap board. Wirewrapping is used for larger, more extensive, projects. Although the distinctions between types of breadboard and the respective supporting hardware are quite flexible, they hold true for a wide range of situations.

Brassboards are full-model shop versions of the circuit that will plug into the actual equipment cabinet where the final circuit will reside. Some brassboards are wirewrapped, but most are printed circuits. The brassboard should be near its final configuration where it can be tested in the actual equipment. In contrast, the breadboard is purely a laboratory model. The brassboard is usually made with model shop handwork methods, not routine automated production methods. The brassboard differs from the final version in that it might contain "dead-bug" and "kluge-card" modifications, in which components are informally mounted on the board to circuit changes.

The full-scale development (FSD) model is the highest preproduction model. It might look very much like the first-article production model that is eventually made, and is intended for field tests of the final product. For example, an FSD model of a two-way landmobile radio transceiver can be mounted in an actual vehicle and used in tests or field trials. For aircraft electronics, the FSD model must be built and tested according to flight worthiness criteria. The FSD model is built as near to regular production methods as possible.

Remember several principles when breadboarding electronic circuits. Although it is possible to ignore these rules, they constitute good practice and ignoring them is risky. The rules are:

- Insert and remove components only with the power turned off
- Wire and/or change the wiring only with the power turned off
- Check all wiring prior to applying power for the first time

First connect the power-distribution and ground wires, then check these wires before you connect any other wires (note: in many schematics, the V− and V+ wiring is omitted, but that doesn't mean that these connections are not to be made).

Use *single-point* (a.k.a. *star*) *grounding* wherever possible. A ground plane or ground bus should not be used unless signal frequencies are low, signal voltage levels are relatively large, and current drain from the dc power supply is low. Otherwise, ground noise and ground-loop voltage drop problems will exist.

If a signal is applied to IC input pins when no dc is applied, the substrate "diode" can become forward biased and possibly damage the device. Therefore: a) do not apply signal until the dc power is turned on, b) turn off the signal source before turning off the dc power supply, and c) never apply a signal with a positive peak that exceeds V+ or a negative peak that exceeds V−.

Make all measurements with respect to common or ground, unless special test equipment is provided. For example, in Fig. 3-4 differential voltage V_d is composed of two ground-referenced voltages, V_1 and V_2. Measure these voltages separately and then take the difference between them (i.e., $V_2 - V_1$).

Although it is relatively easy today to make a small printed-circuit board using "homebrew" techniques, most small projects are constructed (at least initially) on perforated wiring board (usually just called *perf board*).

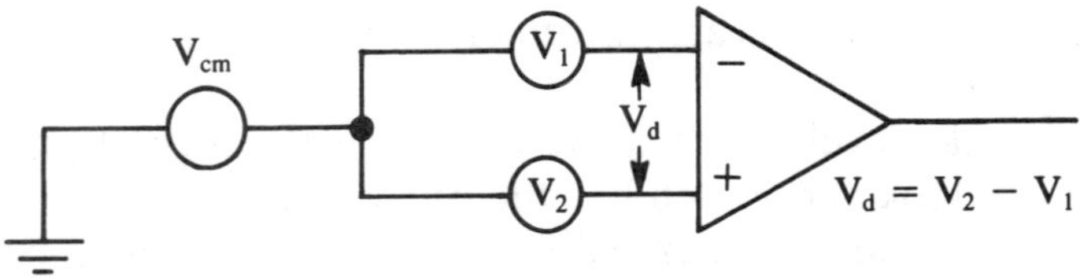

Figure 3-4. Differential voltage V_d is actually the differential voltage $V_2 - V_1$.

Figure 3-5 shows both RF (Fig. 3-5A) and subaudio (Fig. 3-5B) projects that are built on perf board. RF projects should be housed in metal-shielded chassis, such as those shown in Fig. 3-6. The project in Fig. 3-6A is a small HF receiver tuned to 10.7 MHz, and was used as the IF amplifier in a VHF/UHF spectrum analyzer and receiver project. The circuit in Fig. 3-6B was designed as a small RF amplifier circuit.

Designing electronic projects

The process of designing electronic circuits or projects is not an arcane art that is open only to a few initiates. Rather, it is a logical step-by-step process that can be learned. Like any skill, design skill is improved with practice, so don't have excessive expectations for the "first time at bat" and don't be discouraged if the process did not exactly work out as planned the first time.

A

Figure 3-5. Examples of perfboard construction: A) RF projects. B) analog project.

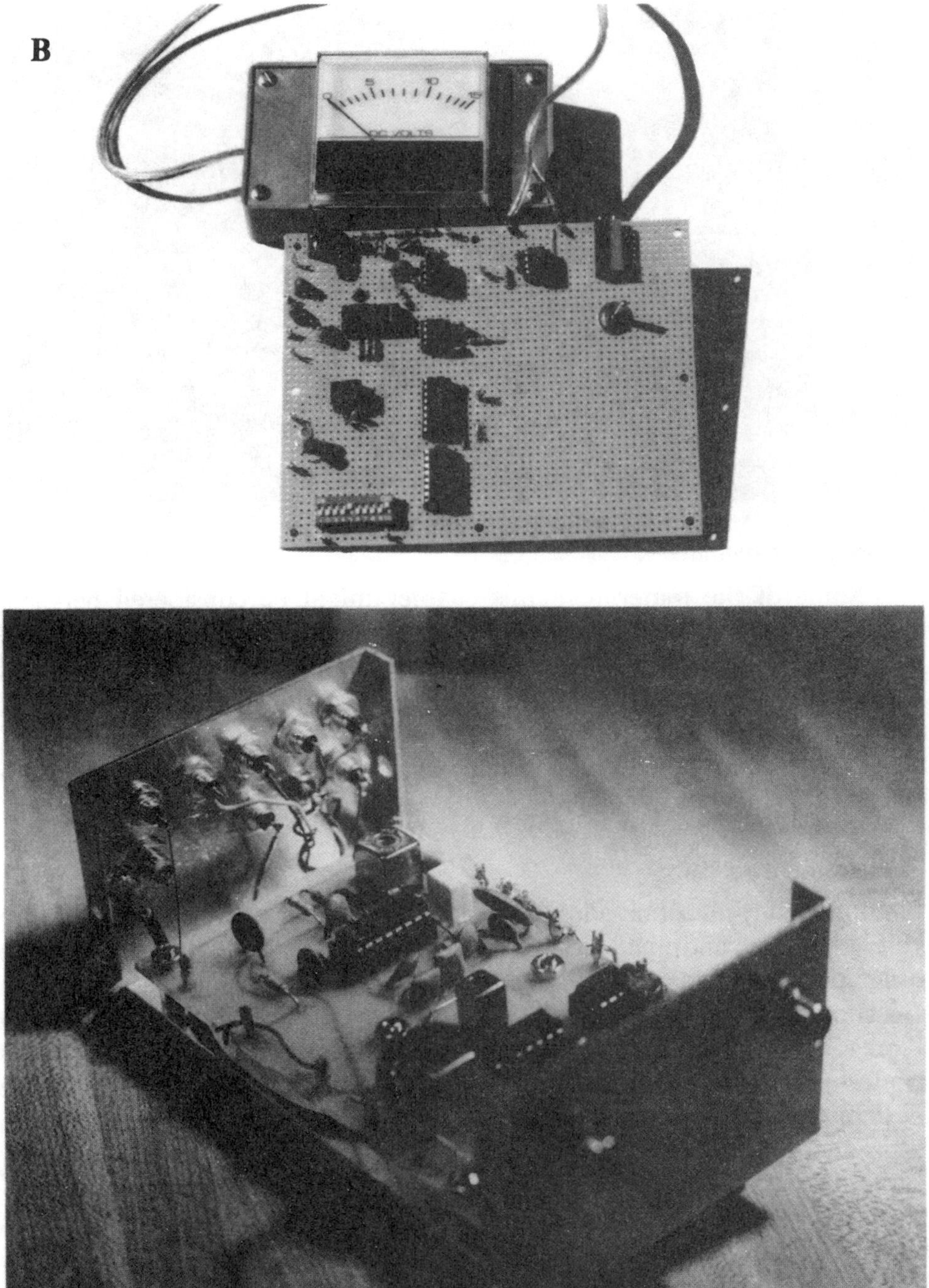

B

A

Figure 3-6. Examples of projects built inside a small aluminum box.

B

Figure 3-6. Continued.

Some of the material in this chapter might be considered personal philosophy, which might be a fair label to attach. Although many technical people claim to disdain "philosophy," we all have it and use it. It's just that some people think about it a lot, others think about it either a little or at only a few times. Still others don't think about it at all—they make decisions that are not based on a logically considered viewpoint, but rather, on dumb luck and pure default.

A design procedure

The procedure that one adopts for designing might well be different from following steps and that's alright. The purpose of offering a procedure is to systematize the process. Although it is conceivable that one can design an instrument by a process similar to Brownian motion (as, theoretically, a Tom Clancy novel can be written by an infinite collection of monkeys pounding randomly on an infinite number of PC keyboards), it is the systematic approach that most often yields success. This procedure assumes a product that is a one-of-a-kind instrument, as in a scientific laboratory or plant. Designing a product for production and sale follows a similar procedure, but it involves marketing and production problems as well. The steps in the procedure, some of which are iterative with respect to each other, follow:

1. Define and tentatively solve the problem.
2. Determine the critical attributes that are required of the final

product; incorporate these into a specification and a test plan, which determines objective criteria for acceptance or rejection.
3. Determine the critical parameters and requirements.
4. Attempt a block-diagram solution.
5. Apportion requirements (e.g., gain, frequency response, etc.) to the various blocks.
6. Perform analysis and simulate the block diagram to test validity of the approach.
7. Design specific circuits to fill in the blocks.
8. Build and test the circuits.
9. Combine the circuits on a breadboard.
10. Test the breadboarded circuit according to a fixed test plan.
11. Build a brassboard that incorporates all changes made in the previous steps.
12. Test the brassboard and correct the problems.
13. Design and construct the final configuration.
14. Test the final configuration.
15. Ship the product.

Solving the (right) problem The purpose of the designer is to solve problems with analog circuits, digital circuits, a computer or whatever else is available in the armamentarium.

Two related problems are often seen in the efforts of novices. First, the designer will often have a tentative favorite approach in mind before the problem is properly understood. The decisions are based on what the designer is most comfortable doing. For example, many younger designers are likely to select the digital solution in a knee-jerk manner, which excludes any consideration of the analog solution. All options should be evaluated and the one that best fits the need should be selected.

Second, be sure you understand the problem to be solved. Although this advice seems trivial, it is also true that failure to understand the problem at hand often sinks designs before they have a chance to manifest themselves. There are several facets to this problem. For example, it is natural for engineers to think that an elegant solution is complex and large-scale. If this mistake is made, then the product will probably be overdesigned and have too many "whistles and bells." It was, after all, designed to solve a much harder problem than was actually presented.

Another aspect is to understand the final customer's *use* for the product. It is all too easy to get caught up in the specifications, or in our own ideas about the job, and to altogether overlook what the user needs to accomplish with the product. An example is derived from biomedical instrumentation. A physiologist requested a pressure amplifier that would measure blood pressures over a range of 0 to 300 mmHg (Torr). What the salesman never told the plant was that it would be used on humans (safety and regulatory issues), that blood would come in contact with the diaphragm (cleaning and/or liquid isolation issues), and that it would occasionally be used for measuring 1 to 5 mmHg central venous pressures (which implies low-end linearity issues).

Part of the problem in determining the level of complexity or the specific design function is miscommunication between the user and the designer. Although miscommunications occur frequently between in-house designers and their clients, it is probably most common between distant customers and engineers in the plant. Of course, marketing people might never let the engineer and customer meet (either from ignorance or a fear that the salesman's little lies will surface: "The reason I hate engineers is that, under duress, they tend to blurt out the truth").

The proper role of the designer is to scope out the problem, understand what the circuit or instrument is supposed to do, how and where the user is going to use it, and exactly what the user wants and expects from the product.

Determine critical attributes This step is basically the fruit of understanding and solving the correct problem. From the solution of the problem one can determine a set of attributes, characteristics, parameters, and other indices of the product's final nature.

At this point, one must write a specification that documents what the final product is supposed to do. The specification must be clearly written so that others can understand it. A concept or idea does not really exist, except in the mind of the originator. One must, according to W. Edwards Deming[1], create *operational definitions* for the attributes of the product. One cannot simply say "it must measure pressure to a linearity of one percent over a range of 0 to 100 psi." Rather, it might be necessary to specify a test method under which this requirement can be met. After all, more than one standard for pressure and measurement might exist, and there is certainly more than one definition of linearity. The operational definition serves the powerful function of providing everyone with the same set of rules; basically, it levels the playing field.

---

[1]*Out of the Crisis*, W. Edwards Deming, MIT Press.

Part of this step, and of making an operational definition, is to write a test plan for the final product. Here, one determines (and often contractually agrees) exactly what the final product will do and defines the objective criteria of goodness or badness that will be used to judge the product.

Determine critical parameters and requirements Once the product is properly scoped out, it is time to determine the critical technical parameters that are needed to meet the test requirements (and hopefully the user's needs—if the test requirements are properly written). Parameters such as frequency response, gain, and so forth tend to vary in multimode instruments, so one must determine the worst case for each specification item and design for it.

Attempt a block-diagram solution The block diagram is a signals-flow or function-diagram that represents stages, or collections of stages, in the final instrument. In large instruments, several indexed levels of block diagram might exist, each one going finer in detail.

Apportion requirements to the blocks Once the block diagram solution is on paper, tentatively apportion system requirements to each block. Distribute gain, frequency response, and other attributes to each block. Remember that factors such as gain distribution can have a profound effect on the dynamic range. Also, the noise factor and drift of any one amplifier can have a tremendous effect on the final performance; added cost and complexity often arises in these types of parameters.

Analyze and simulate Once the block diagram is set and the requirements are apportioned to the various stages, it is time to analyze the circuit and run simulations to see if it will actually work. A little desk checking goes a long way toward eliminating problems later, when the thing is first prototyped. Plug in typical input values and see what happens on a stage-by-stage basis. Check for the reasonable outputs at each stage. For example, if the input signal should drive an output signal to 17 volts, and the operational amplifiers are only operated from 5-volt power supplies, then something is wrong and will have to be corrected.

Design specific circuits for each block At this point, the remainder of this book is of most useful to you, because it is on the specific circuits that we will dwell later. In this step, fill in those blocks with real circuit diagrams.

Build and test the circuits At this point, one must actually construct the individual circuits, and test them to make sure they work as advertised (unless, of course, the circuit is so familiar that no testing is needed). Keep in mind that some of the best ideas for simplified circuits might not actually work . . . and this is the place to find out. Use a benchtop breadboard that allows circuit construction with plug-in stripped-end wires.

Combine the circuits in a formal breadboard Once the validity of the individual circuits is determined, combine them together in a formal breadboard. Whether built on a benchtop breadboard or on a prototyping board, make sure that the layout is similar to what is expected in the final form.

Test the breadboard Test the overall circuit according to a formally established objective criteria. This test plan should be developed earlier in step two. As problems arise and are solved, make changes and/or corrections and *document the results.* Perhaps the main failing of inexperienced designers is that they do not properly document their work, not even in an engineer's or scientist's notebook.

Build and test a brassboard version A brassboard version is made as close as possible to the final configuration. Although breadboarding techniques can be a little sloppy, the brassboard should be a properly printed circuit board. The test criteria should be the same as before, updated only for changes. If problems turn up, they should be corrected before proceeding further. Remember that the most common problems that occur when leaping from breadboard to brassboard are in the layout (e.g., coupling between stages), power distribution, and ground noise (these are the areas of difference between the two configurations).

Design, build and test final version Once all of the problems are known, solved, and the changes have been incorporated, it is time to build the final product as it will be given to the user. At this point the reputation of the designer is made or broken; now the product will finally be evaluated by the client.

Summary

When circuits are designed and tested, it is useful to have a power supply available that is capable of providing all the necessary voltages and signals needed by the circuit. This normally includes a dual dc supply capable of supplying 100 milliamps from each polarity, as well as overload and short-circuit protection. The circuits are usually first constructed on a perf board or breadboard. After the circuit is tested, it is more permanently constructed. Electronic project design should begin with a clear understanding of the problem, followed by a block diagram solution. The blocks are then converted to actual circuits.

Questions

1. What are the necessary characteristics of a bench power supply for IC circuit construction?
2. Name the four levels of test circuits prior to actual production models.
3. List the main steps in circuit design procedure.

4 USING ELECTRONIC INSTRUMENTS

AFTER COMPLETING THIS chapter, you will be able to:

- ☐ Understand what a multimeter does
- ☐ Describe signal generators
- ☐ Explain the oscilloscope

Some particularly skilled electronics enthusiasts seem to be able to divine circuit problems, almost if by magic. Although it seems like magic to the noninitiate, it is really only a reflection of the fact that a large amount of experience is at work. But even the most skilled, seemingly magical, electronics technologist will agree that a good selection of test equipment is the key to wringing out circuit problems and repairing faults.

Figure 4-1 shows my own workbench (although since the photo was taken, I've added a couple additional signal generators, a spectrum analyzer, and a few small devices as well). Notice that the test equipment is a mix of old and new. The large black signal generator in the center is a Measurements Model 80 that was made in 1949 and was refurbished by me. It produces up to 100,000 μV of RF, modulated or unmodulated, at frequencies from 2 to 400 MHz.

I traded a "parts only" sideband rig to a local ham for the signal generator. He rebuilt the SSB rig and I rebuilt the Model 80 (we both knew the condition of the other's offering, lest you wonder). At hamfests, I've seen these same generators (in both military and civilian versions) for as little as $50.

The nature of the test equipment that you select depends on the job that you want to do. It is assumed that the readers of this book are interested in solid-state circuits, integrated circuits, and dc and relatively low-frequency (< 1 MHz) ac circuits. The test equipment discussed in this chapter is aimed at those readers.

Figure 4-1. Workshop layout with bench, power supplies, signal generator, and oscilloscope.

For IC work, you will need several different types of instruments: a multimeter, signal generators, an oscilloscope, power supplies, and certain accessories. Let's look at these classes and some practical examples, each in their turn.

Kit-built instruments are a good idea, especially for newcomers. Besides saving some money, the kit-built instrument provides the owner with project-construction experience under controlled circumstances. I have built literally scores of Heathkits (The Heath Co., Benton Harbor, MI, 49022).

Multimeters

A multimeter is an instrument that combines a multirange voltmeter, a milliammeter (or ammeter), an ohmmeter, and possibly some other functions. This single instrument can measure all of the basic electrical parameters that the technician needs to understand when troubleshooting a receiver.

The earliest multimeters were called *volt-ohm-milliammeters (VOM)*. These instruments were passive devices and the only power that they used was a battery for the ohmmeter function. Because they are passive instruments, they are still preferred for troubleshooting medium- to high-power radio transmitters. They are included here, even though not recommended, because they are inexpensive and many readers will buy them at places like Radio Shack.

The *sensitivity* of the VOM is a function of the full-scale deflection of the meter movement that is used in the instrument. Sensitivity was rated in terms of *ohms-per-volt full scale.* Three common standard sensitivities were found: 1000 ohms/volt, 20,000 ohms/volt, and 100,000 ohms/volt. These measurements represented full-scale meter sensitivities of 1 mA, 100 μA, and 10 μA, respectively.

Voltage was read on the VOM by virtue of a multiplier resistor in series with the current meter, which forms the basic instrument movement. The front panel switch selected the multiplier resistor, which determined the voltage scale that was used.

The earliest form of active multimeter was the *vacuum-tube voltmeter (VTVM).* These instruments used a differential balanced-amplifier circuit that was based on dual triode tubes (e.g., 12AX7) to provide amplification for the instrument. As a result, the VTVM was much more sensitive than the VOM. It typically had an input impedance of 1 megohms, with an additional 10 megohms in the probe, for a total input impedance of 11 megohms. Compare this feature with the VOM, which had a different impedance for each voltage scale.

The next improvement was the solid-state meter. These instruments used a field-effect transistor (FET) in the front end and other transistors in the rest of the circuit. As a result, these instruments were called *field-effect transistor voltmeters* (FETVM). The FETVM has a very high input impedance, typically 10 megohms or more.

Finally, the most modern instruments are various digital multimeters (DMM), in both handheld and bench styles (Fig. 4-2). These instruments are the meter of choice today if you are going to purchase a new model. In fact, it might be a little difficult to find a nondigital model any place except Radio Shack. The DMM is preferred today because it provides the same high input impedance as the FETVM. But, because the display is digital instead of analog, the readout is less ambiguous.

One bit of ambiguity exists, however, and that is *last digit bobble.* The digital meter only allows certain discrete states, so if a minimum value is between two of the allowed states, then the reading might switch back and forth between the two (especially if noise is present). For example, a voltage might be 1.56456 volts, but the instrument is only capable of reading to three decimal places. Thus, the actual reading might bobble between 1.564 and 1.565 volts.

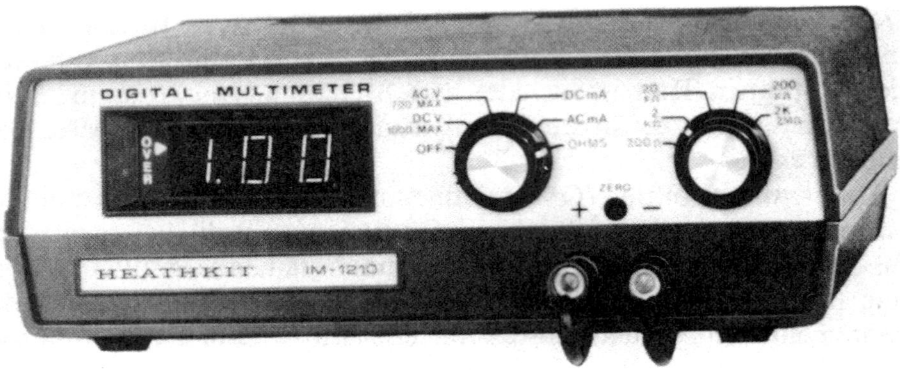

Figure 4-2. Digital multimeter (courtesy The Heath Company).

Last digit bobble is not really a problem, however, unless you bought an instrument that is not good enough. The number of digits on the meter should be one more than is required for the actual service work. Digits are specified in DMMs in terms of the numeric places that the meter can read. For example, the instrument in Fig. 4-2 is considered a 2½ digit model. The "½ digit" refers to the fact that the most significant digit (on the left side) can only be 1 or 0 (in which case it is blanked off). Thus, the full-scale of this meter is always 199. The decimal point location is a function of the range setting.

For many users, the selected DMM can be a 2½ digit model, but if you anticipate a lot of detailed solid-state work, then opt instead for a 3½ digit model. These instruments will read to 1999 on each scale. Oddly, the cost difference between the two is small.

Figure 4-3 shows a handheld digital multimeter that was manufactured by Hewlett-Packard for a professional market. It is extremely useful for portable servicing and other applications where a bench model instrument is not appropriate or convenient.

An interesting aspect of DMMs is that they use low-voltage for the ohmmeter, so the instrument can be used in the circuit on solid-state equipment. The ohmmeter on other forms of instruments will forward-bias the pn junctions of solid-state devices, so it cannot be used for accurate measurements. However, this same feature of the DMM also means that the normal ohmmeter cannot be used to make quick tests of pn junctions on devices such as transistors and diodes. However, some DMMs have a special setting that will allow this operation. In some, it is a "high-power" setting, and in others it is designated by the normal arrow diode symbol.

Another feature of the modern DMM is an aural continuity tester. It will beep anytime the resistance between the probes is low (i.e., when a short circuit exists). This feature is especially useful for testing multiconductor cables for continuity when the actual resistance is not a factor.

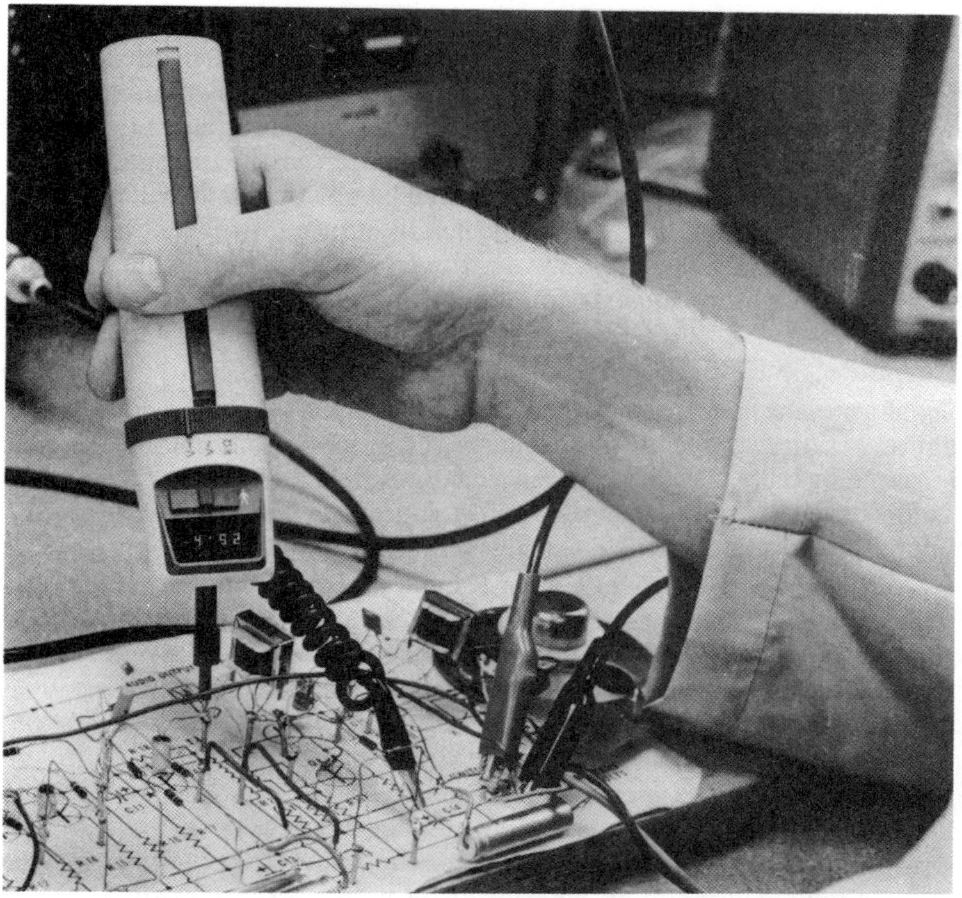

Figure 4-3. Handheld digital multimeter for bench and portable testing.

Older instruments

As is also true with other equipment, the multimeter shows up on the used market (especially hamfests) quite often. It is easy to obtain older workable instruments at real cheap prices. This situation is especially true now that everyone seems to dump old analog instruments in favor of the new digital varieties. However, some pitfalls exist along the pathway.

First, make sure that the instrument isn't so old that it uses a 22.5 volt (or higher) battery in the ohmmeter section. Those instruments were made before about 1956, and will blow most of the transistors or ICs that you measure. Look out especially for the oversized RCA and Hickoc instruments, which were once the mainstay of radio service shops.

Second, make sure that nothing is wrong with the instrument that cannot easily be repaired. VTVM's tend to be in good shape, or at least repairable condition. Unfortunately, many VOMs are in terrible shape. Part of the problem is that the meter will measure current easily enough, but if the operator left the meter on the current setting and then measured a voltage . . . pooff! The instrument burned up. When examining an instrument, try to make a measurement on each scale. Alternatively, remove the case and look for charred resistors and burnt switch contacts.

In any event, when you obtain a used meter, be sure to take it out of the case, clean the switch contacts, and generally spruce it up. The usual forms of contact and tuner cleaner will work wonders on used multimeters as well. Simply cleaning the switch(s) can remove intermittent or erratic operation.

Signal generators

Signal generators produce a signal that can be used to troubleshoot, align, adjust, or simply prove the performance of a piece of electronic equipment. Fortunately, only a few basic signal generators are needed. Signal generators for this market are available in several varieties, including: audio generators, function generators, and RF signal generators.

Audio generators produce at least sine waves on frequencies within the range of human hearing (20 Hz to 20 kHz). Some models produce frequencies over a greater range and others produce square waves as well as sine waves. The standard audio signal generator has a variable *amplitude* or *level* control, and a fixed output impedance of 600 ohms. Audio generators might or might not have an output level meter.

An audio signal generator is the Heath IG-18 (Fig. 4-4). This instrument produces sine- and square-wave outputs which have individually calibratible amplitudes. The meter measures the output level of the sine-wave signal. Frequency is set on this instrument with a series of switches. The

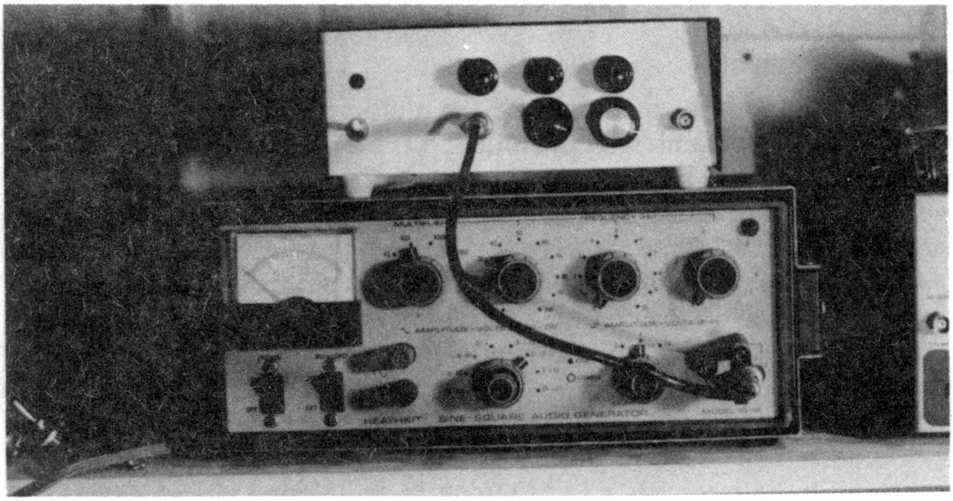

Figure 4-4. Audio signal general generator with add-on adapter.

first switch is a range multiplier and each successive switch is an order of magnitude less (but arranged in decade groupings). The settings shown are X10, 30, 0, 1, so the frequency set is 10 × (30 + 0.1) or 301 Hz. The box on top of the IG-18 signal generator is a homebrew project that will produce sawtooth and triangle waves from the IB-18 outputs. Since this photo was taken, the circuitry in the small box was relocated inside the IG-18 case.

A *function generator* is much like an audio generator but it also outputs a triangle waveform, in addition to the sine- and square-wave signals. Some function generators also produce pulses, sawtooths, and other waveforms as well. Typical function generators operate from less than 1 Hz to more than 100 kHz. Many models exist that have a maximum output frequency of 500 kHz, 1 MHz, 2 MHz, 5 MHz, and in at least one case, 11 MHz. Like the audio generator, the standard output impedance is 600 ohms. However, some instruments also offer 50-ohm outputs (standard for RF circuits), and a TTL-(digital) compatible output. The latter is a digital output that is compatible with the ubiquitous *transistor transistor logic (TTL* or *T^2L)* family of digital logic devices (these have either 74xx- or 54xx-type numbers).

The *sweep function generator* allows the output frequency to sweep back and forth across a range around the set center frequency. As a result, frequency-response evaluations can be made with the sweep function generator and an oscilloscope.

RF signal generators output signals that are generally in the range above 20 kHz and typically have an output impedance of 50 ohms. This value of impedance is common for RF circuits, except in the TV industry (where 75 ohms is the standard). RF signal generators are available in various types, but can easily be grouped into two general categories: *service grade* and *laboratory grade.* For service work, either type is usable.

My own bench has several different signal generator instruments: the Heath IG-18 sweep function generator, a Model 80 (2-400 MHz) RF signal generator, and an elderly Precision Model E-200C, which I refurbished after buying it at a hamfest in nonworking condition.

Oscilloscopes

Probably no other instrument is as useful as the oscilloscope in working on IC circuits. You can look at signals and waveforms instead of just the averaged dc voltages (as are measured on the DMM). Figure 4-5 shows a simple service-grade instrument that can be used for most service jobs on AM, FM, and shortwave receivers, as well as most CB and amateur radio transmitters. The oscilloscope is formally called the *cathode-ray oscillo-*

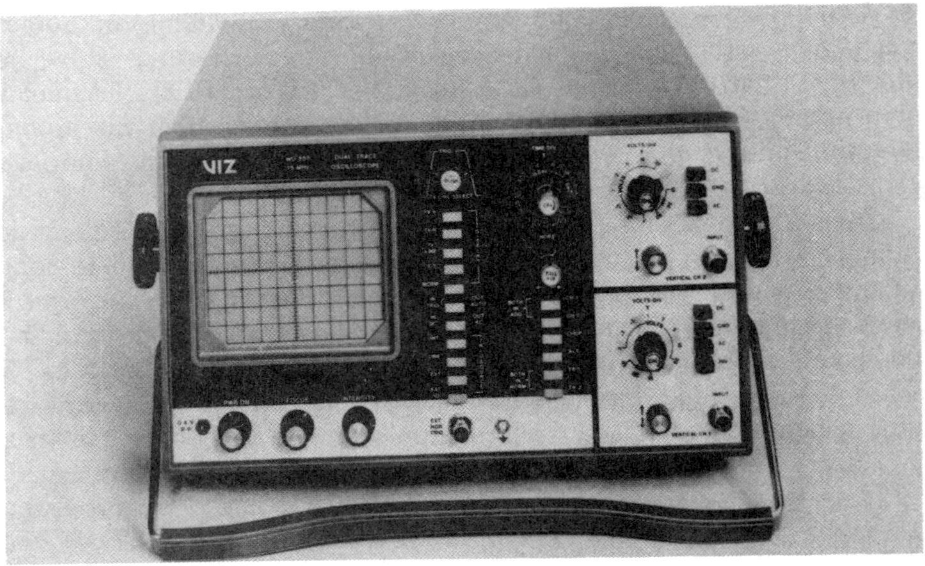

Figure 4-5. Oscilloscope.

graph *(CRO)*. It displays the input signal on a viewing screen that is provided by a *cathode-ray tube (CRT)*.

The light is produced on the CRT viewing screen by deflecting an electron beam vertically (Y-direction) and horizontally (X-direction). When two signals are viewed with respect to each other on an X-Y oscilloscope, a *Lissajous figure* is formed.

The quality of oscilloscopes is often measured in terms of the *vertical bandwidth* of the instrument. This specification refers to the −3 dB frequency of the vertical amplifier (or amplifiers, if it is a dual-beam 'scope). The higher the bandwidth, the higher the frequency that can be displayed and the sharper the rise-time pulse that will be faithfully reproduced. For ordinary service work, a 5-MHz 'scope will suffice. However, there will be times when you will need a high-frequency 'scope, so buy as much bandwidth as you can afford. Although once in the price stratosphere, even new 50-MHz models can be bought relatively cheaply today.

Dc power supplies

Although often overlooked as "test equipment," the simple dc power supply is definitely part of the test bench. In addition to powering units that either lack a dc supply or have a defective dc supply, the bench power supply can be used for a variety of biasing and other troubleshooting functions. Chapter 5 features the basic theory of dc power supplies and chapter 6 features construction projects that you can build.

Other instruments

The *signal tracer* is a high-gain audio amplifier that can be used to examine the signal at various points along the chain of stages in a radio or audio amplifier. A *demodulator probe* will permit the signal tracer to "hear" RF and IF signals. The signal tracer is a back-up replacement for the oscilloscope (if it is absolutely impossible to obtain). However, the signal tracer is an insufficient back-up to the 'scope. But this instrument does have advantages and its low cost makes the signal tracer a worthwhile addition to the bench.

Summary

The basic piece of test equipment is a multimeter, capable of measuring voltage, current, and resistance. The multimeter may be either analog, using a meter movement, or digital, using a numeric display. The digital multimeter generally has the advantage of being easier to read, and it often contains additional test features. Signal generators provide ac signals at different frequencies, and function generators provide different waveforms, such as square waves and triangular waves, as well. These generators have different frequency and amplitude capabilities. Oscilloscopes provide a visual display of voltage, which is especially useful when the voltage varies with time. Oscilloscopes are rated by their maximum amplitude and maximum frequency capabilities.

Questions

1. What three functions do multimeters have?
2. What is the importance of ohms-per-volt rating in a multimeter?
3. How do VOMs, VTVMs, and DMMs differ?
4. What is a signal generator? a function generator? What are the two primary ratings for each of them?
5. What is an oscilloscope? What does it measure? What is the most important rating for the oscilloscope?

5 DC POWER SUPPLY THEORY

AFTER COMPLETING THIS chapter, you will be able to:

- ☐ Explain the different parts of the dc power supply and the function of each
- ☐ Understand the considerations when power supply components are selected

Ac power supplies

The power supply in ac-operated circuits converts the alternating current that is supplied to the residential power system to the various dc voltages that are needed to operate the circuit.

One complete ac sine wave cycle consists of a complete positive excursion and a complete negative excursion. If time t is required to complete one cycle, then the *frequency* is $1/t$. In the United States, the standard ac frequency is 60 cycles per second (60 Hz). Overseas, it is common to find 50 Hz ac.

The nominal ac voltage supplied to residences is 115 volts ac, although the "nominal" voltage actually resides within a normal range of 105 to 125 volts. Also, the term *ac volts* must be defined. The standard ac waveform that is supplied by the power company is the sine wave (Fig. 5-2). Alternating current is bidirectional. By convention, the voltage or current in one direction is plotted above the baseline and is labelled positive. The voltage or current in the opposite direction is plotted below the baseline and is labelled negative. The ac voltage rises in a positive direction to a peak value ($+V_p$), and then falls back to zero. It then reverses direction, and climbs to a negative peak, $-V_p$. Following the negative peak, the ac sine waveform drops back to zero and begins to become positive again on the next cycle.

At least three voltage measurements are typically used to describe the standard sinusoidal waveform: peak, rms, and peak-to-peak. The *peak voltage* (V_p) is measured from the zero-volt baseline to either the positive or negative peak and the *peak-to-peak voltage* ($V_{p\text{-}p}$) is measured from the positive peak to the negative peak, $V_{p\text{-}p} = [(+V_p) - (-V_p)]$. For the standard baseline symmetrical sine wave (Fig. 5-1), $V_{p\text{-}p} = 2V_p$.

The *rms (root mean square) voltage* is the voltage that is normally cited when talking about ac. For example, "115-volts ac" really means "115-volts ac rms." Although the most rigorous definition of rms voltage uses calculus mathematics, for sine waves

$$V_{\mathrm{rms}} = \frac{V_p}{[2]^{1/2}}$$

$$= \frac{V_p}{1.414}$$

$$= 0.707V_p$$

For our purposes, the rms ac voltage is 0.707 times the peak voltage. The rms voltage definition is derived from the amount of equivalent work that the sine-wave power can do. The rms voltage is equal to the dc voltage, which will perform the same amount of heating in a resistance load.

Although the term *voltage* is used here for illustration purposes, the previous discussion also applies to the current in resistive circuits.

Components of a dc power supply

The following sections will look individually at the various components of the electronic dc power supply. Remember that its purpose is to convert ac to nearly pure dc.

Transformers

A *transformer* (Fig. 5-3A) consists of two intimately close wire coils that are wound around a common magnetic core. One coil is called the *primary winding* and the other is the *secondary winding*. The coils are arranged to maximize the magnetic coupling (the extent that the magnetic flux, which surrounds the primary, cuts across the coils of the secondary winding).

Magnetic coupling is enhanced by winding the coils on a suitable magnetic transformer core. The magnetic core of the transformer is made of laminated iron sheets that are clamped together (Fig. 5-3B). The laminated construction prevents eddy-current losses from decreasing the efficiency of the transformer.

Only two windings exist on the simplest transformer: primary and secondary. The primary winding is connected to the ac source and the secondary winding is connected to the load. In most transformers, the primary winding is wound closest to the core.

The primary and secondary voltages of the transformer are related by the ratio of primary to secondary turns:

$$\frac{V_{pri}}{V_{sec}} = \frac{N_{pri}}{N_{sec}} \tag{5.1}$$

Where:

V_{pri} is the voltage applied to the primary winding
V_{sec} is the voltage appearing across the secondary winding
N_{pri} is the number of turns of wire in the primary winding
N_{sec} is the number of turns of wire in the secondary winding

Thus, a 6.3-volt ac transformer will have a turns ratio of 115/6.3, or about 18 : 1. So, the primary will have 18 turns for each turn in the secondary. If the secondary voltage is higher than the primary voltage, then the transformer is called a *step-up transformer*, but if the secondary voltage is lower than the primary voltage, it is a *step-down transformer*. Many transformers are actually multiple secondary transformers, which have both step-up and step-down windings.

The windings of a transformer are color coded for easy identification. In general, for U.S.-made transformers:

Primary winding: black
High-voltage secondary: red
First low-voltage secondary: green
Second low-voltage secondary: brown
5-volt ac (or third low-voltage) secondary: yellow

Some foreign-made transformers do not follow this coding scheme, so be cautious. For example, a particular transformer from Japan used the normal black for the primary windings, but used red for the low-voltage winding.

If the winding is *center-tapped*, then the center-tap wire will be the same color as the main wires for that winding, but it will also have a yellow tracer (except on the yellow 5-volt ac winding, where the tracer is some other color, usually black). The standard for writing the colors on the center-tap wire is to list the main body color first, and then the tracer. Therefore, a wire listed as "green/yellow" has a green body with a yellow tracer; this particular example is a 6.3-voltac (CT) winding.

The voltage rating of the secondary winding is the rms value of the end-to-end voltage. If the transformer secondary is center-tapped, then the secondary rating might be listed in the form "12.6-VCT" (volts center-tapped), or "6.3-0-6.3." Both of these ratings denote exactly the same transformer; however, the voltages are not measured end-to-end in the latter case, but rather from the center tap to each end.

Rectifiers

The purpose of a rectifier in a dc power-supply circuit is to remove the impurities of the ac line current and make it suitable for dc-craving electronic circuits, which require pure, or nearly pure, direct current.

Before discussing the details of solid-state rectifiers, let's review the two basic forms of electrical current in the context of rectification: dc and ac. The key feature of dc is that it is *unidirectional* (i.e., current flows through the circuit in only one direction). Direct current is zero until it is turned on, and it will then rise to a certain level and remain there. Electrons flow from the negative terminal to the positive terminal of the power supply, and that polarity never reverses direction.

Alternating current, on the other hand, is *bidirectional.* On one half-cycle, the current flows in one direction. Then the power-supply polarity reverses, so the current flows in the opposite direction. The electrons still flow from negative to positive, but since the positive and negative poles have switched places, the physical direction of current flow has reversed. In the normal ac-power mains, the voltage and current waveforms vary as a sine wave. By convention, flow in the "positive" direction is graphed above the zero volts (or zero amperes) line and flow in the "negative" direction is graphed below the zero line (Fig. 5-1).

Raw ac is incompatible with nearly all electronic circuits, so it must be changed to dc by a rectifier and a *ripple filter.* The main requirement for a rectifier is that it must convert bidirectional ac into a undirectional form of

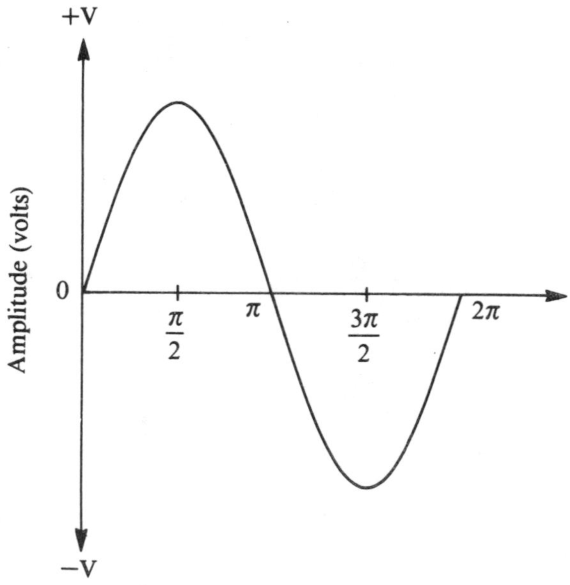

Figure 5-1. Sine wave.

current. Although industry once used rotary mechanical switches, synchronous vibrators, and vacuum tubes to accomplish the rectification job, all modern circuits rely on solid-state pn-junction rectifiers.

Pn-junction diode rectifiers

Although various types of rectifiers were used in the past, all common rectifier diodes in use today are *silicon pn-junction diodes* (shown schematically in Fig. 5-2). The pn-junction diode rectifier consists of a silicon semiconductor material that is doped with impurities to form an n-type material at one end and a p-type material at the other end. The charge carriers (which form the electrical current) in the n-type material are negatively charged electrons and the charge carriers in the p-type material are positively charged "holes."

In the *reverse bias* case (Fig. 5-2A), the negative terminal of the voltage source (V) is connected to the p-type material and the positive terminal is connected to the n-type material. Positive charge carriers are thus attracted away from the pn junction toward the negative voltage terminal, while negative charge carriers are drawn toward the positive terminal. That leaves a charge-free *depletion zone* in the region of the junction that contains no carriers. Under this condition, little or no current is flowing across the junction. Theoretically, the junction current is zero, although in real diodes, a tiny *leakage current* always runs across the junction.

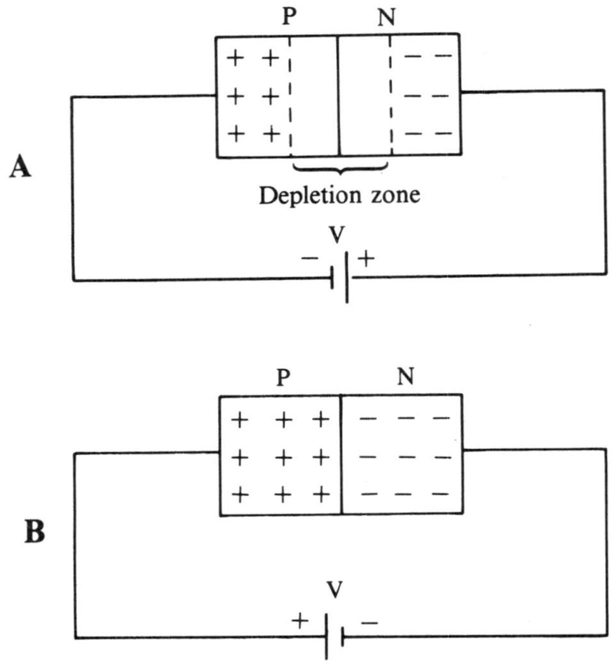

Figure 5-2. Pn-junction diode: A) reverse-biased, B) forward-biased.

In the *forward bias* (Fig. 5-2B), the polarity of voltage source (V) is reversed from Fig. 5-2A. The positive terminal is applied to the p-type material, and the negative terminal is applied to the n-type material. Because similar charges repel, the charge carriers in both p- and n-type materials are driven away from the power-supply terminals, toward the junction. The depletion zone disappears and allows positive and negative charges to get close to the boundary between regions. As these opposite charges attract each other across the junction, a current flows in the circuit.

From this description, it is apparent that a pn-junction diode is able to convert bidirectional ac into unidirectional current because it allows current to flow in only one direction. Thus, it is a rectifier. However, the rectifier output current is not pure dc (as from batteries), but rather it is a form of current, called *pulsating dc.*

Figure 5-3 shows the standard circuit symbol for the solid-state rectifier diode, along with some common shapes of actual diodes. The input side (where ac is applied) is the anode and the "dc" output side is the cathode. The diodes (Figs. 5-3B through 5-3G) are positioned so that the respective anodes and cathodes are aligned with those of the circuit symbol in Fig. 5-3A. Rectifers 5-3B through 5-3E are epoxy-package devices, the type seen most often in small electronics devices. The cathode will be marked either with a rounded end (5-3B), a line (5-3C) a diode arrow (5-3D), or a plus sign (5-3E).

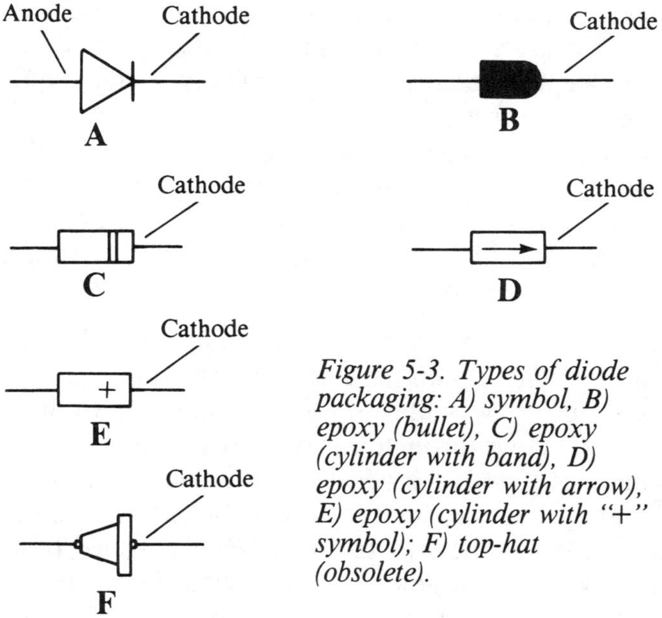

Figure 5-3. Types of diode packaging: A) symbol, B) epoxy (bullet), C) epoxy (cylinder with band), D) epoxy (cylinder with arrow), E) epoxy (cylinder with "+" symbol); F) top-hat (obsolete).

The diode shown in Fig. 5-3F is the old-fashioned (now obsolete) "top-hat" type. Unless otherwise specified, the top-hat type can safely pass a current of 500-milliamperes; those in Figs. 5-3B through 5-3E generally pass 1-ampere (or more, for larger-sized, but similar packages).

Rectifier specifications Several key rectifier specifications exist: forward current, leakage current, surge current, junction temperature, and peak inverse voltage (PIV)—also called peak reverse voltage (PRV).

The *forward current* is the maximum constant current that the diode can pass without damage. For the popular 1N400x series of rectifiers, the forward specified current is one ampere (A). The *leakage current* is the maximum current that will flow through an intact reverse-biased pn junction. In an ideal pn junction diode, the leakage current is zero; in high-quality practical diodes, it is typically very low compared to the forward current. The *surge current* is typically very much larger than the forward current, and inexperienced people sometimes mistake it for the operating current of the diode. Surge current is the maximum short-duration current that will not damage the diode. "Short duration" typically means one ac cycle (1/60 second, or 16.67 mS, in a 60-Hz system). *Do not use the surge current rating as if it is the forward-current rating!*

The specified *junction temperature* is the maximum allowable operating temperature of the pn junction. The actual junction temperature depends on the forward current and how well the package (and environment) dissipates internal heat. Although typical maximum junction temperatures range up to +125 degrees Celsius for silicon devices, a good design requires the lowest possible temperature. One reliability guide requires that the junction temperature must be held to a maximum of +110°C.

The *peak inverse voltage* (*PIV*) is the maximum allowable reverse bias voltage that will not damage the diode. If PIV is exceeded then the reverse current suddenly increases from the leakage current value to a very high, destructive, value. The PIV rating of the diode is the limiting rating in certain power supply designs.

Rectifier circuits Figure 5-4 shows a solid-state rectifier diode (D1) in a simple *halfwave-rectifier circuit.* In Fig. 5-4A the diode is forward-biased: the positive terminal of the voltage source is connected to the anode of the rectifier. Current (*I*) flows through the load resistance (*R*). In Fig. 5-4B, the opposite situation is found: the negative terminal of the voltage source is applied to the anode, so the diode is reverse-biased and no current flows.

The output current through the load (*R*) is graphed (Fig. 5-4C) as a function of time when an ac sine wave is applied. From time T_1 to T_2, the diode is forward-biased, so current flows in the load (also from T_3 to T_4. However, during the period T_2 to T_3, the diode is reverse-biased, so no current flows. Because the entire sine wave takes up the period T_1 to T_3, only half of the input sine wave is used. The output waveform, shown in Fig. 5-4C, is not yet pure dc, and is called *halfwave-rectified pulsating dc.*

The halfwave rectifier is low cost, but wastes energy because it only uses one-half of the input ac waveform. Efficiency is increased by using the entire waveform in a *fullwave-rectifier circuit.*

Figure 5-5A shows the classic two-diode fullwave rectifier. This circuit uses a transformer that has a center-tapped secondary winding. Because the center tap (CT) is used as the zero-volts reference point (and is grounded in most circuits), the polarities at the ends of the secondary are always opposite each other (i.e., 180 degrees out of phase). On one half cycle, point A is positive with respect to the CT, while point B is negative. On the next half cycle, point A is negative and point B is positive, with respect to the CT. This situation makes D1 forward-biased on one half cycle, while D2 is reverse-biased. Alternatively, on the next half cycle, D1 is reverse-biased and D2 is forward-biased.

Follow the circuit of Fig. 5-5A through one complete ac cycle (times T_1 through T_3 in Fig. 5-5B). On the first half cycle ($T_1 - T_2$), point A is positive, so D1 is forward-biased and conducts current and D2 is reverse-biased. current I_1 flows from the CT through load *R*, diode D1, and back to the transformer at point A. On the alternate half cycle, current I_2 flows from the CT, through load *R*, diode D2, and back to the transformer at point B.

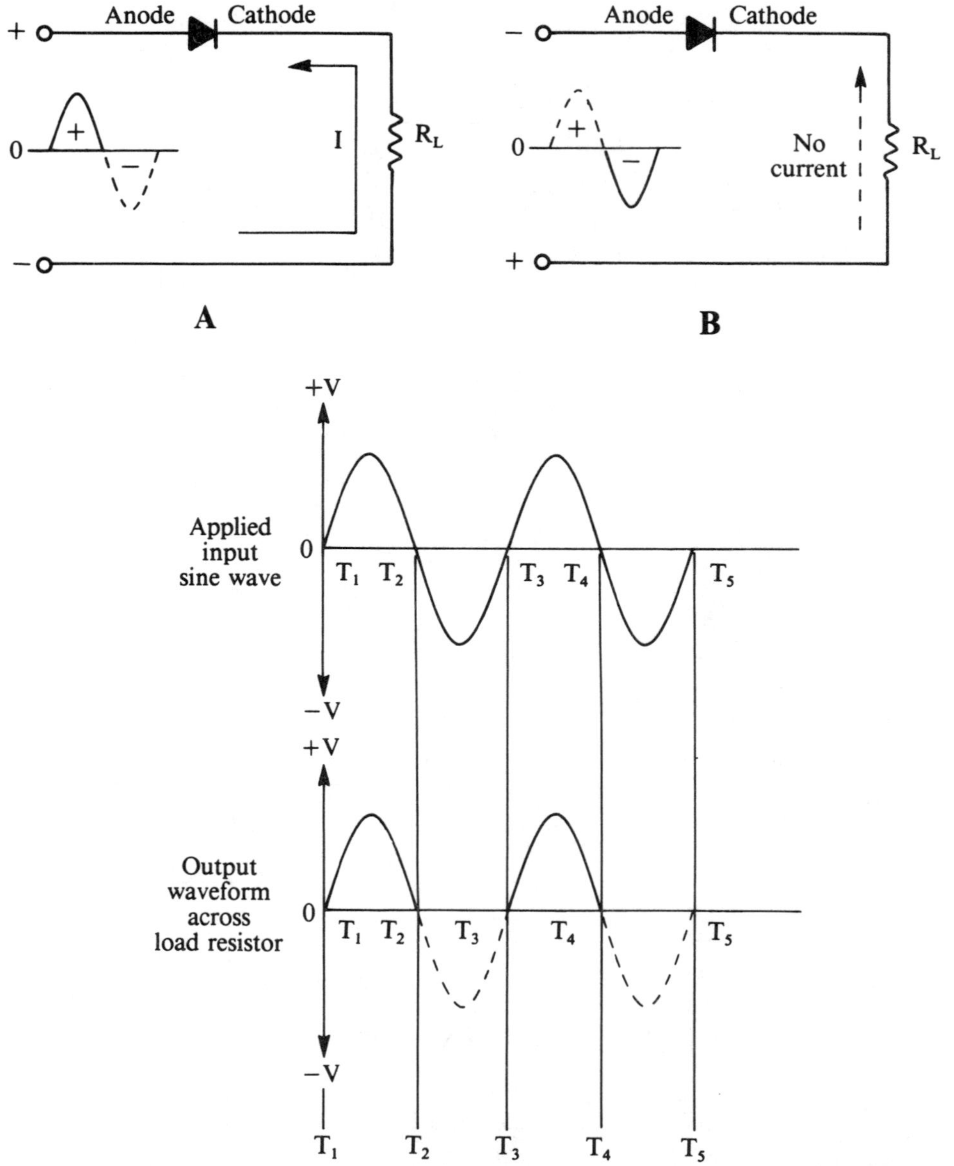

Figure 5-4. Half-wave rectifier circuit: A) positive half cycle, B) negative half cycle, C) half-wave rectification of an ac sine wave.

Now notice what happened: I_1 and I_2 are equal currents that are generated on alternate half cycles, and they *flow in load* R *in the same direction.* Thus, a unidirectional current is flowing through load R on both halves of the ac sine wave. The waveform that results from this action (Fig. 5-5B) is called *full-wave rectified pulsating dc.*

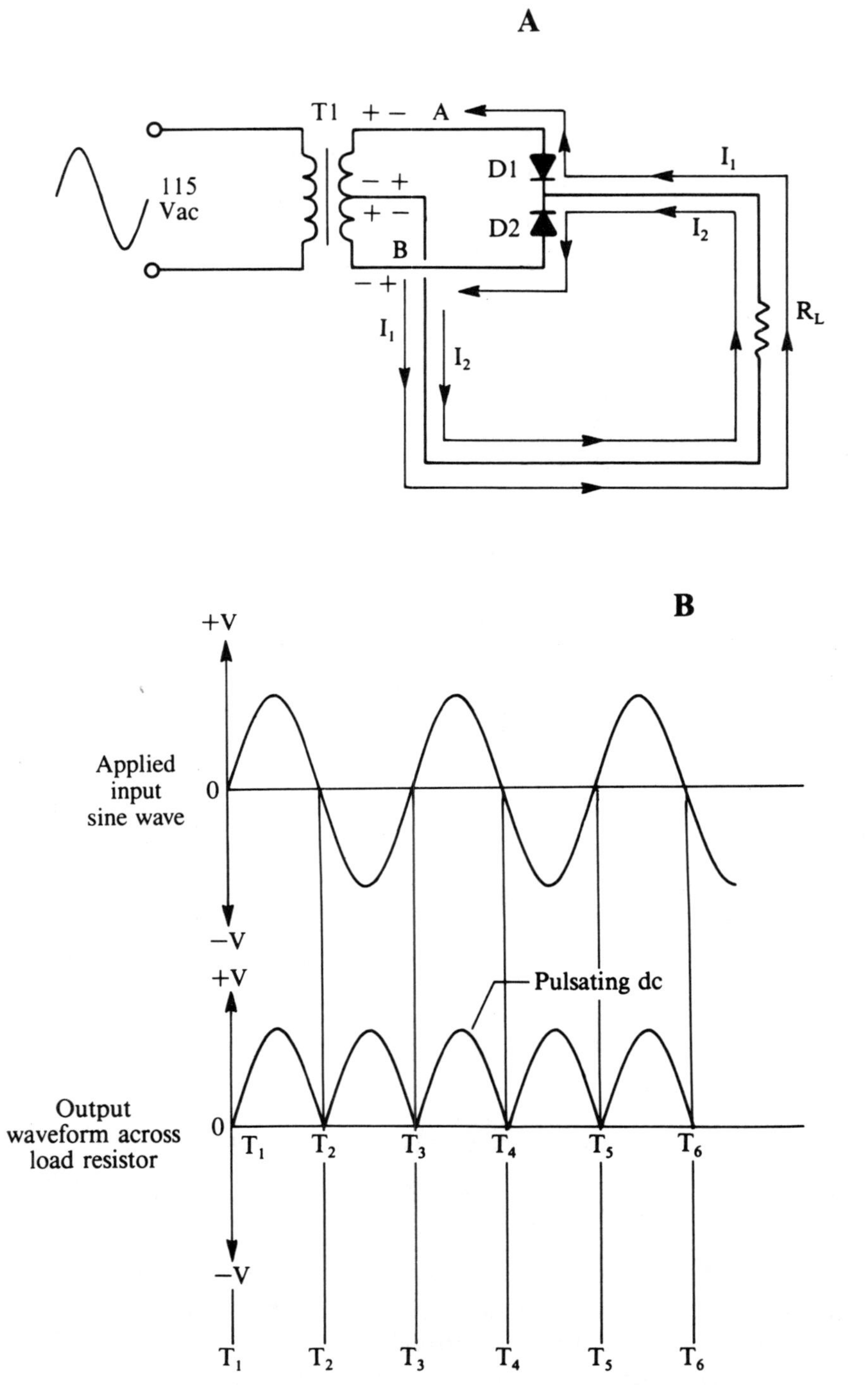

Figure 5-5. A) Full-wave rectifier circuit, B) full-wave rectification waveform.

The center tap on the secondary winding of the transformer can be eliminated by using the *full-wave bridge rectifier* circuit of Fig. 5-6A. This circuit requires twice as many rectifier diodes as the classic full-wave rectifier, but allows the use of a simpler transformer with no center tap. The operation of the circuit is similar to that of the regular full-wave rectifier circuit (see Fig. 5-6B). On one half cycle, point A is positive and point B is negative. Current I_1 flows from the transformer at point B, through D4, load R, diode D1, and back to the transformer at point A. On the alternate half cycle, point A is negative and point B is positive. In this case, current I_2 flows from point A, through diode D3, load R (in the same direction as I_1), diode D2, and back to the transformer at point B. As is true in any full-wave rectifier, *current in the load flows in the same direction on both half cycles.*

A bridge rectifier can be built with four discrete diodes (D1 through D4). In most modern equipment, however, a *bridge-rectifier stack* is used. These components are bridge rectifiers that are built into a single package with four leads. Figure 5-6C shows various alternate schematic diagram circuit symbols for bridge-rectifier stacks.

Figure 5-7 shows a "half-bridge" full-wave rectifier. This circuit is used more today because of the dual-polarity power supplies used in a lot of equipment. Operational amplifiers and some CMOS devices typically require $\pm$ 12-volt or $\pm$ 15-volt bipolar dc power supplies. A full-wave bridge rectifier stack can be coupled to a center-tapped transformer to create a pair of full-wave rectified dc power supplies. The CT is the common (or ground); the bridge positive terminal supplies the positive output voltage and the negative terminal supplies the negative output voltage. Both outputs from the half-bridge rectifier are full-wave rectified pulsating dc.

Selecting rectifier diodes The two most-often-used parameters to specify practical power-supply diodes are: *forward current* and *peak inverse voltage.* The forward-current rating of the diode must be at least equal to the maximum-current load that the power supply must deliver. In practical circuits, however, a safety margin is necessary to account for tolerances in the diodes and variations of the real load (as opposed to the calculated normal load). Also, if the diode rating is somewhat larger than the load current, the circuit will be much more reliable. Select a diode with a forward-current rating of 1.5 to 2 times the calculated (or design goal) load current—or more, if available. Selecting a diode with a very much larger forward-current rating (e.g., 200-A diode for a 100-mA circuit) is both wasteful and likely to make the diode not work exactly like a rectifier diode. The general rule is to make the rating as high as is feasible. The "1.5-to-2" times rule, however, should allow a reasonable margin of safety without wasting money on higher-grade diodes.

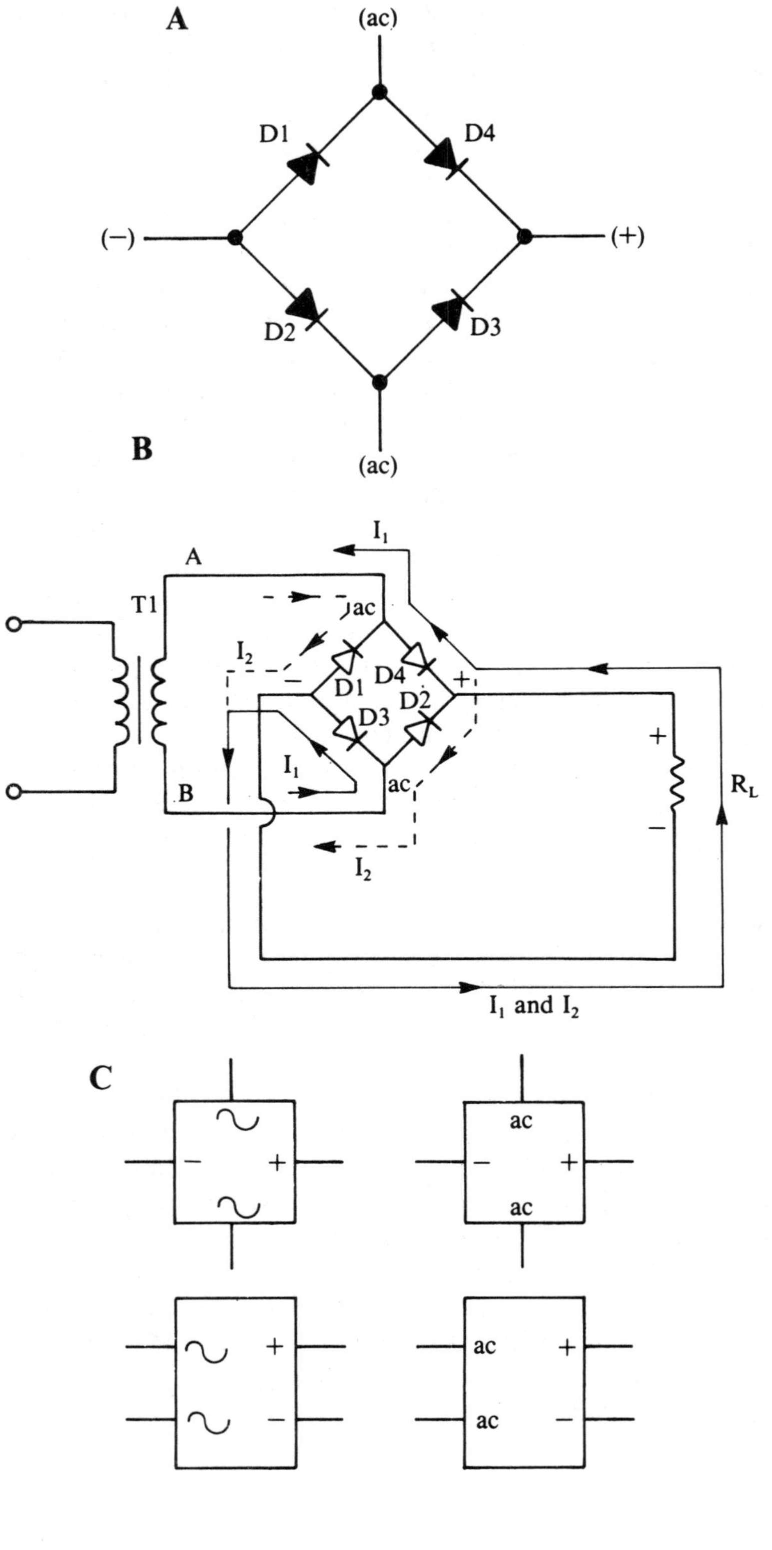

Figure 5-6. Full-wave bridge rectifier: A) circuit, B) in a power-supply circuit, C) circuit diagram symbols.

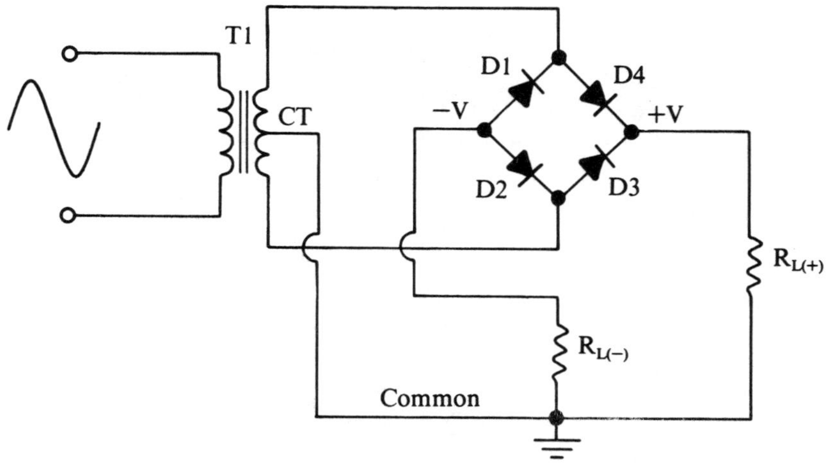

Figure 5-7. Dual-polarity full-wave rectifier with a bridge.

The peak inverse voltage (PIV) rating can be a little more complicated to determine. In unfiltered, purely resistive circuits, the PIV rating needs to only be greater than the maximum peak-applied ac voltage (1.414 times rms). But if a 20 percent safety margin is desired (a good idea), then make it 1.7 times the rms voltage.

Most rectifiers in practical dc power supplies are used in *ripple-filtered circuits* (e.g., Fig. 5-8A), and that makes the PIV selection problem different. Figure 5-8B shows a redrawn simple half-wave rectifier capacitor filtered circuit to better illustrate the circuit action. Keep in mind that capacitor C1 is charged to the peak voltage with the polarity shown. This peak voltage is 1.414 times the rms voltage. The peak voltage across the transformer secondary (V) is in series with the capacitor voltage. When voltage V is positive, the transformer voltage and capacitor voltage cancel out, making the diode reverse voltage nearly zero. However, when the transformer voltage (V) is negative, the two negative voltages (V and V_c) add algebraically to twice the peak voltage, or a total of $2.83 \times V_{ac(rms)}$.

The reverse voltage across the diode is approximately 2.83 times the rms voltage. Therefore, the absolute minimum PIV rating for the diode in a filtered dc supply is 5.83 times the applied rms. If a 20-percent safety margin is preferred, then the diode PIV rating should be 3.4 times the applied rms voltage (or more).

Figure 5-9 shows the proper method for mounting an axial lead rectifier on a prototyping perf board or PC board. In order to improve heat dissipation, allow ⅛ to ¼ inch of space beneath the rectifier diode. The space beneath the diode body allows air to circulate (keeping the diode cooler) and prevents the diode heat from damaging the board. Incidentally, this same tactic is also useful when power resistors or other hot passive

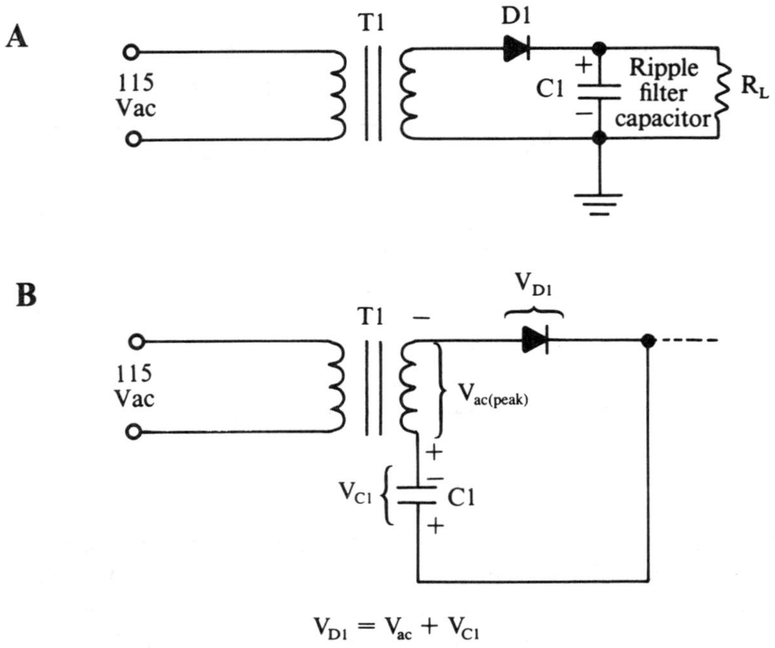

Figure 5-8. "Brute-force" ripple filter: A) circuit, B) redrawn for clarity.

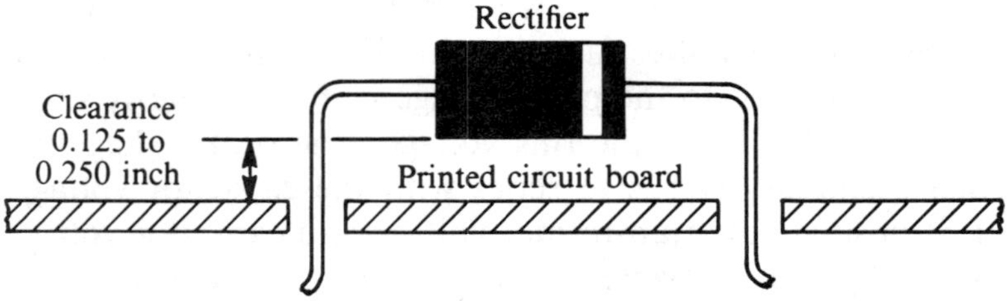

Figure 5-9. Mounting rectifier diode on perfboard or PCB.

components are used. This method is used anytime, *except* where excessive vibration is expected, which includes most sedentary circuits and equipment. Where vibration is a problem, use a Teflon® spacer beneath the diode and board.

Ripple filter circuits

The pulsating dc output from either full-wave or half-wave rectifiers is almost as useless for most electronic circuits as the ac input waveform. A *ripple-filter circuit* is therefore needed to smooth the pulsating dc into a purer dc. Figure 5-10A shows the simplest filter circuit: a single capacitor (C1) that is connected in parallel with the load. Circuit action is shown in

Fig. 5-10B. The capacitor stores electrical charges on voltage peaks, then dumps that charge into the load when the voltage drops between peaks. The shaded area of Fig. 5-10B shows the filling in caused by the capacitor charge. The output voltage is the sum of both rectifier and capacitor contributions, and it is represented by the heavy line in Fig. 5-10B.

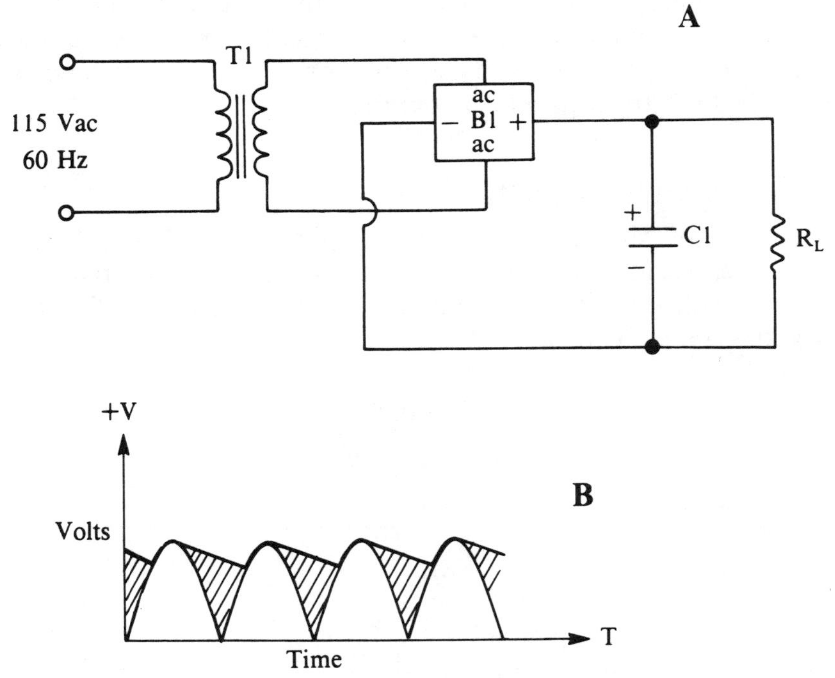

Figure 5-10. Brute-force rectifier: A) circuit, B) waveform.

The value of the filter capacitor is determined by the amount of ripple factor that can be accepted. The *ripple factor (r)* is the ratio of the ripple voltage amplitude to the average voltage of the rectified waveform. Values of r tend to be in the range 0.01 to 1.0 for common electronic circuits. The standard calculation for ripple factor for 60-Hz ac circuits is:

Half-wave rectifier circuits:

$$r = \frac{1{,}000{,}000}{208\ C_1\ R_L} \tag{5.2}$$

Full-wave rectifier circuits:

$$r = \frac{1{,}000{,}000}{416\ C_1\ R_L} \tag{5.3}$$

Where:

C_1 is the capacitance in microfarads.
R_L is the load resistance in ohms.

A *voltage regulator* smooths the ripple considerably. Power-supply makers sometimes claim in advertising that their products have the "equivalent of one farad of output capacitance." What they mean is that the ripple reduction produced by the voltage regulator is the same as having a one farad (1,000,000 μF) capacitor across the load.

Pi (π) network filter circuits

Another type of filter circuit is the RC pi-network (Fig. 5-11). Output voltage V_1 represents a circuit like the previous ones, but output voltage V_2 has a lower voltage and a substantially lower ripple factor. For full-wave circuits, the ripple factor is:

$$r = \frac{2{,}500{,}000}{C_1\, C_2\, R_1\, R_L} \tag{5.4}$$

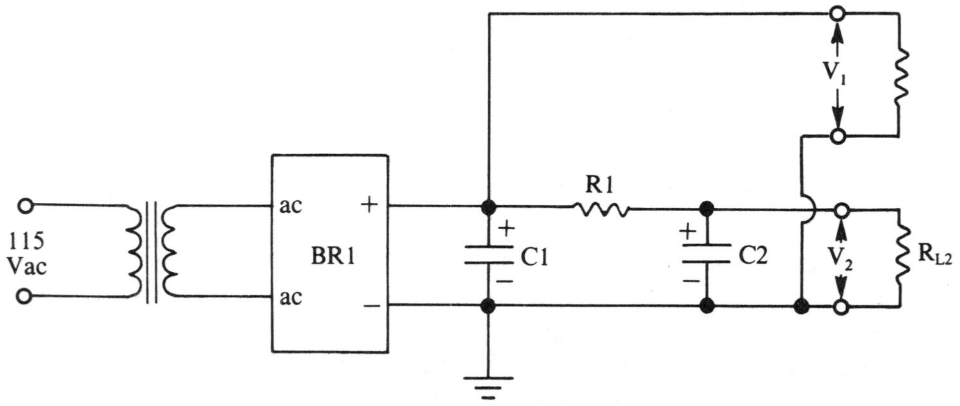

Figure 5-11. RC π-network ripple filter.

Capacitors used in ripple filter circuits A number of different filter capacitors have been used in dc power supplies over the years. However, *aluminum electrolytic capacitors* are now most commonly used in ripple filtering. In those capacitors, the "plates" are aluminum foil that have been covered with an aluminum-oxide compound. One difference between the oil-filled capacitor and the electrolytic capacitor is that the electrolytic is polarity sensitive. The capacitor will be marked so that the positive and negative leads can be identified. The most obvious marking shows the + symbol for the positive lead or terminal, and either −, a black band, black wire, or no marking for the negative lead.

It is very important that you observe the polarity markings on electrolytic capacitors. *If an electrolytic capacitor is wired into the circuit backwards it might explode and throw a lot of fiber material and metal shrapnel.*

Three ratings are used on the electrolytic capacitor: Capacitance, dc working voltage, and temperature range. The *capacitance* is rated in *microfarads* (μF), and in low-voltage dc power supplies, it typically ranges from 500 to 25,000 μF. The general rule for replacement is to select a capacitance that is at least as great as the original. It is generally not a problem to use more capacitance than originally called for in the radio power supply. For example, if you are trying to replace a 400-μF capacitor, it is okay to use any value from 400 μF to about several times the original capacitance. Larger capacitances can also be used, but this practice can cause stress on the rectifier during the initial "cold-charge" period, just after the radio is turned on.

The *dc working voltage* (*wvdc*) is the maximum sustained dc voltage that the capacitor can bear. This voltage is an absolute maximum, and it must not be exceeded. In fact, evidence suggests that it should not even be approached. Because of normal manufacturing tolerances, assume that the real working voltage is about 10 percent lower. For example, if you have a 100-wvdc capacitor, assume that the actual working voltage is 10 volts less, or about 90 wvdc.

The maximum temperature rating will be given in degrees Celsius on the body of the capacitor. This rating is rarely an issue with small projects, but could be in some special cases where poor ventilation exists.

Voltage regulation

The output voltage of ordinary rectifier/filter dc power supplies is not stable, but rather, it might vary considerably over time. Two main sources of variation exist in the output of a dc power supply. First, the ac input voltage always fluctuates. Ordinary commercial power lines vary from 105 to 120 Vac (rms) normally, and it might drop to 95 volts during local power "brownouts."

The second source of variation is created by load variation (see Fig. 5-12), because real dc power supplies are not ideal. The ideal textbook power supply has zero ohms internal resistance. Real power supplies, however, have a certain amount of internal resistance (represented by R_S in Fig. 2-15). When current is drawn from the power supply, voltage (V_1) drops across the internal resistance, and this voltage is subtracted from available voltage (V). In an ideal power supply, output voltage V_O is the same as V,

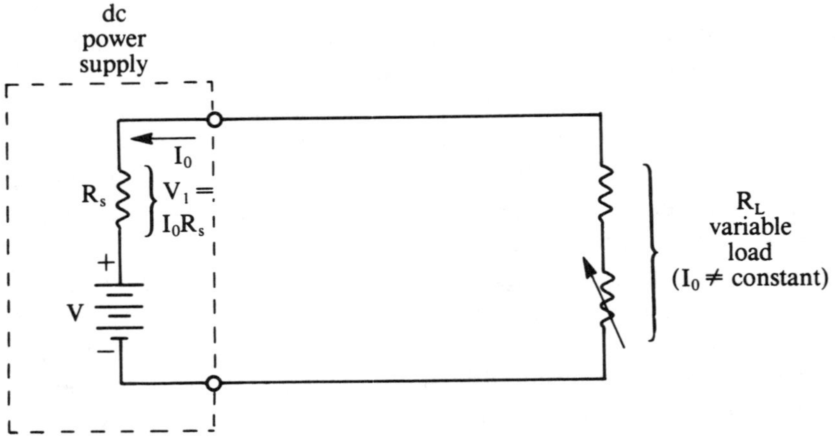

Figure 5-12. Source of voltage regulation problems: internal power-supply resistance.

but in real supplies, V_O is equal to $V - V_I$. Because V_I varies with changes in load current I_O, the output voltage will also vary with changes in current demand.

Many electronic circuits do not work properly under varying supply-voltage conditions. Oscillators and some waveform generators, for example, tend to change frequency if the dc power-supply voltage changes. Obviously, some means must be provided to stabilize the dc voltage for sensitive circuits. The zener diode is perhaps the simplest such voltage-regulator device.

Three-terminal IC voltage regulators At one time, the main voltage regulator was the *zener diode*, although it was not the best solution for many circuits. Fully electronic regulators work better in the face of changing loads and input voltage. Although they were once a bit difficult to design and use, electronic voltage-regulator circuits for low- to medium-current levels (up to 5 amperes) are reasonably simple to build now that simple *three-terminal IC voltage regulators* are available. The circuit with positive three-terminal regulators is shown in Fig. 5-13 and the typical package styles are shown in Fig. 5-14.

Capacitor C1 is the normal ripple filter capacitor, and it should have a minimum value of 1000 μF/ampere of load current (some authorities insist on 2000 μF/ampere). Capacitor C4 is used to improve the transient response to sudden increases in current demand (which happens frequently in digital circuits). Capacitor C4 should have a value of approximately 100-μF/ampere load current. Capacitors C2 and C3 are used to improve the

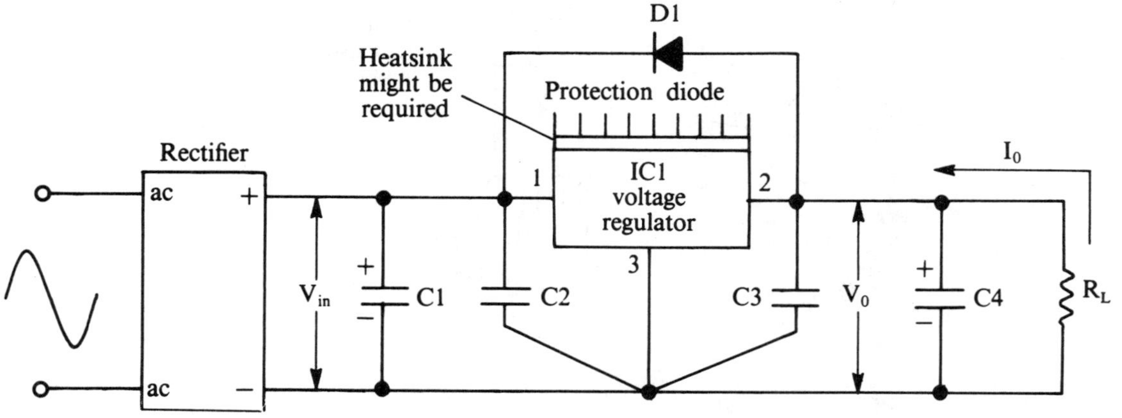

Figure 5-13. Using the 3-terminal IC regulator.

immunity of the voltage regulator to transient noise impulses. These capacitors usually range from 0.1 to 1 μF and they must be mounted as close as possible to the body of voltage regulator IC1.

Diode D1 is not shown in a lot of circuits, but is highly recommended for applications where C4 is used. If the diode is not present, then charge that is stored in C4 would be dumped back into the regulator when the circuit is turned off. That current has been implicated in poor regulator reliability. The mechanism of failure is that the normally reverse-biased pn junction formed by the IC regulator substrate and the circuitry become forward-biased under these conditions. This situation allows a destructive current to flow. The diode should be a one ampere type at power supply output currents up to two amperes, and larger for larger output current levels. For most low voltage, one ampere or less, dc power supplies a 1N400x is sufficient.

IC voltage regulator packages Several three-terminal IC voltage-regulator packages are shown in Fig. 5-14. The "H" package (Fig. 5-14A) is used at currents up to 100 mA, the TO-220 "T" package (Fig. 5-14B) at currents up to 750 mA, and the TO-3 "K" package (Fig. 5-14C) at currents to one A. Some people use these regulators at higher currents by installing the regulators on large heatsinks. However, this is a bad practice and should be avoided. If you need more current, then select a higher-rated three-terminal regulator device . . . don't try to wring more out of a lesser device.

Two general IC regulator families exist. One is designated "78xx," in which the "xx" is replaced with the fixed output voltage rating. Thus, "7805" is a 5-volt regulator and a "7812" is a 12-volt regulator. The "LM-340y-xx" series is also used. The "y" is the package style (H, K or T)

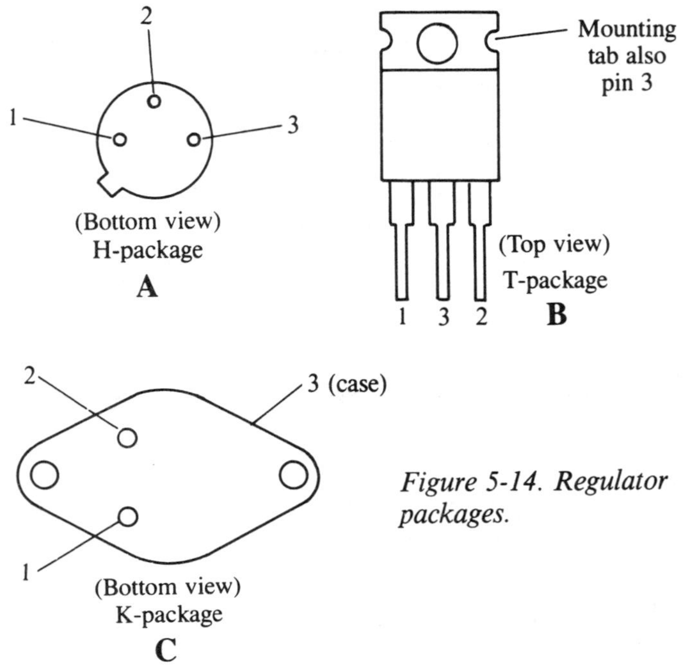

Figure 5-14. Regulator packages.

and the "xx" is the voltage. Thus, the LM-340K-05 is a 1-ampere, 5-volt regulator in a "similar-to-TO3" type-K package; the LM-340T-12 is a 12-volt, 750-mA regulator in a plastic TO-220 power-transistor package. Negative polarity versions of these regulators are available under the "79xx" and LM-320y-xx designations. Figure 5-15 shows the typical circuit symbol. Notice that the pin-outs on the voltage regulator device are different from those of the positive regulator.

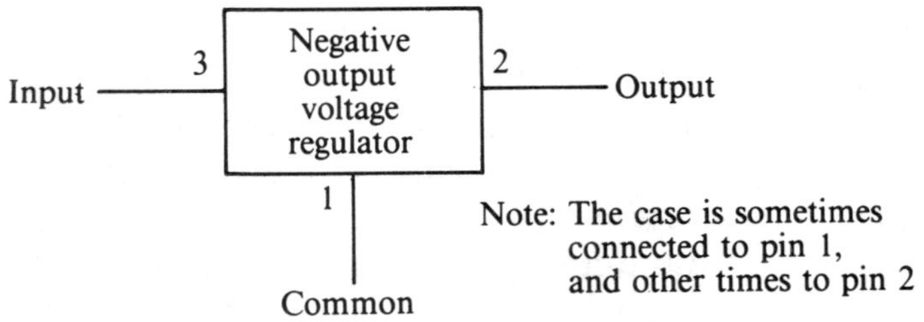

Figure 5-15. Negative voltage regulator.

Minimum dc input voltage The minimum input voltage that is required at the three-terminal IC voltage regulator is usually 2.5 volts higher than the rated output voltage. Thus, for a +5-volt regulator, the minimum allowable input voltage is 7.5 Vdc. The power dissipation is proportional to the voltage difference between this input potential and the rated output potential. For a one-ampere regulator, therefore, the dissipation will be 2.5 watts if the minimum voltage is used and it will be considerably higher if a higher voltage is used. An input voltage should be close to the minimum allowable voltage. For +5-volt supplies that are used in digital projects, a standard 6.3-Vac transformer is sufficient. When full-wave rectified and filtered with 1000 μF/ampere or more, the output voltage will be approximately +8 Vdc.

Summary

The dc power supply is used to convert the bidirectional alternating current (ac), which is available from the power mains, to the unidirectional direct current (dc), which is needed to operate electronic circuits. The typical dc power supply that is used in electronics projects consists of several different components: transformer, rectifier, ripple filter, and (in many cases) a voltage regulator. The transformer scales the 115-volt ac voltage from the power lines up or down, as needed, for the particular project or application. The rectifier converts the bidirectional ac into unidirectional pulsating dc, while the ripple filter smoothes the pulsating dc into nearly pure dc. The voltage regulator stabilizes the voltage in the presence of changing load currents and ac input voltage.

Questions

1. What is the function of a dc power supply?
2. Draw a block diagram of a dc power supply, label the parts, and give the function of each.
3. What is peak inverse voltage?
4. Sketch each of the following: half-wave rectifier, full-wave rectifier using a center-tapped transformer, and bridge rectifier.
5. What considerations should be taken into account when diodes for a power supply are selected?
6. What effect does a filter capacitor have on **a.** dc output? **b.** peak inverse voltage for rectifier diodes?
7. What is voltage regulation? What devices are used to perform this function?

Problems

1. Calculate the peak voltage for a 115-volt ac signal.
2. Calculate the peak voltage for a 220-volt ac signal.
3. A transformer has 500 turns in the primary and 50 turns in the secondary. If the input voltage is 115 volts, what is the output voltage?
4. A transformer will be used to convert 115 volts to 12.6 volts. If the primary has 1000 turns, how many should the secondary have?
5. An unfiltered half-wave rectifier has a 12-volt rms input. What is the peak inverse voltage for the diode?
6. Repeat Problem 5 if a filter capacitor is placed in the circuit.
7. A 60-Hz signal is half-wave rectified and then filtered by a 22-μF capacitor. If the load is 1 kohm, what is the ripple factor?
8. A 60-Hz signal is full-wave rectified and then filtered by a 1000-μF capacitor. If the load is 500 ohms, what is the ripple factor?
9. For the circuit in Fig. 5-11, $C_1 = C_2 = 1000\ \mu$F, $R_1 = 500$ ohms, and $R_L = 1$ kohm. Calculate the ripple factor.
10. Repeat Problem 9 for $R_1 = 5$ kohms.

6 DC POWER-SUPPLY PROJECTS

A number of experiments and projects follow in the upcoming chapters and all of them require a dc power supply to operate properly. Three basic dc power supplies are needed for ordinary workbench electronics. First, for digital projects, the dc power supply should offer +5 Vdc at a current rating of at least 1 ampere (up to 5 amperes is useful). The second necessary form of dc power supply is the dual-polarity ±12-Vdc (or 15 Vdc) power supply. These supplies are used for analog projects, as well as for certain CMOS digital circuits. Finally, the adjustable dc power supply can be set by a front-panel control to any voltage within its range.

Many readers will want to build their own dc power supply, even if only for practice in electronic construction. Others will prefer to buy either a dc power supply or a prototyping board with a built-in power supply. For those who build, some precautions are in order.

1. All of the dc power supplies in this chapter operate at 110 volts ac. This voltage level is potentially very dangerous or fatal. *Do not work on the supply when it is plugged in, and remember that these circuits have deadly points!* Keep the end of the power cord in sight while working on the circuit so that you are sure it is *not* plugged in.
2. Always use a fuse or circuit breaker in the primary winding of the circuit. If a polarized plug is used (highly recommended!), then place the fuse/breaker in series with the hot side of the ac power line.
3. Keep all 110-Vac points inside a grounded metal cabinet.

The basic methods of construction, as well as a description of the function of the components (see also the last chapter) will be given only for the first project. The others are nearly identical, so it would be redundant to repeat the same information. If you plan to build one of these circuits, first read the description of Project 6-1 regardless of whether you plan to build it.

Project 6-1
+5 Vdc power supply for digital projects

The standard voltage for digital circuits is +5 Vdc at 1 ampere for more. The +5 Vdc potential must be tightly regulated between tolerance limits of +4.90 and +5.1 volts. Although modern logic families (such as CMOS and other MOS forms) can operate over a number of different power-supply potentials, the TTL family requires these tight limits—so the TTL limit forms the design basis for most digital circuits. As a result, the manufacturers of microprocessor chips routinely use +5-Vdc potential for their chips. The specified currents range from 1 ampere for very small systems to as much as seven amperes for larger systems.

Figure 6-1 shows the circuit for a simple +5-Vdc power supply that is capable of delivering maximum output currents from one to five amperes, depending on the components selected (those shown are for 1-ampere supplies). The same circuit, with some alteration, would also provide maximum currents of 100 mA or 500 mA depending on the voltage regulator selected (but more of that later). For now, let's look at the various components and how to select some of the options.

Fuse The job of a fuse (F1 in Fig. 6-1) is to interrupt the circuit when excessive electrical current flows. It therefore protects the circuit from further damage when a fault or an output short circuits. The fuse value should be 1.5 to 2 times the normal maximum current flow, according to one popular rule. Place the fuse in the hot lead from the ac power lines, not the neutral. A polarized ac line cord might be necessary to ensure that the fuse remains in the hot line.

Power switch The on/off power switch (s1 in Fig. 6-1) controls the power supply. Be sure to use a switch that is rated at the correct current and voltage for the application.

MOV A *metal-oxide varistor* (*MOV*) is a special device that protects against high-voltage transients that arrive on the ac power lines. According to some studies, a large number of these transients (> 1,500 volts peak) can occur on residential and light industrial power systems. These transients cause seemingly random "bomb-outs" and other odd pathological behavior from some digital computers.

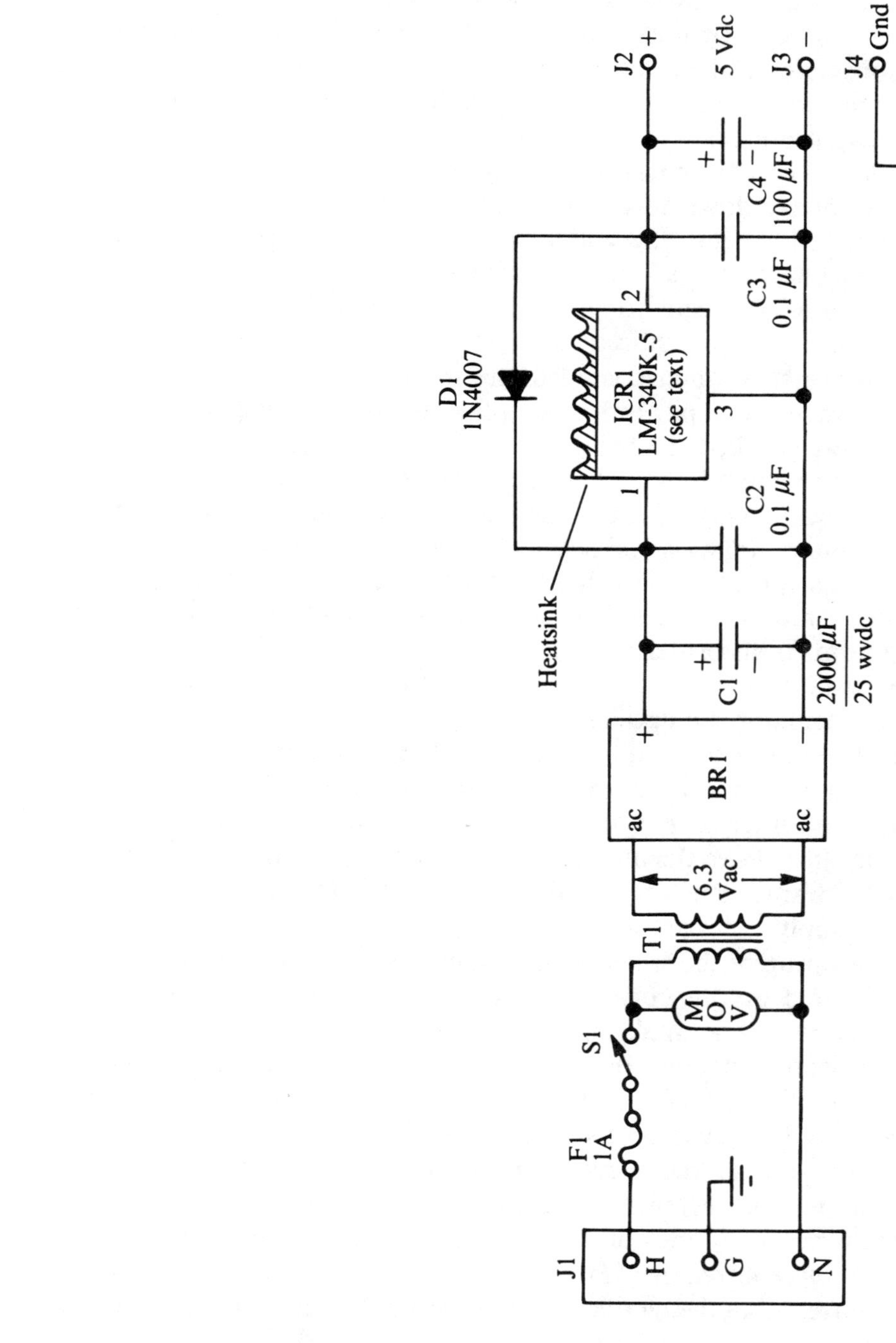

J1: 115-Vac chassis socket and matching line cord

J2, J3, J4: Large size (3/8" base) five-way binding post (+ = red, − = black, Gnd = other color)

T1: 6.3 Vac @ >1 Ampere

Figure 6-1. +5-volt dc power-supply circuit.

The MOV clips those transients to some maximum voltage that is well within the range tolerated by the computer and its power supply. Although the MOV is considered to be optional by some people, I consider it to be essential if proper operation of the digital circuit or computer is a must. Besides, it's cheap protection.

Transformer The power transformer (T1 in Fig. 6-1) steps the 115-Vac power line voltage down to the level where it is most useful for the electronic circuits. In general, the voltage regulators (more in a moment) require an input voltage that is 2.5 or 3 volts higher than the rated output voltage. Thus, a +5-Vdc regulator needs +8 Vdc as a minimum input voltage. This voltage sets the required secondary voltage of the transformer.

When transformer secondary voltages are specified, the *rms voltage* is used. On a sine wave (in ac power line supplies) the peak voltage is 1.414 times the rms voltage ($V_p = 1.414\ V_{rms}$). The output of the transformer should be sufficient to supply a peak voltage that is just above that which will produce the input voltage that the regulator needs. One of the standard transformer secondary voltages is 6.3 Vac, a holdover from vacuum tube days, but it is still useful. The peak voltage of this transformer is 1.414 × 6.3 Vac = 8.9 volts. Thus, for a +5-Vdc power supply, a 6.3-Vac *filament* (vacuum-tube term) transformer is sufficient.

You don't want to use a higher voltage transformer secondary rating, even though the regulator can usually tolerate up to 30, 35, or even 40 volts in some cases. The power dissipation of the regulator is the product of the load current and the difference between input and output voltages. Thus, in order to reduce the internal power dissipation of ICR1, keep the input voltage to a point just above the minimum value. That's why it's preferable to use a 6.3-Vac transformer, rather than a 12.6-Vac transformer for a +5-Vdc power supply.

The current rating of the transformer secondary must be sufficient to produce the required output current under constant conditions without burning up, plus reserve a small amount for reliability purposes. Some people keep a reserve capability of 20 percent, and others prefer at least 40-percent reserve. Thus, a 1-ampere power supply should have a transformer with a secondary current of 1.2 amperes or more.

Rectifier The job of the rectifier (BR1 in Fig. 6-1) is to convert bidirectional alternating current into a form of unidirectional current, called *pulsating dc*. Two ratings of rectifiers must be considered: *forward current* and *peak inverse voltage* (*PIV*). The forward current rating tells us the maximum current that the rectifier can pass without harm. Again, apply a 20 percent or more tolerance for reliability's sake. The PIV rating is the maximum reverse voltage that the rectifier can tolerate without harm. For a filtered dc power supply, such as this one, the PIV voltage of the rectifier should be not less than 2.83 times the applied rms (i.e., twice the applied peak voltage). For +5-Vdc power supplies, this requirement is not inordinately difficult to meet because the lowest common value is 50-volts PIV.

Filter capacitor The job of the ripple filter capacitor (C1 in Fig. 6-1) is to smooth the pulsating dc into nearly pure dc. For nonregulated power supplies, use the maximum ripple factor that the circuits being served can tolerate to select the capacitance. In regulated power supplies (such as Fig. 6-1), the manufacturers of regulators (ICR1) will recommend a value of at least 1,000 μF per ampere of output current. Other authorities ask for 2,000 μF/ampere.

The working voltage rating, usually called *working voltage dc* or *wvdc*, is the maximum constant voltage that the capacitor will accommodate. Again, apply at least a 20-percent tolerance. Again, some people prefer a wider tolerance than the standard. For a 5-Vdc power supply that uses a 6.3-Vac transformer, the output of the rectifier will be close to 8.9 volts. With a reasonable safety factor, the wvdc rating of C1 will be 15 wvdc or more. Therefore, select a 2,000 μF/15 wvdc or greater capacitor for C1. Note: reasonably sized 2,000 μF capacitors are easily obtained at ratings greater than 25 wvdc.

Noise capacitors Capacitors C2 and C3 in Fig. 6-1 are used to suppress the transient noise that sometimes enters the dc power supply from external sources. These noise pulses can interrupt the operation of the regulator (ICR1) and also affect its reliability. To prevent this problem, mount C2 and C3 (0.1 to 1.0 μF) as close as possible to the pins of the regulator device. These capacitors are often mounted directly to the device.

IC voltage regulator The job of the voltage regulator (ICR1 in Fig. 6-1) is to keep the output voltage constant despite: changes in input voltage, and changes in output load current requirements. At one time, voltage regulators were somewhat more complex and difficult to design, so they were used only for the most critical applications. Today, however, modern three-terminal IC regulators are used extensively. These components are easy to use and are quite cheap. The complexity is still present, but it is internal to the IC and is therefore "transparent" to the user.

Heatsinks The voltage-regulator device in Fig. 6-1 is heatsinked, as is the rectifier-bridge stack. These parts are often overlooked, but are still quite important. The heatsink is designed to carry away the heat that is generated inside the regulator, so it affects reliability. In general, a 10°C decrease in a device's junction temperature doubles the mean time between failures. The heatsink is (in my opinion) not optional. For the regulator in Fig. 6-1, select a heatsink that matches the package style. For the bridge rectifier, use a heatsink on rectifiers if the power supply is rated at 3 amperes or greater.

Output capacitor Capacitor C4 in Fig. 6-1 is used to smooth out variations in the output voltage that are caused by sudden changes in the load current demand. The regulator responds after a very short time to the change, but that time is long enough to create a "glitch" in the power-supply voltage. Capacitor C4 smooths the glitch by dumping a small reservoir of charge into the circuit. Select a value for C4 according to the rule 100 μF/ampere of output current. Any working voltage over 10 wvdc is usable (for other than 5-Vdc power supplies, use a capacitor that is rated at least two times the output voltage).

Protection diode Some power supplies use a reverse-polarity protection diode (D1 in Fig. 6-1) to prevent damage to the regulator in case the output voltage becomes larger than the input voltage. This situation can occur at turn/off if capacitor C4 is used or if the load circuit contains a large capacitance. For 1-ampere and 3-ampere dc supplies, the 1N4007 is sufficient (1,000 volts PIV, at 1 ampere). For larger supplies, use a 1000-volt PIV 3-ampere diode for D1.

Parts for larger +5-Vdc power supplies

The values of components specified in the previous project assume that the output voltage is 1 ampere. The usable 1-ampere regulators are LM-309K, 7805 if in a K-package (if the T package is selected, limit output current to about 750 mA), or LM-340K-05. If you want a power supply that outputs up to 3 amperes at 5 Vdc, select the LM-323K regulator. For the +5-Vdc 5-ampere supply, look for Lambda Electronics type LS-1905.

In addition to the regulator, the other components also need to be beefed up. The power transformer must have a secondary-current rating that is equal to the desired maximum output current plus a reasonable safety margin (20 percent by some opinion, 50 percent by others). The rectifier current rating must also be increased. Attempting to draw 3 amperes from a 1-ampere rectifier will have the predictable result: failure.

Project 6-2
±12 Vdc dual polarity power supply

The other dc power supply that is often needed in small electronic projects is the ±12-Vdc 1-ampere bipolar dc power supply. This supply (Fig. 6-2) provides two output voltages (measured with respect to ground): +12 volts and −12 volts. The current rating should be at least 500 mA, and possibly as high as 3 amperes, per output.

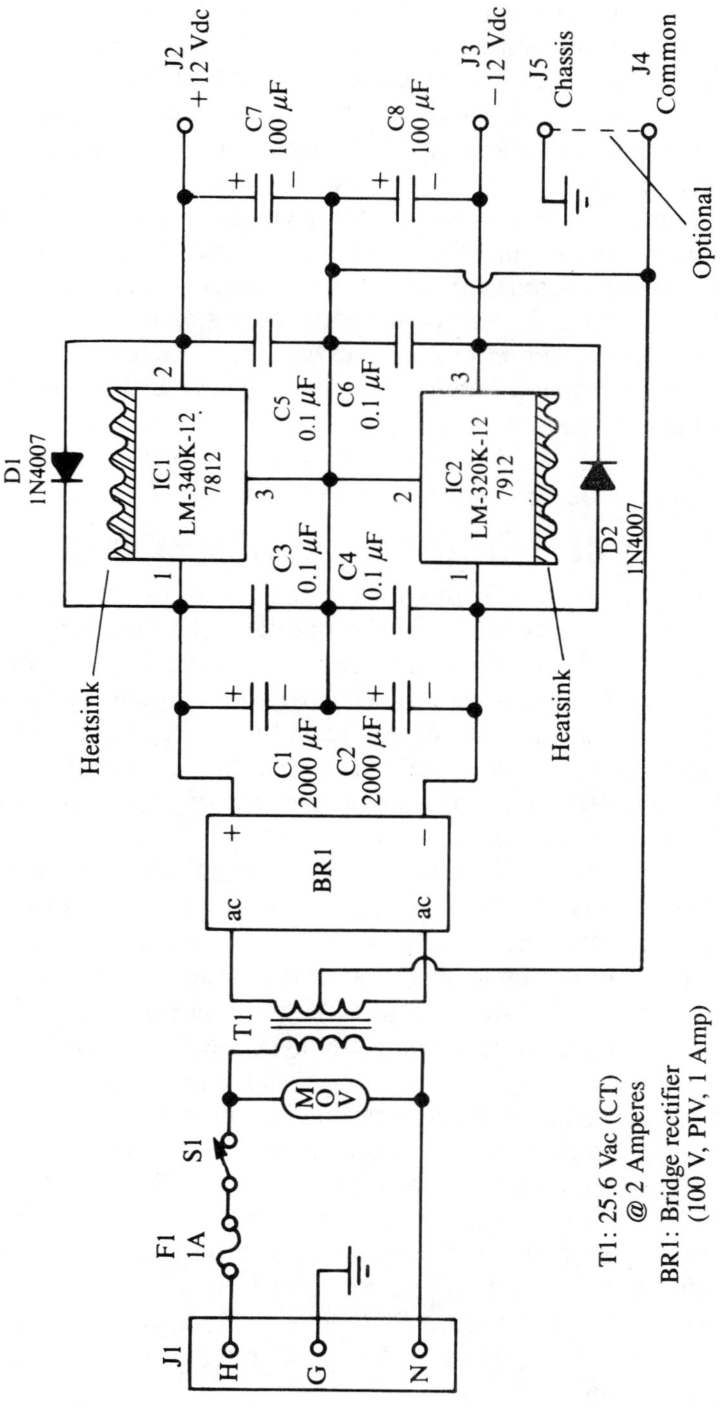

Figure 6-2. Bipolar ±12-Vdc power-supply circuit.

The rectifier circuit of Fig. 6-2 is similar to that of Fig. 6-1, except that it uses a center-tapped transformer. Although a standard bridge-rectifier stack (BR2) is used, the rectifier is actually a pair of regular (nonbridge) full-wave rectifiers in tandem. Thus, the full current rating of the rectifier is set to at least the maximum output current of the supply plus a 20-percent safety margin.

For the popular version in which each output provides up to 1 ampere, the transformer must be rated for at least 2 amperes and (with safety margins considered) up to 2.5 or 3 amperes is desirable. The secondary voltage rating is 25 Vac center-tapped or greater.

The dc power supply is relatively easy to design and build, at least for the purposes stated herein. You should have little or no trouble applying this material to actual projects.

Project 6-3
+12-volt 1-ampere dc power supply

This project (Fig. 6-3) provides a single, monopolar dc power supply that will deliver 1 ampere at a rated potential of 12 Vdc. Most users will need to build two of these circuits in order to provide the ±12 volts that are needed for linear IC and other projects. An advantage to using separate 12-volt supplies is that they can be connected in series to provide 24 Vdc. Parallel connection to increase the current rating to 2 amperes is possible, but isolation diodes must be used to prevent the two regulators from driving each other on and off.

The bridge rectifier in this circuit must have a forward current rating that is at least as high as the load current. It's a good idea to have a 50-percent margin for reliability. For example, if you plan to make a 1-ampere dc power supply that will be run continuously at or near the full-rated output current, select a 1.5-ampere or higher-rated bridge rectifier. Continuous-service rules are always tighter than intermittant-service rules, so if you use the supply for only very short periods, it is possible to get away with the higher current rating. In any event, no one has ever regretted following the more conservative approach to ratings.

The peak inverse voltage (PIV) rating of the rectifier diode must be at least 2.83 times the rms secondary voltage of the transformer. This criteria is rarely a problem when designing and building very low voltage power supplies (e.g., 5 Vdc), but as the voltage rises, the selection of rectifier PIV becomes critical. Standard PIV values include: 50, 100, 200, 400, 600, and 1000 volts. I prefer to use the 1000-volt PIV diodes and bridges for the safety margin.

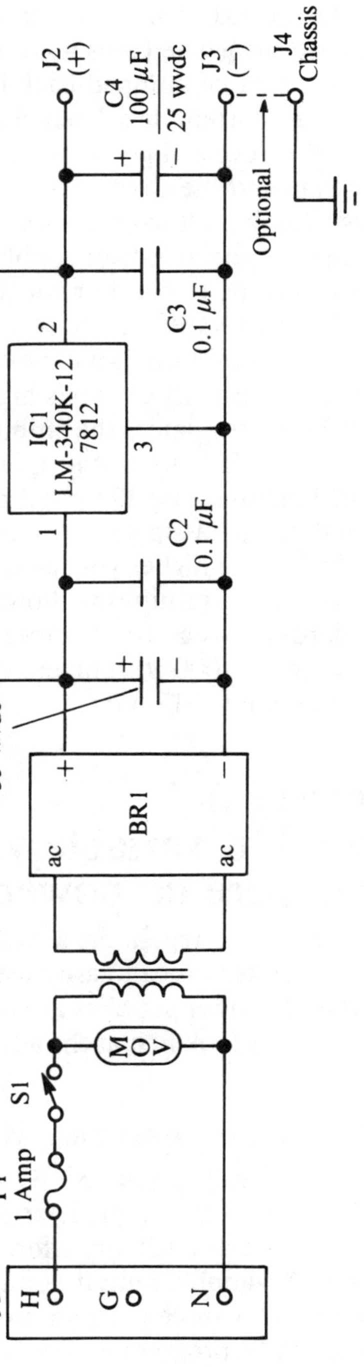

Figure 6-3. + 12-Vdc power-supply circuit.

Diode D1 is used in many circuits, but is highly recommended for applications where C4 is used. If the diode is not present, the charge stored in C4 would be dumped back into the regulator when the circuit is turned off. That current has been implicated in poor regulator reliability. The normally reverse-biased pn junction that is formed by the IC regulator substrate and the circuitry becomes forward-biased under these conditions. This situation allows a destructive current to flow. The diode should be a 1-ampere type at power-supply currents up to 2 amperes, and larger for larger current levels. For most low-voltage (1 ampere or less) supplies a 1N4004 to 1N4007 diode is satisfactory.

The minimum input voltage (V_{in}) to the three-terminal IC voltage regulator is usually 2.5 volts higher than the rated output voltage. Thus, for a +12-volt regulator, the minimum allowable input voltage is 14.5 Vdc. The power dissipation is proportional to the voltage difference between this input potential and the rated output potential. Therefore for a 1-ampere regulator, the dissipation will be 2.5 watts if the minimum voltage is used and higher if a higher voltage is used. For example, an input voltage should be close to the minimum allowable voltage be used. For 12-Vdc supplies, a standard 12.6-Vac transformer is sufficient. When full-wave rectified and filtered with 1000 μF/ampere or more, the rectifier output voltage will be approximately +17 Vdc.

Project 6-4
LM-338 variable voltage, 5-ampere dc power supply

Although fixed regulators are convenient for dc power supplies (indeed, for most circuits you can easily use one of the fixed voltage values), multiple-voltage dc power supplies are occasionally necessary. Before describing the project, let's look first at the adjustable three-terminal IC voltage regulators.

Adjustable three-terminal IC voltage regulators

The previously discussed IC voltage regulators are fixed-output-voltage types. They offer a predetermined output voltage that is unchangeable without an extraordinary effort, which usually deteriorates the voltage regulation. A variable-output-voltage three-terminal voltage regulator, on the other hand, can be programmed by selection of two resistors, on the other hand, can be programmed by selection of two resistors to any voltage that is desired within its range. These devices can be used for either variable dc power supplies or to supply custom output voltages other than the standard fixed voltages. For example, the LM-317 and LM-338 devices are variable dc voltage regulators that can deliver up to 1.5 amperes and 5 amperes, respectively, at voltages up to +32 Vdc.

The project Figure 6-4 shows a typical power-supply circuit for these regulators. In the version shown, the LM-338 regulator is selected, so the current rating is 5 amperes. The input voltage must be 3 volts higher than the maximum output voltage. The output voltage is set by the ratio of two resistors, R1 and R2, according to the equation:

$$V_o = 1.25 \text{ volts } \left(\frac{R_2}{R_1} + 1\right) \tag{6.1}$$

An example from the National Semiconductor *Linear Databook* shows 120 ohms for R1, and a 5 kohms for potentiometer R2. This combination produces a variable output voltage of 1.2 to 25 Vdc, when V_{in} is greater than 28 Vdc. Diode D1 can be any of the series 1N4002 through 1N4007 for LM-317 supplies and any 3-ampere type for LM-338 supplies.

In order to make this power supply work, select a power transformer that offers not less than 5 amperes (or 1.5 amperes if the LM-317 is used) and a secondary voltage rating of 25.6 Vac or more. These transformers are easily available and are sometimes seen in catalogs from Radio Shack and mail-order sources.

When building this power supply, remember that the current rating is higher than in the 1-ampere supplies. The wiring in this supply should be at least #14, and using #12 wire would not be a waste. In lower current supplies, #16 or even #18 wire could be used, but not at 5 amperes. The IR drop losses would create an quasiregulated situation because of the wire resistance.

Project 6-5
LM-338 variable voltage 10-ampere dc power supply

Figure 6-5 shows a pair of LM-338 devices (IC1 and IC2) that are used in parallel to produce a variable voltage 10-ampere dc power supply. The regulators cannot be simply connected in parallel, but rather they require supporting circuitry to keep the power supply running evenly. Otherwise, the two regulators would take turns shutting each other down.

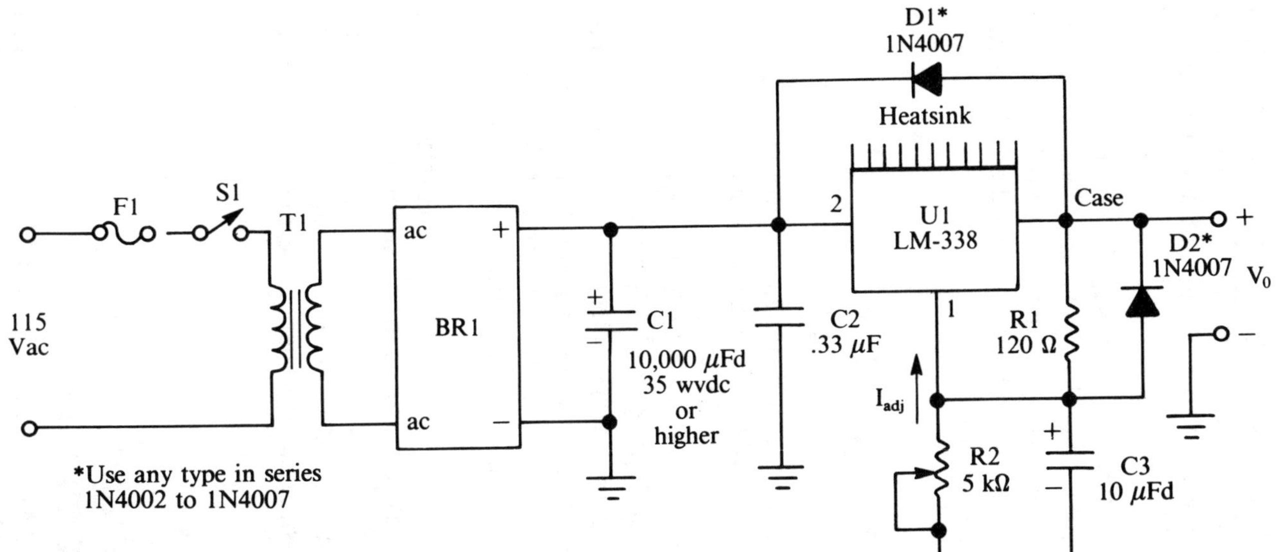

$$V_0 = 1.25 \times \left[\frac{R_2}{R_1} + 1\right] + R_2 I_{adj}$$

Figure 6-4. Variable voltage, 3-ampere dc power-supply circuit.

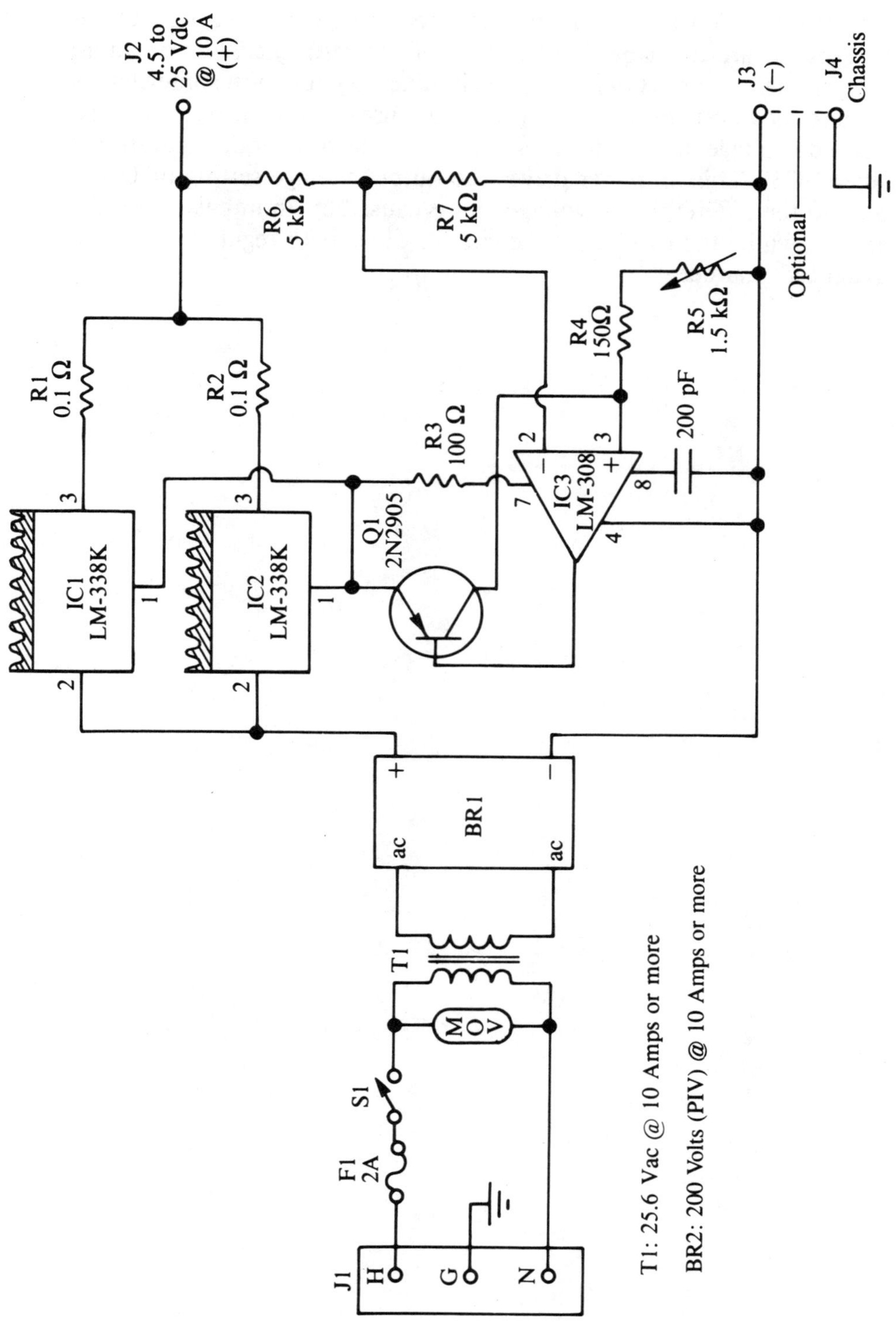

Figure 6-5. High-current variable-voltage dc power supply.

The 0.1-ohm 5-watt resistors that are used in series with each regulator input terminal are available in the form of car radio *fusistors* or audio power-amplifier *bias resistors.* In those circuits, they function somewhat as fuses. Here, however, they not only serve as fuses of a sort, but they also provide a dc voltage drop to develop a signal for the differential operational amplifier (IC3). This amplifier drives the output voltage setting of U1 in response to the differential dc voltage that is caused by an unbalance in the currents drawn by the two legs of the circuit. Thus, both regulators operate in a balanced manner.

7 INTRODUCTION TO OP AMPS

AFTER COMPLETING THIS chapter, you will be able to:

- ☐ Describe the different types of op amps
- ☐ Understand the characteristics of op amps
- ☐ Discuss standard op amp circuits

Op-amp circuit symbols

Figure 7-1 shows the two basic symbols that are used to represent operational amplifiers in schematic circuit diagrams. The version in Fig. 7-1A is the IEEE standard symbol; the symbol of Fig. 7-1B is, however, the most commonly used, so it is used in this book. The difference is that the IEEE version in Fig. 7-1A is used *only* for operational amplifiers, but the symbol in Fig. 7-1B can be used for both operational amplifiers and other linear differential amplifiers.

Categories of operational amplifiers

A number of different types of op amp are on the market. Some of these devices have been available for a long time and others are relatively new. Let's take a look at a few types (the following categories are intended to be representative, rather than exhaustive).

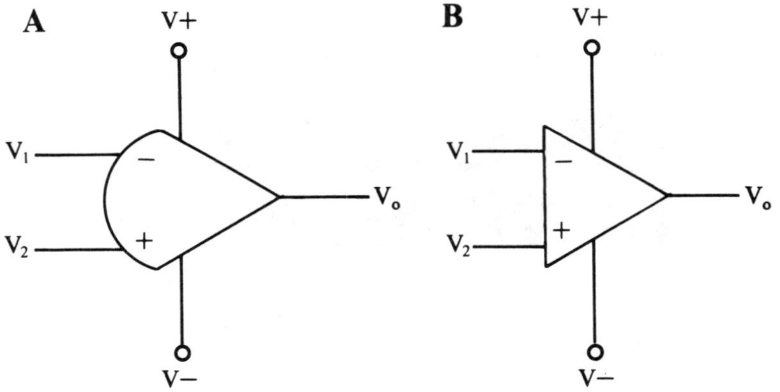

Signals:
V_o = Output voltage
$V_1 + V_2$ = Input voltages

Figure 7-1. Operational amplifier symbols: A) proper symbol, B) most commonly used symbol.

General-purpose op-amps General-purpose op amps are neither special-purpose nor premium devices. Most of these devices are said to be "frequency compensated," so designers trade bandwidth for inherent high (but not absolute) stability. As such, the general-purpose devices can be used in a wide range of applications with few external components. Op amps are usually selected from this category, unless some property of another class gives it a unique advantage.

Voltage comparators Voltage comparators are not true operational amplifiers, but they are based on op-amp circuitry. Although all op amps can be used as voltage comparators, the reverse is not true: IC comparators (e.g., LM-311) cannot usually be used as op amps. The voltage comparator examines two input voltages (V_1 and V_2) and issues an output that reports whether $V_1 > V_2$, $V_1 < V_2$, or $V_1 = V_2$.

Low-input-current op amps Although ideal op amps have zero input bias current, real devices have a small bias current as a result of input transistor biasing or leakage. This class of devices offers very low levels of input bias current because MOSFET, JFET, or superbeta Darlington transistors are used in the input stage instead of npn/pnp bipolar devices. Such transistors produce extremely low input current values. The manufacturer might also elect to use a nulling method in the process that further reduces input bias current. Low-input-current devices typically have picoampere level currents, rather than the microampere or milliampere input currents that are found in other devices.

Low noise op amps Low to reduce internally generated noise. They are generally used in the first stage or two of a cascade chain to improve the overall noise performance of amplifier circuits.

Low-power op amps Low-power op amps optimize internal circuitry to reduce power consumption. Many of these devices also operate at very low dc power-supply potentials (e.g., ±1.5 Vdc).

Low drift op amps Dc amplifier circuits can experience an erroneous change of output voltage as a function of the temperature. Low-drift devices are internally compensated to minimize the temperature drift. These devices are typically used in instrumentation circuits where drift is an important concern—especially when handling low-level input signals.

Wide-bandwidth op amps Wide-bandwidth op amps, called *video op amps* in some literature, have a very high gain-bandwidth product (GBW). One high-performance device, for example, has a GBW of 100 MHz, compared with 300 kHz to 1.2 MHz for various 741-family devices.

Single dc supply op amps Single dc supply op amps provide op-amp-like behavior from a monopolar (typically V+) dc power supply. However, complete op-amp performance is not available from some of these devices because the output voltage cannot assume negative values.

High-voltage op amps Most op amps operate at dc power supply potentials of ±6 to ±22 Vdc. A few high-voltage devices operate from ±44-Vdc supplies, and at least one proprietary hybrid model operates from ±100-Vdc power supplies.

Multiple devices In this category, more than one op amp must be included in the same package. Multiple devices are available with two, three, or four operational amplifiers in a single package. The 1458 device, for example, contains two 741-family devices in either 8-pin metal cans or miniDIP packages.

Limited- or special-purpose op amps The devices in this category are designed for either limited or highly specialized uses. The type LM-302 buffer, for example, is an op amp that is connected internally into the noninverting unity-gain follower configuration. Some consumer audio devices also fit into this category. For example, several dual op-amp devices (used for stereo audio) are optimized for audio frequency applications.

IC instrumentation amplifiers Although the instrumentation amplifier is arguably a special-purpose device, it is sufficiently universal to warrant a class of its own. The IA is a dc differential amplifier that is made of either two or three internal op amps. The voltage gain can be set by either one or two external resistors, depending on the design.

Op-amp input circuits

The input stage of the typical op amp is a dc differential amplifier. A true operational amplifier must be able to perform a wide range of mathematical operations with all combinations of input and output signal polarity; this capability requires both inverting and noninverting inputs. For this same reason, bipolar dc power supplies must be used in op-amp circuits: the results (i.e., output voltages) can be either zero, positive, or negative . . . and the device must accommodate these three possibilities. An op amp must, therefore, have differential inputs and bipolar output polarity.

Dc differential amplifiers

Figure 7-2A shows a simplified dc differential amplifier circuit, which might be found in the input stage of an op amp. Two bipolar npn transistors (Q1 and Q2) are connected from their emitters to a single *constant current source* (*CCS*), I_3. Because current I_3 cannot vary (it is a CCS), changes in either I_1 or I_2 will also affect the other current. For example, an increase in current I_1 means a necessary decrease in current I_2 in order to satisfy Eq. 7-1:

$$I_3 = I_1 + I_2 = \text{constant} \tag{7-1}$$

Where:

I_3 is a constant current from a CCS
I_1 is the C-E current of transistor Q1
I_2 is the C-E current of transistor Q2

Although the collector and emitter currents for each transistor are not actually equal (they differ by the amount of the base current), we can conveniently neglect I_{B1} and I_{B2} for now. For purposes of this discussion, we can assume that $I_1 = I_{C1} = I_{E1}$, and $I_2 = I_{C2} = I_{E2}$ (even though they are not equal in real circuits).

Consider two voltage drops created by current I_2 (which is the C-E current flowing in transistor Q2): V_2 is the voltage drop I_2R_2 and V_{CE2} is the collector to emitter voltage drop across Q2. Voltage V_{CE2} depends on the conduction of Q2, which in turn is determined by the signal applied to the noninverting input (+In).

If the signal voltages applied to both −In and +In inputs are equal, then I_1 and I_2 are equal. In that case, the normal static values of V_2 and V_{CE2} are approximately equal (as determined by internal bias networks), so the relative contributions of V− and V+ to output potential V_o are equal. How does this work?

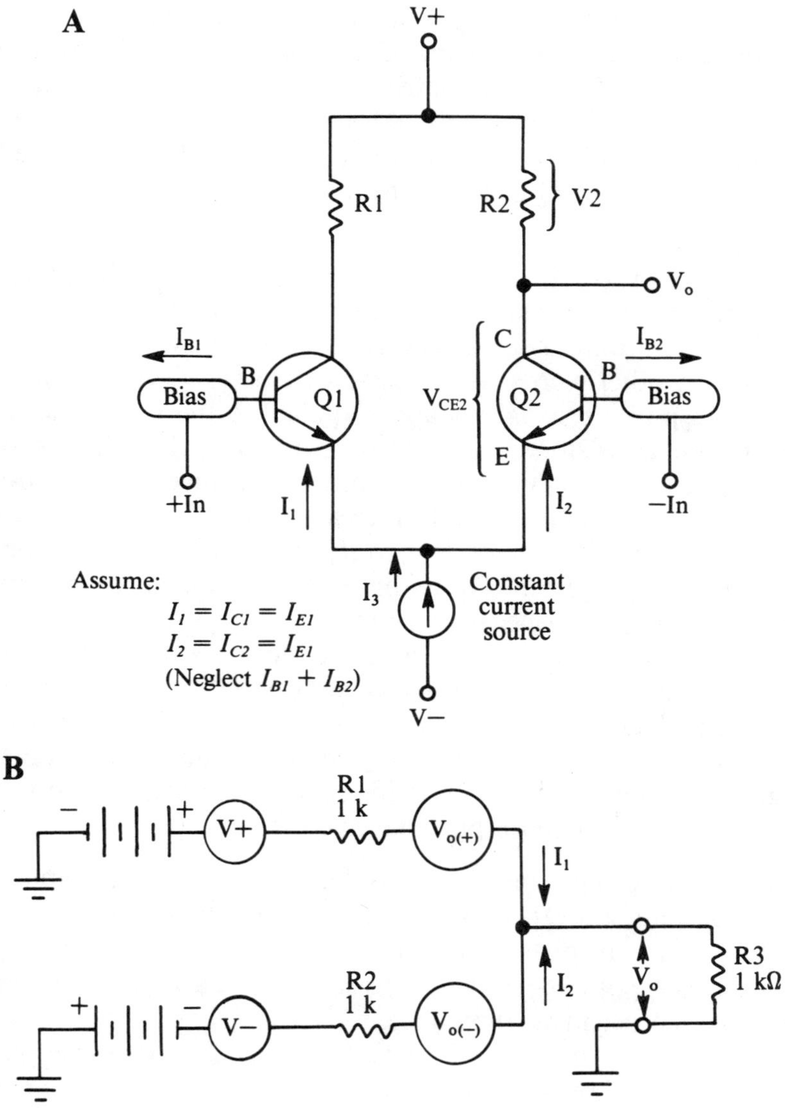

Figure 7-2. A) Bipolar transistor dc differential amplifier, B) signals present in a dc differential amplifier.

Assume:

$V+ = +10$ Vdc
$V- = -10$ Vdc
$V_o = V_{o(-)} + V_{o(+)}$
$R = R_2 = R_3 = 1\ \text{k}\Omega$

A) $$V_{o(-)} = \frac{(V-)(R_3)}{(R_2 + R_3)} = \frac{(-10\ \text{Vdc})(1\ \text{k}\Omega)}{(1\ \text{k}\Omega + 1\ \text{k}\Omega)} = \frac{(-10\ \text{Vdc})(1)}{2} = -5\ \text{Vdc}$$

B) $$V_{o(+)} = \frac{(V+)(R_3)}{(R_1 + R_3)} = \frac{(+10\ \text{Vdc})(1\ \text{K}\Omega)}{(1\ \text{k}\Omega + 1\ \text{k}\Omega)} = \frac{(+10\ \text{Vdc})(1)}{2} = +5\ \text{Vdc}$$

C) $V_o = V_{o(-)} + V_{o(+)} = (-5\ \text{Vdc}) + (+5\ \text{Vdc}) = 0\ \text{Vdc}$

A model circuit is shown in Fig. 7-2B to demonstrate how V_o is formed. Output voltage V_o is the sum of two contributors: $V_{o(-)}$ is the contribution from V−, and $V_{o(+)}$ is the contribution from V+. These voltages are derived from the voltage drops across R1 and R2, respectively; in our model, they represent voltage drops V_2 and V_{CE2}. As long as R1 and R2 are balanced, the sum $V_o = V_{o(-)} + V_{o(+)}$ is equal to zero. However, if either R1 or R2 change, V_o will be nonzero.

Inverting input (−In) Consider the operation of the inverting input (−In), which is the base terminal of transistor Q2. Remember, an inverting input produces an output signal that is 180 degrees out of phase with the input signal. In other words, as V_{-IN} goes negative, V_o goes positive.

If a positive signal voltage is applied to the −In input, and +In = 0, then npn transistor Q2 is turned on harder. The effect is an increase in current I_2 because the collector-emitter resistance of Q2 drops. Now V_2 and V_{CE2} are not equal: voltage V_2 increases, while V_{CE2} decreases. The result is that the contribution of V− to V_o is greater, so V_o becomes negative. Therefore, the base of Q2 is the inverting input because a positive input voltage produced a negative output voltage.

If the signal voltage that is applied to the −In input is negative instead of positive, then the situation changes. In that case, Q2 starts to turn off, so I_2 drops. Voltage V_{CE2}, therefore, increases and V_2 decreases. The relative contribution of V+ to V_o is now greater than the contribution of V−, so V_o becomes positive. Again, inverting behavior is seen: a negative input voltage produced a positive output voltage.

Noninverting input (+In) Consider the noninverting input, which is the base of transistor Q1. Remember, a noninverting input produces an output signal that is in phase with the input signal. A positive-going input signal produces a positive-going output signal, and a negative-going input signal produces a negative-going output.

Suppose −In = 0 and a positive = signal voltage is applied to +In. In this case, I_1 increases. Because Eq (7-1) must be satisfied, an increase in I1 results in a decrease of I_2 in order to keep $I_3 = I_1 + I_2$ constant. Reducing I_2 reduces V_2 (which is I_2R_2), so the contribution of V+ to V_o goes up: V_o goes positive in response to a positive input voltage. This behavior is expected from a noninverting input.

Now suppose that a negative signal voltage is applied to +In. The conduction of Q1 decreases, so I_2 drops. Again, to satisfy Eq. (7-1), current I_2 increases. This condition increases V_2, reducing the contribution of V+ to V_o, forcing V_o negative. Because a negative input voltage produced a negative output voltage, we can again confirm that +In is a noninverting input.

The "ideal" operational amplifier

When you study any type of electronic device for the first time, it is wise to start with an ideal model of that device and later shift to less-than-ideal practical devices. In some cases, the practical and ideal devices are so different that you might wonder about the wisdom of this approach. However, IC operational amplifiers, even low-cost models, so nearly approximate the ideal textbook op amps that the design equations actually work. This analysis method thus becomes extremely useful for understanding the technology, learning to design new circuits, or figuring out how someone else's circuit works.

Later, this book features inverting and noninverting amplifier op-amp configurations, respectively. These sections will derive design equations that describe the operation of real circuits from the ideal op-amp model. The usefulness of this simplified approach proceeds directly from the closeness of the ideal and practical operational amplifier IC devices.

Seven properties of the ideal operational amplifier

The ideal op-amp is characterized by seven basic properties. From this short list of properties circuit operation and design equations can be deduced. Also, the list gives us a basis for examining nonideal operational amplifiers and their defects (plus solutions to the problems that are caused by those defects). The seven basic properties of the ideal op amp are:

1. Infinite open-loop voltage gain
2. Infinite input impedance
3. Zero output impedance
4. Zero noise contribution
5. Zero dc output offset
6. Infinite bandwidth
7. Both differential inputs (−In and +In) stick together

You will find that some cheap op amps only approximate some of the following ideal properties, but the approximation is extremely good for others on the list.

Property 1: infinite open-loop gain (A_{vol}) The *open-loop gain* of any amplifier is its gain without either negative or positive feedback. *Negative feedback* is a signal fed back to the input 180 degrees out of phase. In operational amplifier terms, this is feedback between the output and the inverting input (−In).

Negative feedback reduces the open-loop gain (A_{vol}) by a certain factor (B) which depends on the properties of the feedback network. Figure 7-3 shows the basic configuration for any negative-feedback amplifier. The *transfer equation* for any linear electronic circuit (whether or not it is an amplifier) is the ratio of the output function to the input function. The transfer equation for an amplifier is also called its *gain*, so for a voltage amplifier it is: gain $= A_{vol} = V_o/V_{in}$. In Fig. 7-3, A_{vol} represents the gain of the amplifier element only (i.e., the gain with the feedback network disconnected). The overall gain of this circuit (i.e., the total gain with both amplifier element and feedback network β considered) is defined as:

$$A = \frac{A_{vol}}{1 + A_{vol}\beta} \tag{7-2}$$

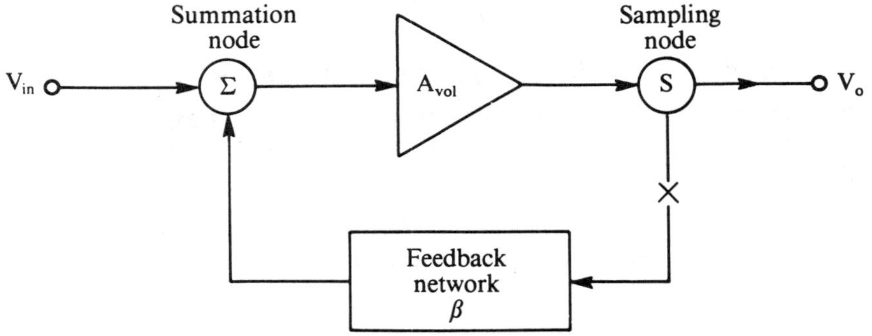

$$A_v = \frac{V_o}{V_{in}} = \frac{A_{vol}}{1 + A_{vol}\beta}$$

Figure 7-3. Feedback amplifier block diagram.

Where:

A_v is the closed-loop gain
A_{vol} is the open-loop gain
β is the transfer equation for the isolated feedback network

In the ideal op amp, A_{vol} is infinite, so the voltage is only a function of the feedback network. In real op amps, the value of the open-loop gain is not quite infinite, but it is quite high. Typical values range from 20,000 in low-grade consumer audio models to more than 2,000,000 in premium units (typically it is from 200,000 to 300,000 in general-purpose op amps).

Property 2: infinite input impedance This property implies that the op-amp input will not load down the signal source (which can be an error source). The input impedance of any amplifier is the ratio of the input voltage to the input current: $Z_{in} = V_{in}/I_{in}$. When the input impedance is infinite, the input current must be zero. An important implication of this property is that the operational amplifier inputs neither sink nor source current. In other words, it will neither supply current to an external circuit, nor accept current from an external circuit. An implication of this property (i.e., $I_{in} = 0$) is important to perform the simplified circuit analysis that is used in chapters 8 and 9 for the gain-setting equations.

Real operational amplifiers have some relatively small input current other than zero. In low-grade devices, this current can be substantial (more than one mA), which will cause a large output offset-voltage error in high-gain circuits. The primary source of this current is the base bias current from the npn and pnp bipolar transistors that are used in the input circuits. Certain premium-grade op amps that feature bipolar inputs reduce this current to nanoamperes or picoamperes. In op amps that use field-effect transistors (FETs) in the input circuits, the input impedance is quite high as a result of the very low leakage currents that are normally found in FETs. The JFET input devices are typically called *BiFET op amps* and the MOSFET input models are called *BiMOS devices*. The RCA/GE CA-3140 device is a BiMOS op amp in which the input impedance approaches 1.5 terraohms (i.e., 1.5×10^{12} ohms). This impedance is near enough to infinite to make the input circuits of those devices approach the ideal.

Property 3: zero output impedance A voltage amplifier (the op amp is a member of this class) ideally has a zero output impedance. All real voltage amplifiers, however, have a nonzero (but low) impedance. Figure 7-4 is a

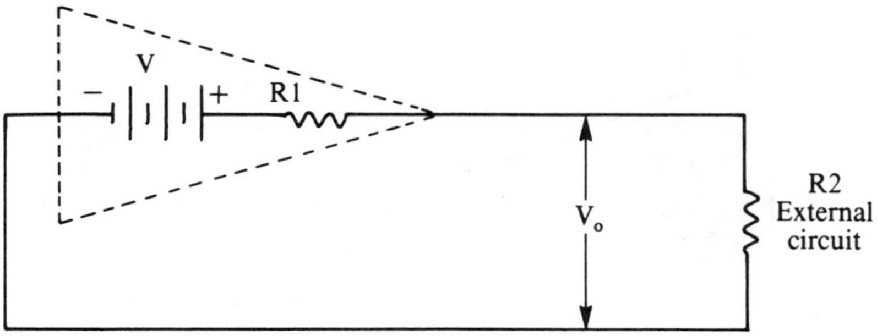

$$V_o = \frac{(V)(R_2)}{(R_1 + R_2)}$$

Figure 7-4. Equivalent circuit for an op amp.

circuit model that represents any voltage source (including amplifier outputs) and its external "load" circuit. Potential V is a perfect internal voltage source with no internal resistance; resistor R1 represents the internal resistance of the source and R2 is the external load. Because the internal resistance (which is usually called *output resistance* in amplifiers) is in series with the load resistance, the actual output voltage (V_o) that is available to the load is reduced by the voltage drop across R1. Thus, the output voltage is:

$$V_o = \frac{V R_2}{R_1 + R_2} \tag{7-3}$$

It is clear from the above equation that the output voltage is equal to the internal source voltage only when the output resistance of the amplifier (R_1) is zero. In that case, $V_o = V(R_2/R_2) = V$. Thus, in the ideal voltage source, the maximum possible output voltage (and the least error) is obtained because no voltage is dropped across the internal resistance of the amplifier.

Real operational amplifiers do not have a zero output impedance. The actual value is less than 100 ohms, and many are typically in the neighborhood of 30 ohms. Thus, in most cases, the op-amp output can be treated as if it is ideal.

Circuit designers set the input resistance of any circuit that is driven by a nonideal voltage source output to be at least 10 times the previous stage output impedance (Fig. 7-5). Amplifier A1 is a voltage source that drives the input of amplifier A2. R_1 represents the output resistance of A1 and R_2 represents the input resistance of amplifier A2. In practical terms, the circuit where $R_2 > 10R_1$ will yield results that are acceptably close to ideal for many purposes. In some cases, however, the rule $R_2 > 100R_1$ or even $R_2 > 1800R_1$ must be followed for greater accuracy.

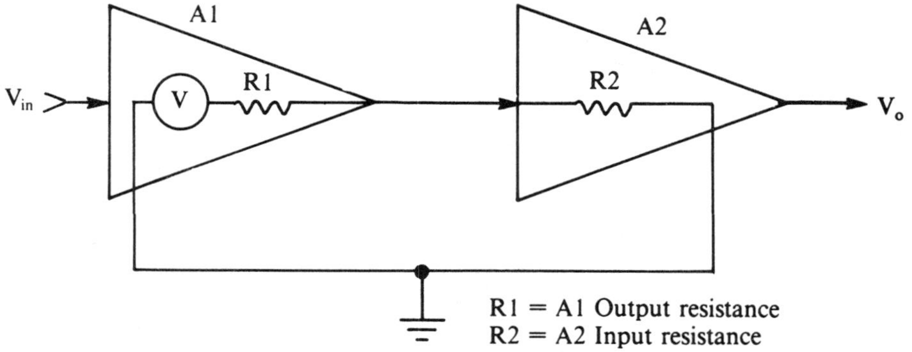

Figure 7-5. Equivalent circuit for driving another op amp.

Property 4: zero noise contribution All electronic circuits, even simple resistor networks, produce unwanted noise. A resistor creates noise as a result of electron movement in its internal-resistance element material. In the ideal operational amplifier, zero noise voltage is produced internally. Thus, any noise in the output signal must be present in the input signal as well. Except for amplification, the output noise voltage will be exactly the same as the input noise voltage. In other words, the op amp contributes nothing extra to the output noise. In this area, practical devices depart quite a bit from the ideal, except for certain premium low-noise models.

Amplifiers use semiconductor devices that create resistive noise (as previously described) and also a special noise of their own. A number of internal noise sources exist in semiconductor devices, and any good text on transistor theory will list more information on them. For now, however, assume that the noise contribution of the op amp can be considerable in low-signal-level situations. Premium op amps are available with very low noise contribution; these devices are usually advertised as premium low-noise types. Others, such as the RCA/GE metal-can CA-3140 device, will offer relatively low-noise performance when the dc supply voltages are limited to +/−5 Vdc, and the metal package of the op amp is fitted with a flexible TO-5-style heatsink.

Property 5: zero output offset The output offset voltage of any amplifier is the output voltage that exists when it should be zero. The voltage amplifier "sees" a zero input voltage when both inputs are grounded. This connection should produce a zero output voltage. If the output voltage is not zero, then an output offset voltage is present. In the ideal op amp, this offset voltage is zero; however, real op amps exhibit at least some output offset voltage. In the real IC op amp, the output offset voltage is not zero, although it can be quite low in some cases.

Property 6: infinite bandwidth The "ideal" op amp will amplify all signals —from dc to the highest ac frequencies. In real op amps, however, the bandwidth is sharply limited. A specification called the *gain-bandwidth product* (*GBW*) is sometimes symbolized by F_t. This specification is the frequency at which the voltage gain drops to unity (1). The maximum-available gain at any frequency is found by dividing the maximum-required frequency into the gain-bandwidth product. If the value of F_t is not sufficiently high, the circuit will not behave in classic op-amp fashion at some higher frequencies that are within the range of interest.

Some op amps have GBW values in the 10- to 20-MHz range, but many others are more limited. The 741 family of devices is very limited; the device will perform as an op amp only within frequencies of a few kilohertz. Above that range, the gain drops considerably. However, in return for this apparent limitation, we obtain nearly unconditional stability; such op amps are said to be *frequency compensated.* The frequency compensation of these devices both reduces the GBW and provides the inherent stability.

Noncompensated op amps yield wider frequency response, but they are more apt to oscillate. These op amps can spontaneously oscillate if certain precautions are not taken in the circuit design and physical layout.

Property 7: differential inputs stick together Most operational amplifiers have two inputs: an inverting (−In) input and a noninverting (+In) input. *Sticking together* means that a voltage that is applied to one of these inputs also appears at the other input. This voltage is real—it is not merely some theoretical device that is used to evaluate circuits. If you apply a voltage to the inverting input and then connect a voltmeter between the noninverting input and the power supply common, the voltmeter will read the same potential on the noninverting input as it did on the inverting input. The implication of this property is that both inputs must be treated the same mathematically. This fact is more noticeable in the concept of virtual as opposed to actual grounds and in the noninverting follower-circuit configuration.

Experiment 7-1

1. Connect Experiment 7-1.
2. Set potentiometer R4 so that the voltage at point A is 1 Vdc.
3. Measure the voltages that appear at points B and C.
4. Compare the voltage at point A with the voltage at point B. Are they the same? Close? Explain any differences.
5. Note the voltage at point C. Can you determine why it has the value that is measured?
6. Change the voltage at point A by varying the potentiometer to several different voltages between 0 and 5 volts, then repeat the experiment at each setting.

The inverting follower produces an output signal that is 180 degrees out of phase with its input signal. The noninverting follower, as you might expect, produces an output signal that is in-phase with its input signal.

Almost all other op-amp circuits are variations on either inverting or noninverting follower circuits. If you understand these two configurations, you can understand (and either design or modify) a wide variety of different circuits that use IC op amp.

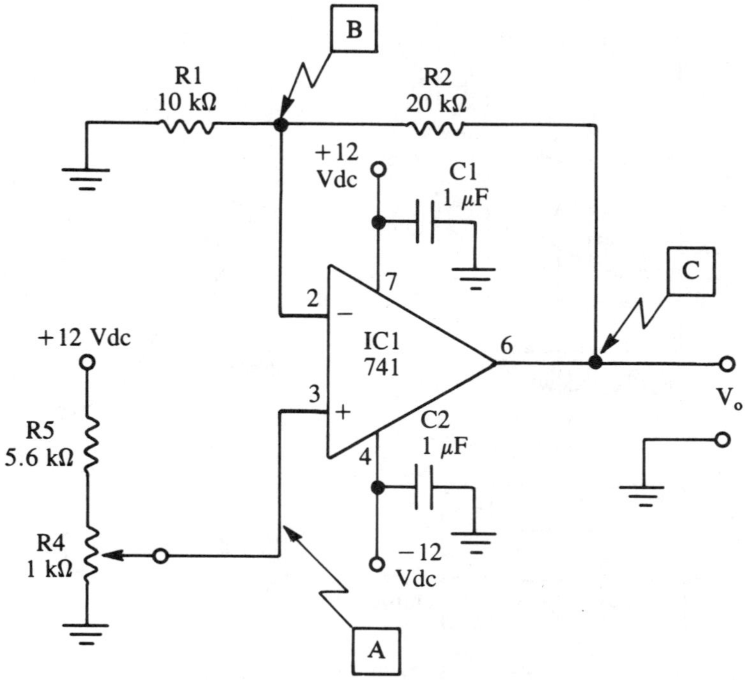

Experiment 7-1. Circuit for Experiment 7-1.

Practical operational amplifiers

Now that we have examined the ideal operational amplifier and some typical device specifications, let's look at practical devices. Because of popularity and low cost, this section features the 741 device. The 741 family also includes the 747 and 1458 "dual-741" devices. Although many higher-grade operational amplifiers are on the market, the 741 (and its close family members) is the industry-standard general-purpose op amp.

Figure 7-6 shows the two most popular packages that are used for the 741. Figure 7-6A is the 8-pin miniDIP package and Fig. 7-6B is the 8-pin metal-can package. The miniDIP pin-outs for a 1458 dual op amp are shown in Fig. 7-6C. The 741 has the following pins:

Inverting input (−In) pin 2 The output signals produced from this input are 180 degrees out of phase with the input signal applied to −In.

Noninverting input (+In) pin 3 Output signals are in-phase with signals applied to the +In input terminal.

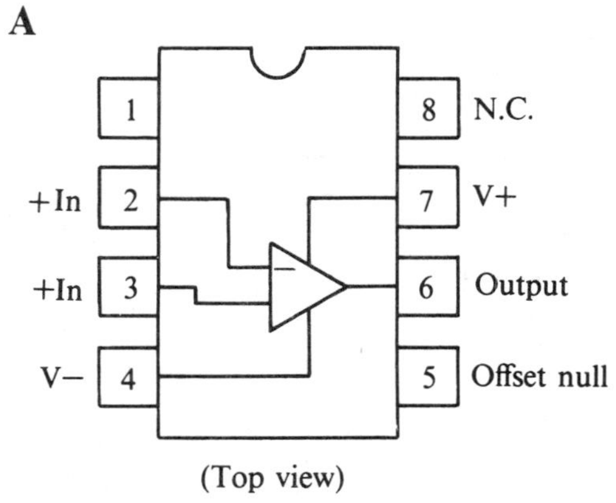

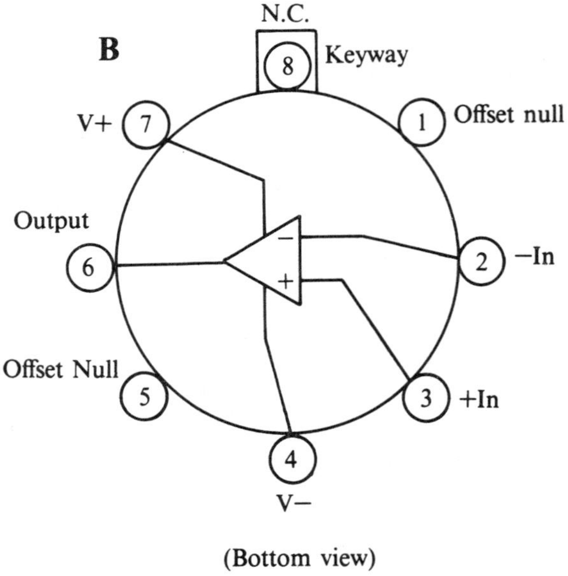

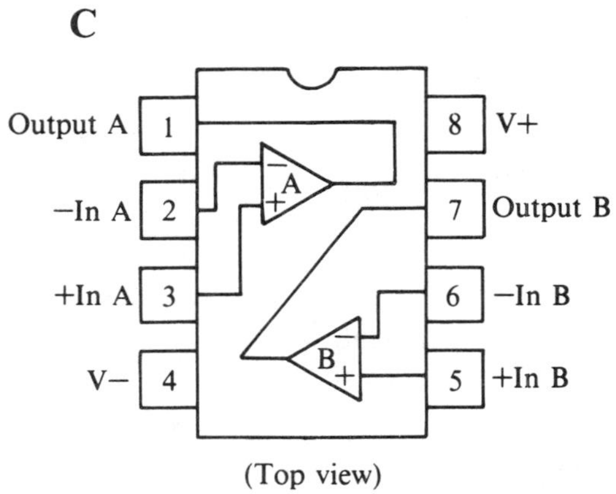

Figure 7-6. Typical op-amp packages: A) mini-DIP industry standard (741), B) metal-package industry-standard (741), C) dual op amp in mini-DIP package (1458, CA-3240).

Output, pin 6 On most op amps, the 741 included, the output is single-ended—output signals are taken between this terminal and the power supply common (see Fig. 7-7). The output of the 741 is said to be short-circuit proof because it can be shorted to common indefinitely, without damaging the IC.

V+ dc power supply, pin 7 The positive dc power-supply terminal.

V− dc power supply, pin 4 The negative dc power-supply terminal.

Offset null, pins 1 and 5 These two terminals accommodate external circuitry that compensates for offset error

The pin-out scheme (Fig. 7-6) is considered to be the industry standard for generic single operational amplifiers. Although many amplifiers use different pin-outs than Fig. 7-6, most of the available devices use this scheme.

Standard circuit configuration

The standard circuit configuration for 741-family op amps is shown in Fig. 7-7. These pin-outs are the industry standard. The output signal voltage is impressed across load resistor R_L which is connected between the output terminal (pin 6) and the power supply common. Most manufacturers recommend a 2-kohm minimum value for R_L. Also, note that some operational-amplifier parameters are based on a 10-kohm load resistance. Because it is referenced to common, the output is single-ended.

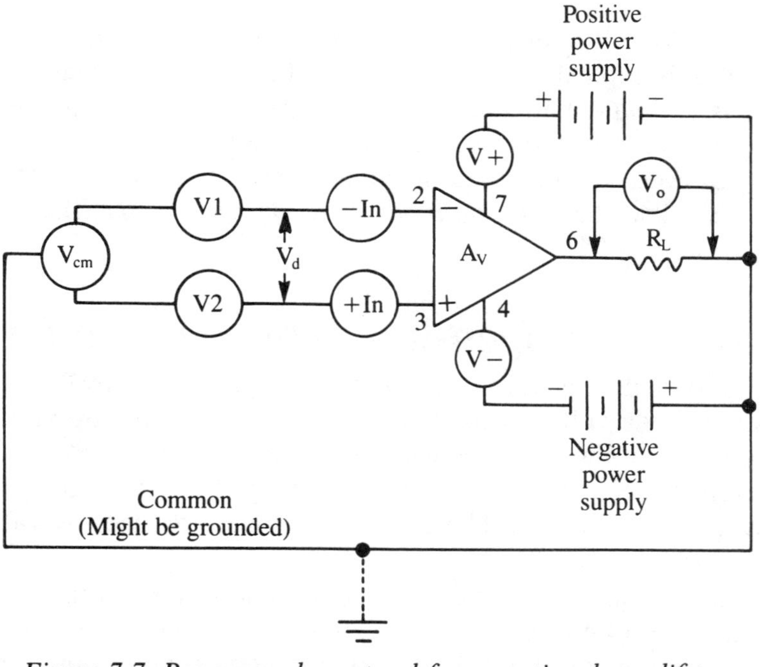

Figure 7-7. Power-supply protocol for operational amplifiers.

The ground symbol shown in Fig. 7-7 indicates (by dotted line) that it is optional. The point of reference for all dc- and signal-voltage measurements is the common connection between the two dc power supplies (V− and V+). Whether or not this point is physically connected to either a ground, the equipment chassis, or a dedicated system ground bus is purely optional. However, this determination is made on the basis of circuit factors other than the basic nature of the op amp.

The V− and V+ dc power supplies are independent. Do not assume that these terminals are merely different ends of the same dc power supply. In fact, V− is negative with respect to common and V− is positive with respect to common.

The two input signals in Fig. 7-7 are labelled V_1 and V_2. Signal voltage V_1 is the single-ended potential between common and the inverting input (−In), and V_2 is the single-ended potential between common and the non-inverting input (+In). The 741 operational amplifier has differential inputs, as is indicated by the fact that both −In and +In are present. Any differential amplifier produces an output that is proportional to the difference between the two input potentials. In Fig. 7-7, the *differential input potential* V_d is the difference between V_1 and V_2:

$$V_d = V_2 - V_1 \tag{7-4}$$

Signal voltage V_{cm} in Fig. 7-7 is the *common-mode signal*, that is, a potential that is applied to both −In and +In inputs simultaneously. This potential is equivalent to the situation $V_1 = V_2$. In an ideal operational amplifier, a common-mode signal will not cause any response. In real devices, however, V_{cm} causes a small response. The freedom from such responses is called the *common-mode rejection ratio* (*CMRR*).

Caution: high open-loop voltage gain (A_{vol})

The flexibility of the operational amplifier is largely caused by the extremely high open-loop dc voltage gain of the device. The open-loop voltage gain (A_{vol}) is the gain of the amplifier without feedback. If the feedback network (shown earlier in Fig. 7-3) had been interrupted at point X, the gain of the circuit would become A_{vol}. The negative feedback reduces overall circuit gain to less than A_{vol}.

The open-loop voltage gain of operational amplifiers is always very high. Some audio amplifier devices that are intended for consumer electronics applications offer A_{vol} of 20,000, and certain premium operational amplifiers offer gains of 1,000,000 and more. Depending on the specific device surveyed, the 741 op amp typically exhibits A_{vol} values in the 200,000 to 300,000 range.

Because of the high A_{vol} values, very small differential-input signal voltages will cause the output to saturate. On the 741 device, the value of the maximum permissible output voltage ($\pm V_{sat}$) is typically about 1 volt (or a little less) below the power-supply potential of the same polarity (certain BiMOS devices, such as the CA-3130 or CA-3140, operate within a few tenths of a volt of the supply voltage). For ±15-Vdc supplies, the maximum 741 output potential is ±14 Vdc. Consider the maximum input potential that will not saturate the op amp at four popular values of power supply potential, assuming the "1-volt-less" rule.

$$V_{inp(max)} = \frac{\pm V_{sat}}{A_{vol}} \tag{7-5}$$

The results for several different cases follow.

$A_{vol} = 300{,}000$		
Power supply	$\pm V_{sat}$	$\pm V_{in(max)}$
±6 Vdc	±5 Vdc	17 μV
±10 Vdc	±9 Vdc	30 μV
±12 Vdc	±11 Vdc	37 μV
±15 Vdc	±14 Vdc	47 μV

One of the consequences of high A_{vol} is that op amps usually saturate at either the V− or V+ power-supply rails when the input lines are either shorted to common or floating. This phenomena is caused by tiny imbalances in the input bias conditions (internal to the device), which are random in nature. Accordingly, you might expect half of a group of op amps to saturate at $+V_{sat}$, and half to saturate at $-V_{sat}$. This situation is probably true for a very large number of devices that are procured from various manufacturers. However, a collection of, say, 100 devices that are purchased at the same time from the same manufacturer will show a marked tendency toward either $-V_{sat}$ or $+V_{sat}$, not the expected random distribution.

The input bias current imbalances tend to be design and process related, so are generally uniform within a given lot from the same source. For example, in a lot of 20 741 devices of the same brand, 19 flipped to $+V_{sat}$ and only one flipped to $-V_{sat}$ at turn-on. According to some authorities, we should expect 10 to fall into each group, but that's not what usually happens.

The behavior of operational amplifiers in the open-loop configuration leads to one category of applications that takes advantage of the very high values of A_{vol}: *voltage comparators*. The first basic op-amp circuit configuration in chapter 8 is called the *inverting follower*.

Summary

The operational amplifier (op amp), originally designed to perform arithmetic operations, is the basis of many electronic circuits. It is basically a differential amplifier, with properties such as a very high open-loop gain, very high input impedance, and very low output impedance. Most op amps have positive and negative power supply inputs, inverting and noninverting signal inputs, and an output lead. Op amps are also available with specific characteristics, such as low noise, single supply, and wide bandwidth.

Questions

1. What is an op amp?
2. List four special-purpose op amps.
3. What are the properties of an ideal op amp?
4. What is the gain bandwidth product?
5. What is an inverting amplifier? a noninverting amplifier?
6. An op amp has supply voltages of ±9 volts. What is the maximum output normally available?

8 INVERTING FOLLOWER CIRCUITS

AFTER COMPLETING THIS chapter, you will be able to:

- ☐ Discuss the basic op amp inverter
- ☐ Explain the inverting summing amplifier
- ☐ Understand what a clipping circuit is

Two basic configurations exist for operational amplifier voltage amplifier circuits: *inverting* and *noninverting*. For some reason that is lost in antiquity, these circuits are usually called *followers*. This chapter examines the basic inverting follower circuit and chapter 9 covers the noninverting follower circuit.

The *inverting follower* is an operational amplifier circuit configuration in which the output signal is 180 degrees out of phase, with respect to the input signal. Figure 8-1A is a cathode-ray oscilloscope (CRO) presentation that shows the relationship between input and output signals for an inverting follower with a gain of approximately −2. Notice the phase reversal present in the output signal, with respect to the input signal. In order to achieve this inversion, the inverting input (−In) of the operational amplifier is active and the noninverting input (+In) is grounded.

Figure 8-1B shows the basic configuration for the inverting-follower (also called *inverting amplifier*) circuits. The noninverting input is not used, so it is grounded. Two resistors are in this circuit: resistor R_f is the negative feedback path from the output to the inverting input and R_{in} is the input resistor. The R_f/R_{in} relationship will be examined in order to determine how the gain is fixed in this type of circuit. But first, let's look at the implications of grounding the noninverting input in this type of circuit.

A

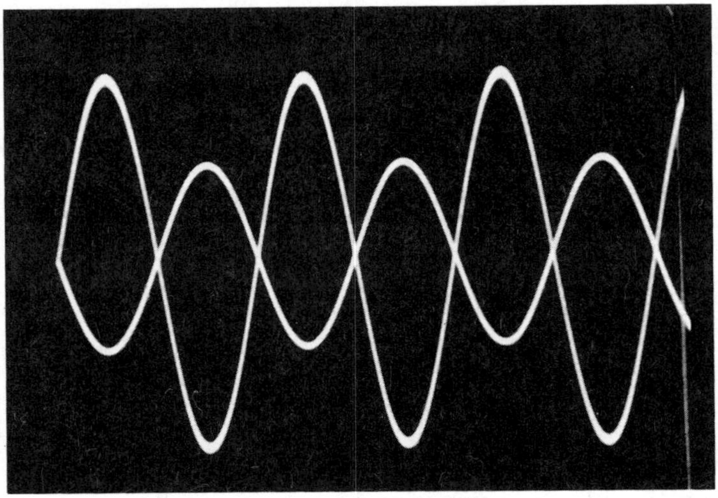

B

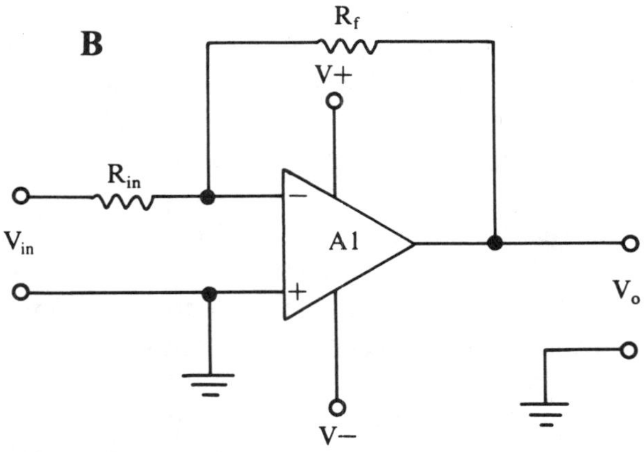

Figure 8-1. A) Input and output waveforms for a gain-of-2 inverting-follower op amp, B) inverting-follower circuit.

What's a virtual ground?

A *virtual ground* is a connection or circuit point that acts like a ground, even though it is not physically connected to either a truly grounded point or to the circuit common point. Although this definition sounds strange at first, it's not an unreasonable description. Unfortunately, this terminology is confusing and therefore leads to an erroneous implication that the virtual ground somehow doesn't really function as a ground.

Chapter 7 features the properties of the ideal operational amplifier. One of those properties states that differential inputs "stick together." Put another way, this property means that a voltage that is applied to one input appears on the other input also.

In the arithmetic of op amps, therefore, both inputs must be treated as if they are both at the same potential. This fact is not merely a theoretical device. For if you actually apply a potential, say 1 Vdc, to the noninverting input, the same 1-Vdc potential can also be measured at the inverting input. Experiment 7-1 addressed this fact.

In Fig. 8-1B, the noninverting input is grounded, so it is at 0-Vdc potential. This, by the properties of the ideal op amp, means that the inverting input of the op amp is also at the same 0-Vdc ground potential. Because the inverting input is at ground potential, but has no physical ground connection, it is said to be at virtual (as opposed to physical) ground. A virtual ground is, therefore, a point that is fixed at ground potential (0 Vdc), even though it is not physically connected to the actual ground or to the common of the circuit. The choice of the term *virtual ground* is unfortunate. The concept is actually quite simple even though the terminology makes it sound a lot more abstract than it really is.

Developing the gain setting equation for the inverting follower circuit

The transfer equation of any circuit is the output function divided by the input function. For an operational amplifier used as a voltage amplifier, the transfer function describes the voltage gain:

$$A_v = \frac{V_o}{V_{in}} \tag{8-1}$$

Where:

A_v is the voltage gain (dimensionless)
V_o is the output signal potential
V_{in} is the input signal potential (V_o and V_{in} are in the same units)

In the inverting follower circuit (Fig. 8-2A), the gain is set by the ratio of two resistors, R_f and R_{in}. Consider the currents flowing in the circuit of Fig. 8-2A. The input bias currents, I_3 and I_4, are assumed to be zero for purposes of analysis. This assumption is reasonable because our model is an ideal operational amplifier. In a real op amp, these currents are not zero and must be accounted for. But in the analysis, we use the ideal model. Thus, in the following analysis, you can ignore the bias currents (assume that $I_3 = I_4 = 0$).

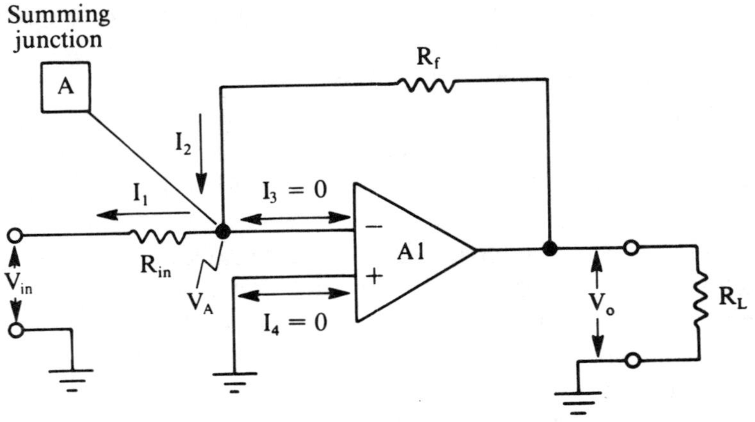

A

Relationships:

$$I_1 + I_2 + I_3 + I_4 = 0$$
If $I_3 + I_4 = 0$, then
$$I_1 + I_2 = 0$$
Kirchoff's current law

$$I_1 = V_{in}/R_{in}$$
$$I_2 = V_o/R_f$$
Ohm's Law

$V_A = 0$ Seventh op-amp property (+In is grounded)

$$A_V = \frac{V_o}{V_{in}} = \frac{A_{vol}}{1 + A_{vol}\beta}$$

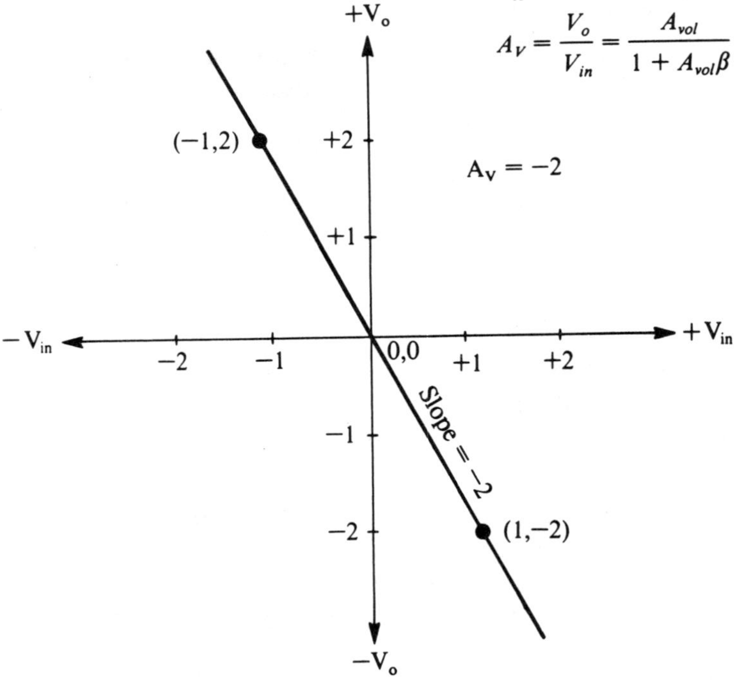

B

Figure 8-2. A) Inverting amplifier that shows currents and voltages, B) transfer function.

Remember that the *summing junction* (point A) is a virtual ground and is at ground potential because the noninverting input is grounded. Current I_1 is a function of applied input voltage V_{in} and input resistance R_{in}. By Ohm's law, then, the value of I_1 is:

$$I_1 = \frac{V_{in}}{R_{in}} \tag{8-2}$$

Further, current I_2 is also related by Ohm's law to output voltage V_o and to feedback resistor R_f (again, because the summing junction is at 0 Vdc):

$$I_2 = \frac{V_o}{R_f} \tag{8-3}$$

How are I_1 and I_2 related? These two currents are the only currents that enter or leave the summing junction (recall that $I_3 = 0$), so by Kirchhoff's Current Law (KCL):

$$I_1 + I_2 = 0 \tag{8-4}$$

so,

$$I_2 = -I_1 \tag{8-5}$$

We can arrive at the gain equation by substituting Eqs. (8-2) and (8-3) into Eq. (8-5):

$$I_2 = -I_1$$

$$\frac{V_o}{R_f} = \frac{-V_o}{R_{in}} \tag{8-6}$$

Algebraically rearranging Eq. (8-6) yields the standard gain equation for inverting-follower op-amp circuits, in standard format:

$$\frac{V_o}{V_{in}} = \frac{-R_f}{R_{in}} \tag{8-7}$$

According to Eq. (8-1), gain (A_v) of the circuit is V_o/V_{in}, so you can also write Eq. (8-7) in the form:

$$A_v = \frac{-R_f}{R_{in}} \tag{8-8}$$

The previous equations show that the voltage gain of an op-amp inverting follower is merely the ratio of the feedback resistance to the input resistance ($-R_f/R_{in}$). The minus sign indicates that a 180-degree phase reversal occurs. Thus, a negative input voltage produces a positive output voltage, and vice versa.

The transfer equation [Eq. (8-14)] is often written to express output voltage in terms of gain and input-signal voltage. The two expressions are:

$$V_o = -A_v V_{in}$$

and,

$$V_o = -V_{in}\left(\frac{R_f}{R_1}\right) \tag{8-9}$$

The transfer function ($A_v = V_o/V_{in}$) can be plotted on graph paper in terms of input and output voltage. Figure 8-2B shows the plot (V_o-vs.-V_{in}) for an inverting amplifier with a gain of -2. In the case of a perfect amplifier, the Y-intercept is 0 volts. Given the nature of Fig. 8-2B, the basic form becomes $V_o = A_v V_{in} \pm V_{offset}$.

Experiment 8-1

In this experiment, an inverting follower op-amp circuit is constructed and tested. For part A of the experiment, the input signal (V_{in}) is a dc voltage from a 1.5-volt dry-cell battery (any size from AAA on up will work). The output voltage (V_o) is found at point C. Point B is the op amp's summing junction. In part B of the experiment, a potentiometer network provides the dc input signal to the op amp.

Part A

1. Connect the circuit of Experiment 8-1 using the battery and double-pole double-throw (DPDT) switch (ignore the potentiometer circuit for now).
2. Turn on the circuit and measure the voltage at point A.
3. Now measure the voltage at point C.

4. Compare the two readings and reconcile them with Eq. (8-10).
5. Set switch S1 to the opposite setting and repeat steps 1-4.
6. Did you notice anything about the polarity of output voltage V_o at point C? Its polarity should have reversed, but the magnitude should have remained the same.
7. Measure the voltage at point B. It should be zero because of what op-amp property?

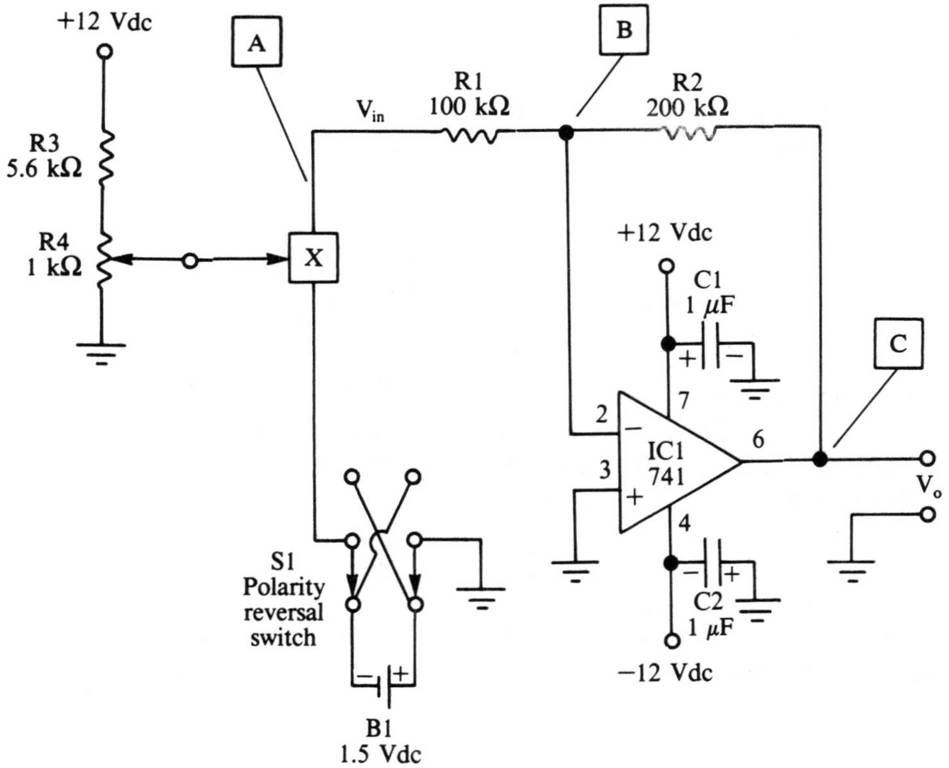

Experiment 8-1. Circuit for Experiment 8-1.

Part B

8. Disconnect the switch and battery at point X and replace it with the potentiometer circuit shown in the inset to Experiment 801.
9. Set the voltage at point A to various potentials by varying potentiometer R4. Repeat steps 1 through 4 at several discrete potentials.
10. Calculate voltage gain A_v from the data that you collected, then calculate the gain from the feedback resistors used. How do they compare? Explain any discrepancy between them.

Multiple input inverting follower circuits

We can accommodate multiple signal inputs on an inverting follower by using a circuit such as Fig. 8-3. A number of applications exist for such

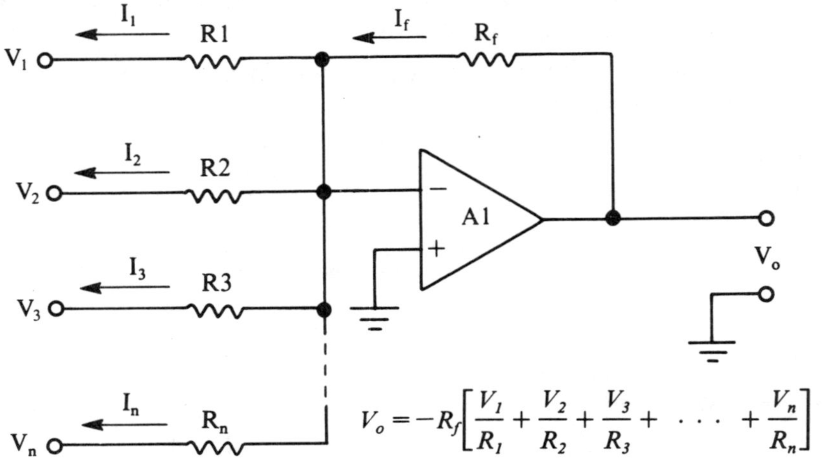

$$V_o = -R_f\left[\frac{V_1}{R_1} + \frac{V_2}{R_2} + \frac{V_3}{R_3} + \cdots + \frac{V_n}{R_n}\right]$$

Figure 8-3. Multi-input inverting-follower op amp.

circuits: summers, audio mixers, instrumentations, and so forth. The multiple input inverter of Fig. 8-3 can be evaluated exactly like Fig. 8-2B, except that you must account for more than one input. Again appealing to KCL, you know that:

$$I_1 + I_2 + I_3 + \ldots + I_n = I_f \tag{8-9}$$

Also by Ohm's law, considering that summing junction A is virtually grounded, you know that:

$$I_1 = \frac{V_1}{R_1} \tag{8-10}$$

$$I_2 = \frac{V_2}{R_2} \tag{8-11}$$

$$I_3 = \frac{V_3}{R_3} \tag{8-12}$$

$$I_n = \frac{V_n}{R_n} \tag{8-13}$$

$$I_f = \frac{V_o}{R_f} \tag{8-14}$$

Substituting Eqs. (8-10) through (8-14) into Eq. (8-9):

$$-\left(\frac{V_1}{R_1} + \frac{V_2}{R_2} + \frac{V_3}{R_3} + \ldots + \frac{V_n}{R_n}\right) = \frac{V_o}{R_f} \tag{8-15}$$

or, algebraically re-arranging Eq. (8-36) to solve for V_o:

$$V_o = -R_f\left(\frac{V_1}{R_1} + \frac{V_2}{R_2} + \frac{V_3}{R_3} + \ldots + \frac{V_n}{R_n}\right) \tag{8-16}$$

Equation (8-16) is the transfer equation for the multiple-input inverting follower.

Experiment 8-2

This experiment extends Experiment 8-1 to include more than one input. Because the circuit includes two independent input signal sources (V_1 and V_2) that are found at points A and B, output signal V_o at point C will be the algebraic sum of the two input signals multiplied by the voltage gain.

Part A

1. Connect the circuit of Experiment 8-2. Make sure that batteries B1 and B2 are connected to the opposite polarity, with respect to each other (e.g., ground the negative end of B1 and the positive end of B2).
2. Measure and record the voltages at points A and B (labelled V_1 and V_2, respectively).
3. Measure and record the output voltage at point C.
4. Calculate the predicted output voltage [Eq. (8-16)] and compare the result with the result of this experiment.

Part B

5. Reverse the polarity of either B1 or B2, but not both.
6. Repeat part A (steps 1 through 4).

Part C

7. Now reverse the other battery (not reversed in part B), and repeat part A.

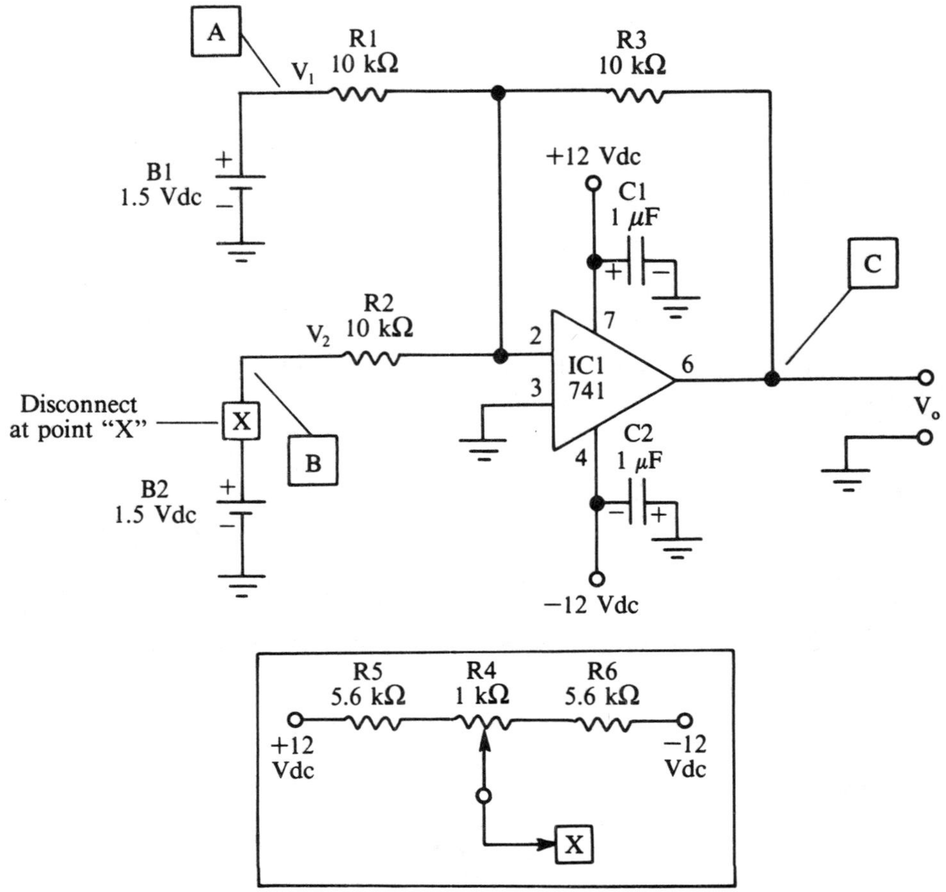

Experiment 8-2. Circuit for Experiment 8-2.

Part D

8. Replace battery B2 with the potentiometer circuit shown in the inset. Break the original circuit at point X and connect the wiper output arm of the potentiometer.

Note: The output of the potentiometer circuit now replaces V_2 in the circuit and can provide both positive- and negative-polarity input signals.

9. Make sure that battery B1 (i.e., voltage V_1) is positive (i.e., the negative terminal of the battery is grounded).
10. Set V_1 to a positive voltage by adjusting R4 while monitoring point B in the circuit.
11. Measure the voltage at point A.

12. Calculate the correct output voltage.
13. Measure the output voltage at point C and compare the readings with the predicted results.

Note: Now you might want to select another value for R3 and re-do this experiment. A 22-kohm resistor for R3 would make the circuit gain −2.2, for example. Alternatively, use a 20-kohm potentiometer in place of R3. Make a cardboard dial and calibrate it in kohms with an ohmmeter. Experiment with the values of R3—from 0 to 20 kohms. Can you predict what will happen when $R_3 < R_1$ or R_2? Trust the equations!

This experiment used dc sources for the op-amp input signal. The next experiment features the same circuit, but with ac signal sources.

Experiment 8-3

In this experiment, a multiple-input op-amp circuit is connected to ac sources. The experiment requires an oscilloscope and an audio-frequency generator. If you do not have an audio oscillator, then you might want to jump ahead to Section III of this book for a project to build. The experiment requires two signal sources, but these can be simulated either by using a single generator with two outputs (e.g., sine wave and square wave) or by using two circuits, such as those shown in the inset of Experiment 8-3A.

1. Connect the circuit of Experiment 8-3. If you use the inset circuit to provide multiple sources, set each potentiometer to full-scale (maximum output voltage).
2. Adjust the signal-generator output level so that V_1 and V_2 (at points A and B, respectively) are each 1-volt peak-to-peak (p-p).
3. Examine the output signal at point C. What do you see? Figure Experiment 8-3B is an example.
4. Change either input level and compare the results.

Project 8-1
Audio microphone/source mixer

An *audio mixer* is a device that linearly combines two or more audio signal sources into one channel. This particular project shows two inputs, but more can be accommodated. These devices are used to provide multiple inputs to public address (PA) and high-fidelity (hi-fi) audio systems. For example, the PA system might take inputs from two different microphones, or a microphone and a tape player. The circuit of Project 8-1 combines these inputs into a single channel for input to the power amplifier.

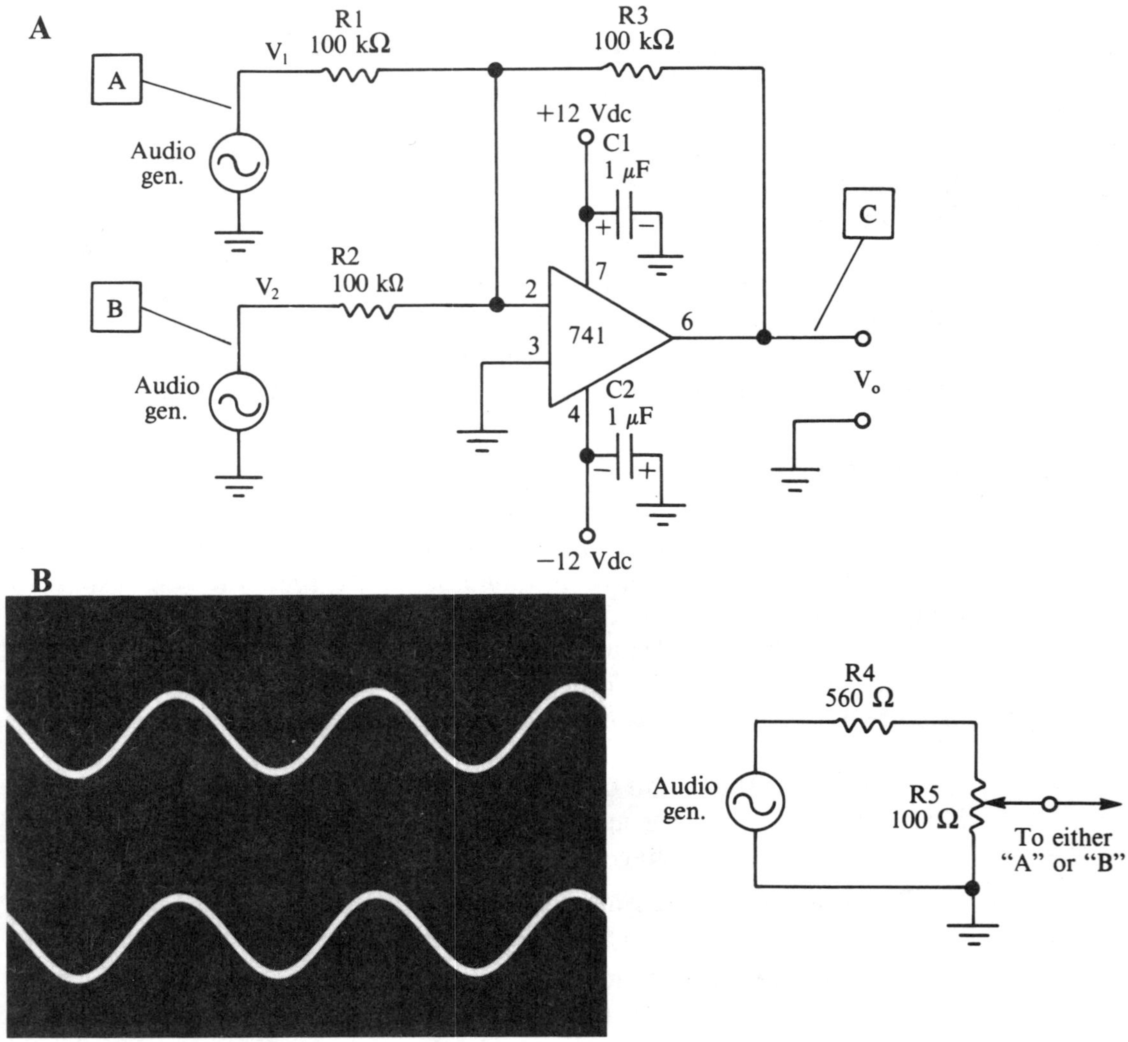

Experiment 8-3. A) Circuit for Experiment 8-3, B) waveforms.

The circuit of Project 8-1 is a *high-level mixer* because it assumes that the signal levels are ready to input to the power amplifier; i.e., they are high-level signals (which are usually 100 mV to 1 volt or so). The gain of this mixer circuit is approximately one. If you wish to use low-level signal sources, increase the gain by increasing the value of R3, R4, or R7 as is appropriate.

Additional inputs can be added by repeating the input circuitry (IC1A and IC1B). Using CA-3240 devices allows two inputs to be provided per added eight-pin miniDIP chip.

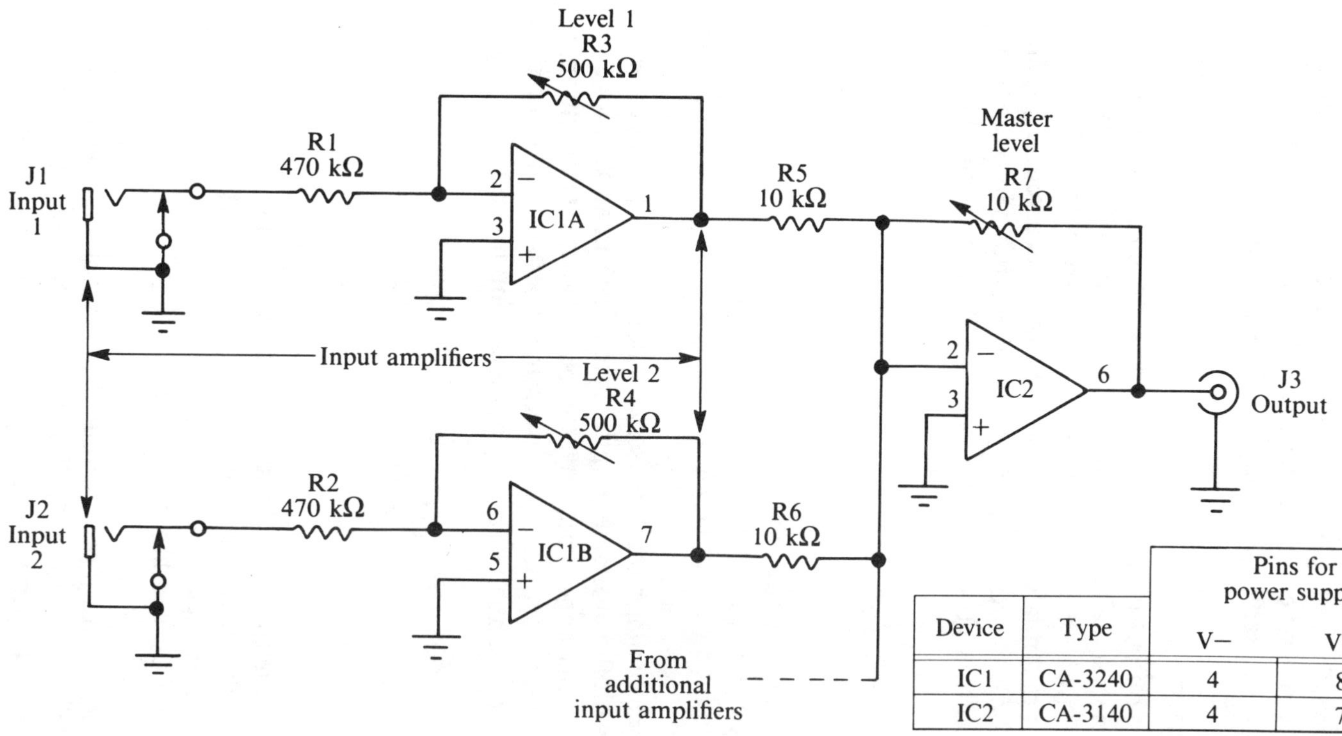

Device	Type	Pins for power supply V−	V+
IC1	CA-3240	4	8
IC2	CA-3140	4	7

V− = −12 Vdc
V+ = +12 Vdc

Note: Bypass V− and V+ on each IC with 0.1 μF tantalum capacitors

Project 8-1. Circuit for Project 8-1.

The input jacks shown here are closed-circuit phone jacks, which can be either the ¼-inch or the 3⁄16-inch standard, depending on your application. If in doubt, use the ¼-inch size, then provide external adapters for 3⁄16-inch source devices. The reason for the use of closed-circuit devices is to ground the input when it is not in use. This tactic results in less noise or hum because it disables unused input lines.

Output current

The op-amp output current (I_o) must be supplied by the output terminal. Typically small-signal op amps supply 5 to 25 mA of current, depending on the device. Power op amps, such as the Burr-Brown OPA-511, supply up to 5 amperes at potentials of ±30 volts. The output current (I_o) splits into two paths (Fig. 8-4); a portion of the output current flows into the feedback path (I_f) and a portion flows into the load (I_L). The total current is:

$$I_o = I_f + I_L \tag{8-17}$$

Where:

I_o is the output current
I_f is the feedback current (V_o/R_f)
I_L is the load current (V_o/R_L)

In normal voltage amplifier service, both I_f and I_L tend to be very small compared with the available output current. But in applications where load and feedback resistances are low, the output currents might approach the maximum specified value. To determine whether this limit is exceeded, divide output potential V_o by the parallel combination of R_f and R_L, or:

$$\frac{V_{o(max)}\,(R_f + R_L)}{R_f RL} < I_{o(max)} \tag{8-18}$$

Where:

$V_{o(max)}$ is the maximum expected output voltage
$I_{o(max)}$ is the maximum allowable output current
R_f is the feedback resistance (in *ohms*)
R_L is the load resistance (in *ohms*)

In general, the output current limit is not approached on ordinary devices unless load and/or feedback resistances are less than 1000 ohms. Power devices, of course, can drive a lower load and/or feedback-resistance combination.

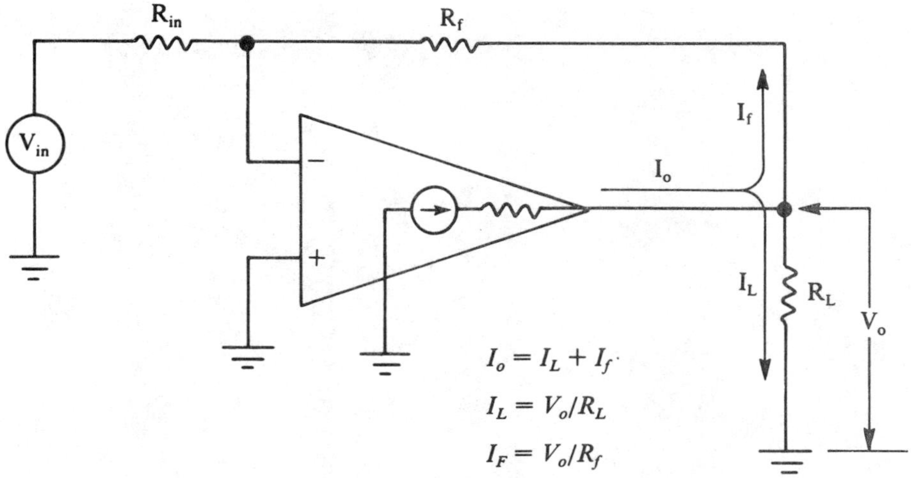

Figure 8-4. Inverting amplifier with signals.

Response to ac signals

Thus far this section on inverting amplifiers has assumed a dc input-signal voltage. The behavior of the circuit in response to ac signals (e.g., sine waves, square waves, and triangle waves) is similar. Recall the rules for the inverter: positive input signals produce negative output signals and negative input signals produce positive output signals. These relationships mean that a 180-degree phase shift occurs between input and output. The relationship is shown in Fig. 8-5A.

Although the dc-coupled op amp will respond to ac signals, a limit must be recognized. If the peak value of the input signal becomes too great, then output clipping (Fig. 8-5B) will occur. The peak output voltage will be:

$$V_{o(peak)} + -A_v V_{in(peak)} \tag{8-19}$$

Where:

$V_{o(peak)}$ is the peak output voltage
$V_{in(peak}$ is the peak input voltage
A_v is the voltage gain

For every value of V− and V+ power supply potentials, a maximum attainable output voltage, $V_{o(max)}$ exists. As long as the peak voltage is less than this maximum allowable output potential, the input waveform will be faithfully reproduced in the output (except amplified and inverted). However, if the value of $V_{o(peak)}$ that is determined by Eq. (8-19) is greater than $V_{o(max)}$, then clipping will occur.

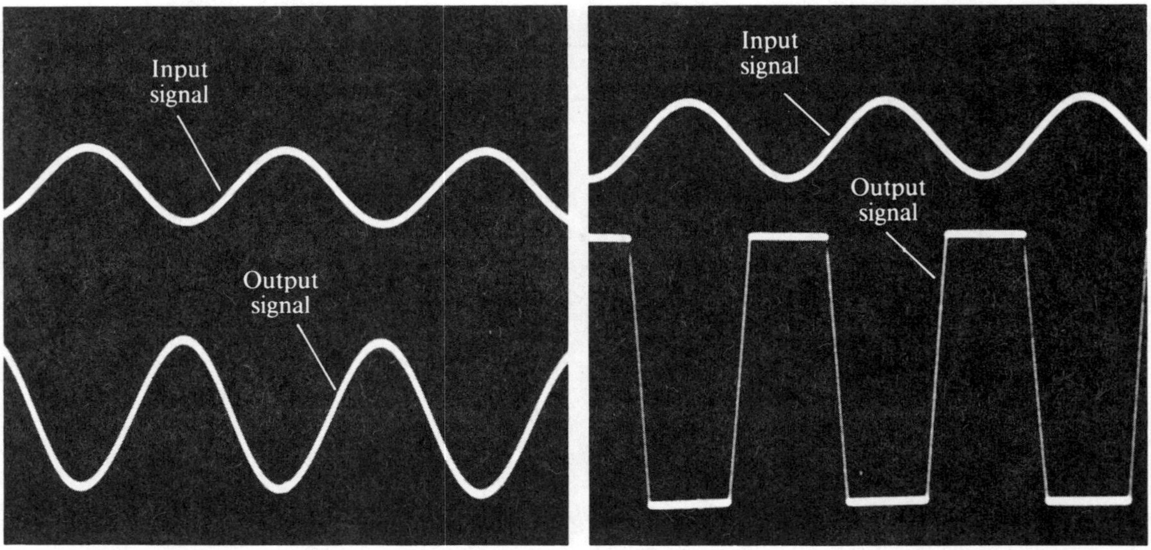

Figure 8-5. A) Inverting-follower input and output signals, B) overdriven amplifier input and output signals.

In a linear voltage amplifier, clipping is undesirable. The maximum output voltage can be used to calculate the maximum input signal voltage:

$$V_{in(max)} + \frac{V_{o(max)}}{A_v} \tag{8-20}$$

Sometimes clipping is desired. For example, in radio transmitter circuits, called *modulation limiters*, simple clippers are often followed by an audio low-pass filter that removes the harmonic distortion that is created by clipping. Another case where clipping is desired is when generating square waves from sine waves. The goal in this case is to drive the input so hard that sharp clipping occurs. Although better ways exist to realize this goal, the overdriven-clipper square wave generator does work.

Experiment 8-4

This circuit illustrates the effects of overdriving an operational amplifier inverting follower circuit. You will need an oscilloscope and an audio frequency sine wave generator for this experiment. If you lack a sine wave generator, then skip ahead to chapter 20 for a possible project. Experiment 8-4 shows an inverting follower that has a gain of -10.

1. Connect the circuit of Experiment 8-4.
2. Adjust the output of the audio generator for a frequency of about 400 Hz and a level of 10 volts p-p.
3. Examine the output voltage on the oscilloscope.
4. Now adjust the output of the signal generator for 20 volts p-p and re-examine the output signal. Notice any clipping?
5. Now back off on the signal level until the clipping completely disappears. Re-examine the output signal.

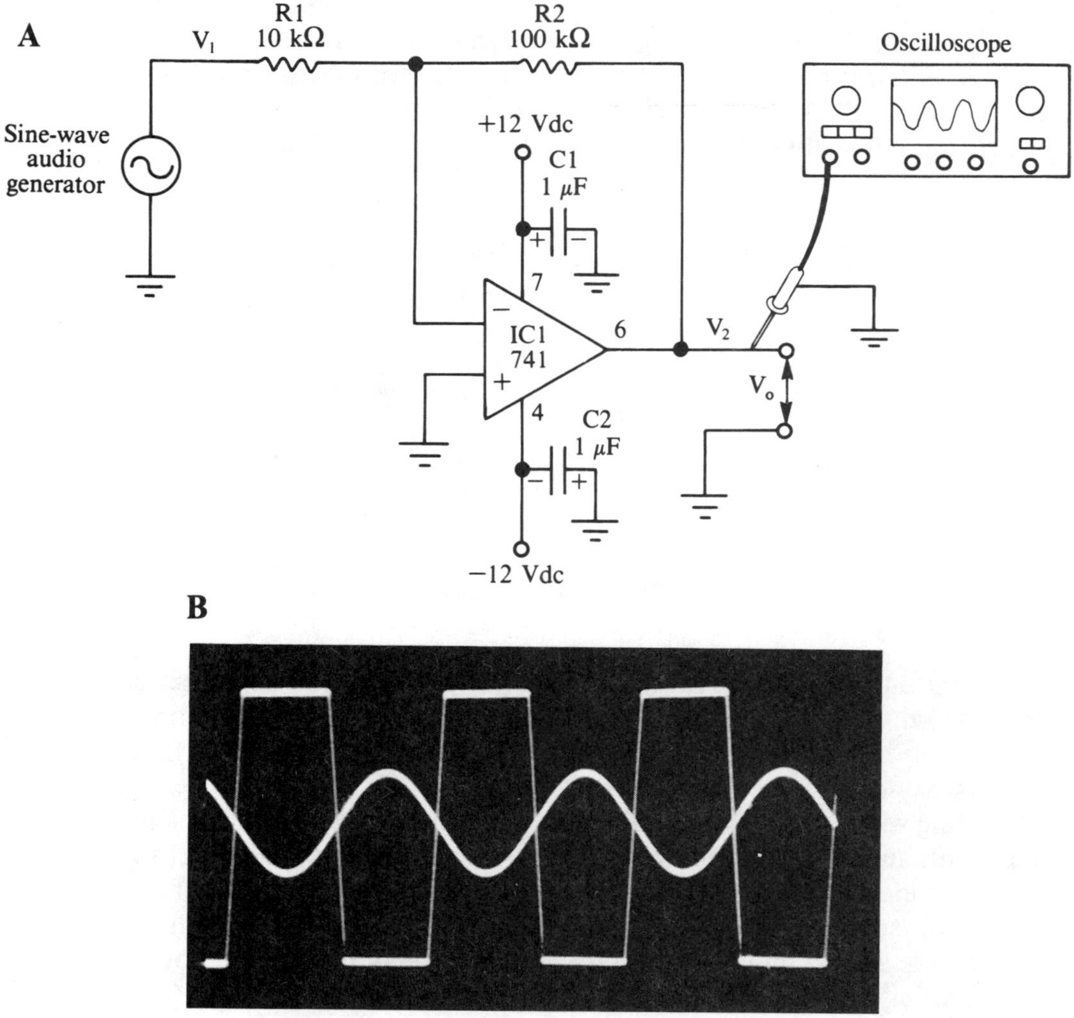

Experiment 8-4. A) Circuit for Experiment 8-4, B) waveforms.

Note: Experiment 8-4B shows an example of a clipped waveform that was taken from this experiment.

If your audio signal generator will not produce the required output levels, then increase the gain of the circuit to −100: make R2 = 1 MΩ and attempt the experiment at lower-signal levels.

Response to ac input signals with dc offset

The case in the previous section assumed a waveform that is symmetrical about the zero-volts baseline. This section features a case where an ac waveform is superimposed on a dc voltage. Figure 8-6 shows an inverting amplifier circuit that has an ac signal source in series with a dc source. In Fig. 8-7A, a 4-volt peak-to-peak square wave is superimposed on a 1-Vdc fixed potential. Thus, because the ac and dc output signals are added together, the nonsymmetrical signal will swing between +3 and −1 volts.

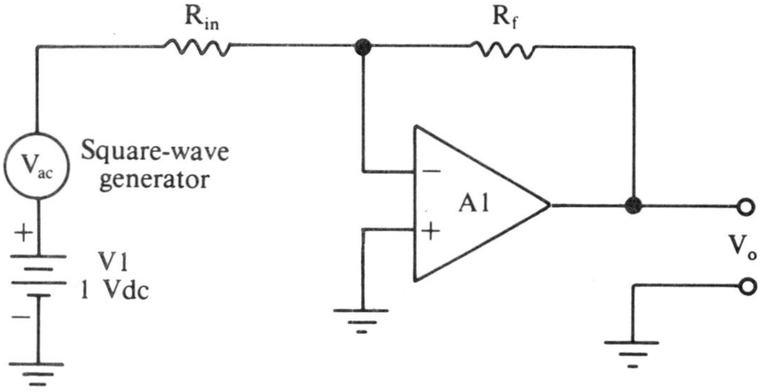

Figure 8-6. Inverting amplifier that has an ac signal superimposed on a dc component.

The output waveform is shown in Fig. 8-7B. With the 180-degree phase inversion and the gain of −2 (Fig. 8-6), the waveform will be a nonsymmetrical oscillation between −6 and +2 volts. Because of gain ($A_v = -2$), the degree of asymmetry also doubles to 2 Vdc.

Dealing with ac signals that have a dc component can lead to problems at high gain and/or high input-signal levels. As was true in the case of the high amplitude symmetrical signal, the output might saturate at either V− or V+ power-supply rails. If this limit is reached, clipping will occur. The dc component is a valid input signal, so it will drive the output to one power-supply limit or the other. For example, if the op amp in Fig. 8-6 has a $V_{o(max)}$ value of ±10 volts, then (with $A_v + -2$) a +4-volt positive input signal will saturate the output and the negative excursion can each −7 volts before causing output saturation.

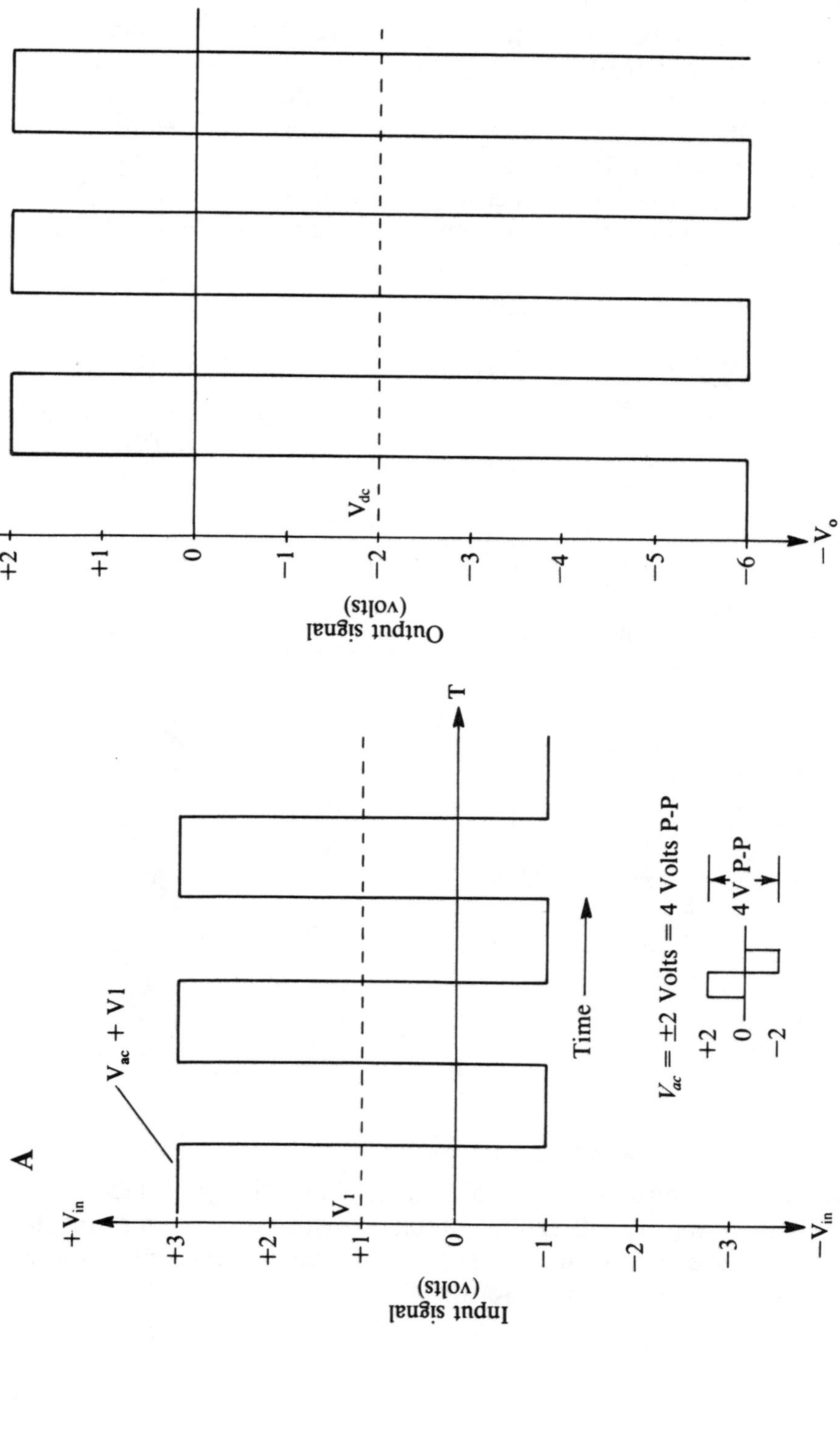

Figure 8-7. A) Square-wave input signal with a 1-Vdc signal, B) output signal.

Experiment 8-5

Perform Experiment 8-4 again, but use the circuit of Experiment 8-5 as the signal source, in lieu of the simple signal generator used previously. This signal source provides a ±1.5-Vdc offset to the input signal. Be sure to mark the zero-volts baseline on the screen of the oscilloscope before proceeding with this experiment. If you have a dual-beam (two-channel) oscilloscope, then use input 1 for the experiment and set input 2 to GND. Adjust position 2 to the zero point of channel 1 and use it as a zero-volts reference throughout the experiment.

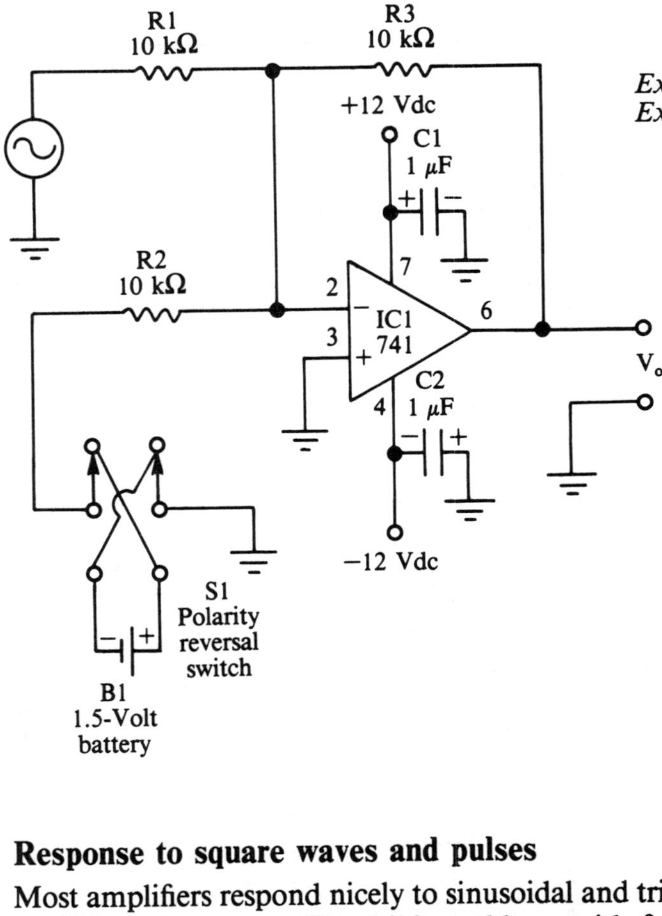

Experiment 8-5. Circuit for Experiment 8-5.

Response to square waves and pulses

Most amplifiers respond nicely to sinusoidal and triangle waveforms. Some amplifiers, however, will exhibit problems with fast rise-time waveforms, such as square waves and pulses. The source of these problems is the high-frequency content of these waveforms.

All continuous mathematical functions (including electronic signal waveforms) are made of a series of harmonically related sine and cosine constituent waves (and possibly also a dc component). The sine wave consists of a single frequency or fundamental sinusoidal wave.

All nonsinusoidal waveforms, however, consist of a fundamental sinewave and its harmonics. The actual waveshape is determined by the number of harmonics present, which particular harmonics are present (i.e., odd or even), the relative amplitudes of those harmonics, and their phase relationship (with respect to the fundamental). These factors can be deduced from the quarter-wave and/or half-wave symmetry of the wave.

The listing of the constituent frequencies (and their amplitudes) forms a *Fourier series*, which determines the bandwidth of the system that is required to process the signal. For example, the symmetrical square wave consists of a fundamental-frequency sine wave (F), plus odd harmonics ($3F$, $5F$, $7F$ . . .) up to (theoretically) infinity (as a practical matter, most square waves are "square" if the first 100 harmonics are present). Furthermore, if the square wave is truly symmetrical, then all of the harmonics are in-phase with the fundamental. Other waveshapes have different Fourier spectrums.

In general, the rise time of a pulse is related to the highest significant frequency in the Fourier spectrum by the approximation:

$$F = \frac{0.35}{T_r} \qquad (8\text{-}21)$$

Where:

F is the highest Fourier frequency (in *hertz*)
T_r is the pulse rise time (in *seconds*)

Because pulse shape is a function of the Fourier spectrum for that wave, the frequency-response characteristic of the amplifier has an effect on the waveshape of the reproduced signal. Figure 8-8 shows an input pulse signal (Fig. 8-8A) and two possible responses. The response shown in Fig. 8-8B is caused by high-frequency attenuation. The rounding shown will be either moderate or severe depending on the bandwidth of the amplifier. In other words, rounding is determined by how many harmonics are attenuated by the amplifier frequency-response characteristic, and to what degree. This problem becomes especially severe when the fundamental frequency (or pulse-repetition rate) is high, the rise time is very fast, and the amplifier bandwidth is low.

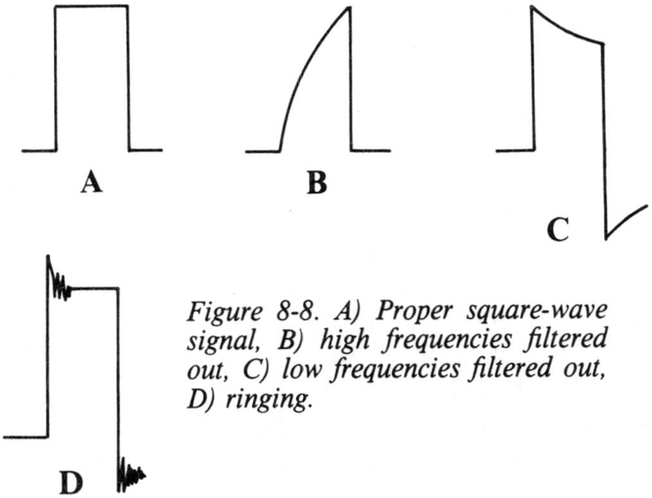

Figure 8-8. A) Proper square-wave signal, B) high frequencies filtered out, C) low frequencies filtered out, D) ringing.

Frequency-compensated operational amplifiers achieve their claimed "unconditional stability" by drastically rolling off the high-frequency response above a few kilohertz. A type 741AE op amp is a frequency-compensated device with a gain-bandwidth product of 1.25 MHz and an open-loop gain of 250,000. Frequency response at maximum gain is 1.25 MHz/250,000 or 5 kHz. Thus, you can expect good square-wave response only at relatively low frequencies. A rule for square waves is to make the amplifier bandwidth at least 100 times the fundamental frequency. As with all such rules, however, this one should be applied with caution, even though it is assimilated into your collection of "standard electronics wisdom."

The other class of problems, *peaking* and *ringing* of the pulse, is shown in Figs. 8-8C and 8-8D. Three principal causes of these phenomena are found. First, a skewed bandpass characteristic is where either the low frequencies are attenuated (or amplified less) or the high frequencies are amplified more. Second, LC resonances in the circuit give rise to ringing. Although not generally a problem at low frequencies, video operational amplifiers might have this problem. Third, both significant harmonics are present at frequencies where circuit phase shifts add up to 180 degrees and the loop gain is unity or greater. When combined with the 180-degree phase shift that is inherent in inverting followers, Barkhausen's criteria for oscillation is fulfilled. Under some conditions, the device will break into sustained oscillation. In other cases, however, oscillation will occur only on fast rise-time signal peaks, as shown in Fig. 8-8D.

Some basic rules

Consider several factors when designing inverting follower amplifiers. First, you must obviously consider the voltage gain that is required by the application. Second, you must consider the input impedance of the circuit. This specification is needed in order to prevent the amplifier input from loading down the driving circuit. In the case of the inverting follower, the input impedance is the value of the input resistor (R_{in}); a simple design rule is in effect:

> The input resistor (hence the input impedance) should be equal or greater than 10 times the source resistance of the previous (driving) circuit.

The implication of this rule is that you must determine the source resistance of the driving circuit and then make the input impedance of the op-amp inverting follower at lest 10 times larger. When the driving source is another operational amplifier, you can assume that the source impedance (i.e., the output impedance of the driving op amp) is 100 ohms or less (it's actually much less). For these cases, make the value of R_{in} at least 1000 ohms (i.e., 10 × 100 ohms = 1000 ohms). This value is based on consideration of available output load current.

In other cases, however, a slightly different problem exists. Some sensors (for example, a thermistor for measuring temperature) have a much higher source resistance. One thermistor has an advertised resistance that varies from 10 to 100 kohms over the temperature range of interest, so a minimum input impedance of 1 megohm (i.e., 10 × 100 kohms) is required. When the input impedance gets this high, the designer might want to consider the noninverting follower, rather than the inverting follower configuration.

In the inverting follower circuit the choice of input impedance drives the design, so it is part of the design procedure:

1. Determine the minimum allowable input resistance (i.e., 10 or more times the source impedance).
2. If the source resistance is 1000 ohms or less, try 10 kohms as an initial trial input resistance (R_{in}).
3. Resistance R_{in} might be lowered if the feedback resistor (R_f) becomes too high for the required gain. The value of R_{in} is the input resistance or 10 kohm, whichever is higher.

4. Determine the amount of gain that is required. In general, the closed-loop gain of a single inverting follower should be less than 500. For gains higher than that figure, use a multiple op-amp cascade circuit. Some low-cost op amps should not be operated with closed-loop gains greater than 200. The reason for this rule is the problems that are found in real (versus ideal) devices. In those cases, the distributed gain of a cascade amplifier might be easier to tame in practical situations.
5. Determine the frequency response (i.e., the frequency at which the gain drops to unity). From steps 3 and 4, you can calculate the minimum gain bandwidth product of the required op amp.
6. Select the operational amplifier. If the gain is high (e.g., over 100), you might want to select a BiMOS or BiFET operational amplifier to limit the output offset voltage that is caused by the input bias currents. Select a 741-family device if you don't need more than a few kilohertz frequency response and if the unconditionally stable characteristics of the 741 is valuable for the application.
7. Also look at the package style. For most applications, the 8-pin miniDIP package is probably the easiest to handle. The 8-pin metal can is also useful and it can be made to fit 8-pin miniDIP positions by correctly bending the leads.
8. Select the value of the feedback resistor:

$$R_f + ABS(A_v R_{in}) = A_v R_{in} \tag{8-22}$$

9. If the value of the feedback resistor is too high (i.e., beyond the range of standard values—about 20 megohms or so) or too high for the input bias currents, try a lower input resistance.

Altering ac frequency response

The natural bandwidth of an amplifier is sometimes too great for certain specific applications. Noise power, for example, is a function of bandwidth, as indicated by the expression: $P_n = \text{KTBR}$. Thus, it is possible that the signal-to-noise ratio will suffer in some applications if the bandwidth is not limited to that which is actually needed to process the expected waveform. In other cases, the rejection of spurious signals suffers if the bandwidth of an amplifier circuit is not tailored to that which is required by the bandwidth of the applied input signal.

Amplifier stability is improved if the loop gain of the circuit is reduced to less than one at the frequency at which the circuit phase shifts (including internal amplifier phase shift) reach 180 degrees. When the distributed phase shift is added to the 180-degree phase shift, seen normally on inverting amplifiers, Barkhausen's criteria for oscillation is satisfied and the amplifier will oscillate. Those criteria are:

- Total phase shift of 360 degrees at the frequency of oscillation
- Output-to-input coupling (can be accidental)
- Loop gain of unity or greater

If these criteria are satisfied at any frequency, the operational amplifier will oscillate at that frequency. For the present, just one technique is featured in case you need to know the method in performing laboratory exercises.

The design goal when tailoring the ac frequency response is to roll off the voltage gain at the frequencies above a certain critical frequency, F_c. This frequency is determined by evaluating the application, and is defined as the frequency at which the gain of the circuit drops off -3 dB from its in-band voltage gain. The response of the amplifier should look like Fig. 8-9A, which is shown in normalized form where the maximum in-band gain is assumed to be 0 dB. Above the critical frequency, the gain drops -6 dB/octave (an octave is a 2:1 change in frequency) by shunting a capacitor across the feedback resistor (Fig. 8-9B). The reactance of the capacitor is shunted across the resistance of R_f, the gain reduces. The low-pass filter characteristic is achieved because the capacitive reactance lowers as frequency increases. The value of the capacitor is found from:

$$C = \frac{1}{2\pi R_f F_c} \qquad (8\text{-}23)$$

Where:

C is the capacitance (in *farads*)
R_f is the feedback resistance (in *ohms*)
F_c is the -3-dB frequency (in *hertz*)

Alternatively, to calculate the capacitance of C in microfarads (μF), use Eq. (8-24):

$$C_{\mu F} = \frac{1{,}000{,}000}{2\pi R_f F_c} \qquad (8\text{-}24)$$

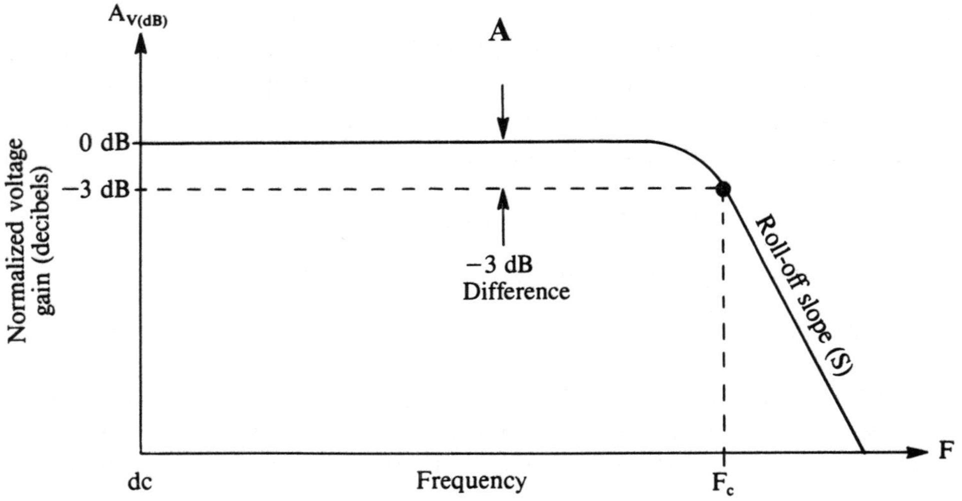

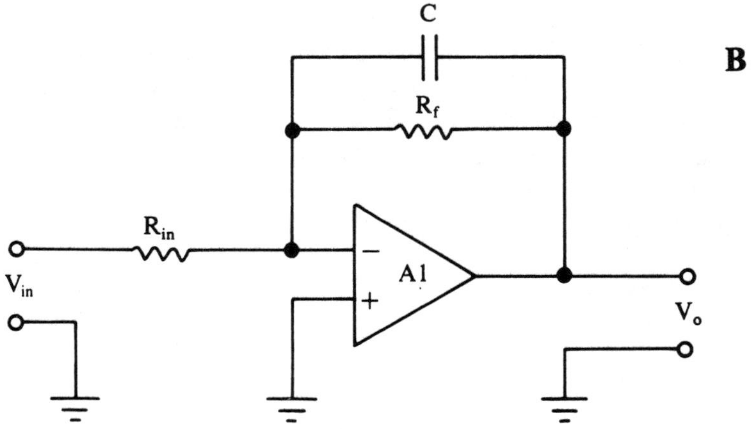

$$C = \frac{1}{2\pi R_f F_c}$$

C in farads
R_f in ohms
F_c in Hertz

or, where *C* is in microfarads (μF):

$$C_{\mu f} = \frac{1{,}000{,}000}{2\pi R_f F_c}$$

Figure 8-9. A) Frequency response of amplifier, B) inverting follower with capacitor for high-frequency roll-off.

Experiment 8-6

This circuit investigates the effect of a feedback capacitor on the frequency response of the op-amp circuit. A square-wave input signal is used because these waveforms are made up of a fundamental frequency (*F*) with large number of harmonics (*3F, 5F, 7F . . .*) added to the fundamental signal. The shape of the square wave depends on how many harmonics are present.

1. Connect the circuit of Experiment 8.6 without the capacitor. This circuit has a gain of −1.
2. Adjust a 400-Hz input square-wave to about 2 volts p-p.
3. Examine the output signal on an oscilloscope.
4. Next, add capacitor c1 in parallel with R2 and re-check the output signal. Select a value of about 0.01 μF to start.
5. Select other values of C1 from 100 pF to 10 μF and examine the response on an oscilloscope.

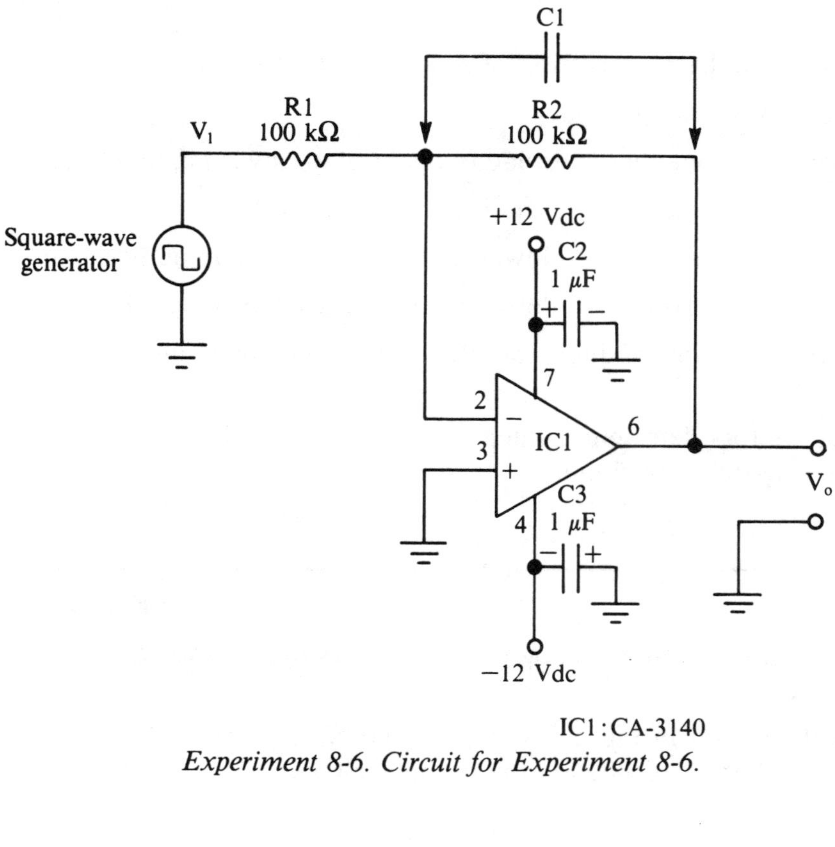

IC1:CA-3140

Experiment 8-6. Circuit for Experiment 8-6.

Note: This experiment can be performed with a sine-wave source and an ac voltmeter. Check the response at a large number of frequencies from 100 Hz to 10 kHz without the capacitor, then redo the measurements at capacitances of 0.001 to 10μF.

Summary

The inverting amplifier generally takes a single input signal and creates an output which is the negative of a dc input, or 180 degrees out of phase for an ac input. Amplification takes place as well. When more than one input appears on the inverting side of the op amp, these signals are added, each having a gain equal to the ratio of the feedback to the input resistor. When the expected output signal exceeds the maximum available output, which is based on the supply voltages, clipping occurs. Frequency response of the op amp often reduces the gain of the amplifier at high frequencies.

Questions

1. Explain the basic operation of an op amp as an inverting amplifier.
2. What is a virtual ground?
3. Which components determine the gain of an inverting amplifier?
4. What is the function of a multi-input inverting-follower circuit?
5. How can an inverting-follower circuit be used as an audio mixer?
6. How can an inverting-follower circuit be used as a sign detector?
7. What effect does adding a dc offset to a signal have on a clipping circuit?
8. What factors determine which resistors are used in an inverting-follower circuit?

Problems

Note: Assume supply voltages of ± 12 volts unless indicated otherwise.

1. For the circuit in Fig. 8-2A, $V_{in} = 2$ volts, $R_{in} = 10$ kohms, and $R_f = 20$ kohms. Calculate the output voltage.

2. Repeat Problem 1 for V_{in} = 0.5 volts, R_{in} = 20 kohms, R_f = 100 kohms.
3. For the circuit in Fig. 8-3, V_1 = 0.1 volt, V_2 = 0.2 volt, V_3 = 0.3 volt, $R_1 = R_2 = R_3$ = 10 kohms and R_f = 20 kohms. Calculate the output voltage.
4. For the circuit in Fig. 8-3, $V_1 = V_2 = V_3$ = 5 volts, R_1 = 10 kohms, R_2 = 20 kohms, R_3 = 40 kohms, and R_f = 10 kohms. Calculate the output voltage.
5. An inverting follower is amplifying a signal coming from a source with an output impedance of 500 ohms. What is the minimum input resistance recommended?
6. An inverting follower circuit has an input resistor of 100 kohms. What is the maximum output resistance of the input signal?
7. For the circuit in Fig. 8-9B, a breakpoint frequency of 1600 Hz is required. If R_f = 20 kohms, what capacitor should be used?
8. Repeat Problem 7 for a breakpoint frequency of 5 kHz and R_f = 50 kohms.

9

NONINVERTING FOLLOWER CIRCUITS

AFTER COMPLETING THIS chapter, you will be able to:

- ☐ Explain what a unity-gain follower is
- ☐ Describe the dc and ac response of noninverting followers
- ☐ Deal with single dc power-supply circuits
- ☐ Understand how to compensate for differences between ideal and practical op amps

Chapter 8 covered the basic inverting-follower circuit. The other standard op-amp circuit configuration is the *noninverting follower*. In this type of amplifier, the signal is applied to the noninverting input of the op amp and the output signal is in-phase with the input signal (Fig. 9-1). The two basic noninverting configurations are: *unity gain* (unity = 1) and *greater than unity gain*. No less-than-unity-gain case exists for noninverting followers, even though such a case does exist for inverting followers ($A_v < -1$ when $R_f < R_{in}$ for inverting followers).

Figure 9-2 shows the circuit for the unity-gain noninverting follower. The output terminal is connected directly to the inverting input, which results in 100-percent negative feedback. The universal voltage gain expression for all feedback amplifiers is:

$$A_v = \frac{A_{vol}C}{1 + A_{vol}\beta} \tag{9-1}$$

Where:

A_v is the closed-loop voltage gain (i.e., gain with feedback)
A_{vol} is the open-loop gain (i.e., gain without feedback)
β is the feedback factor
C is the input attenuation factor

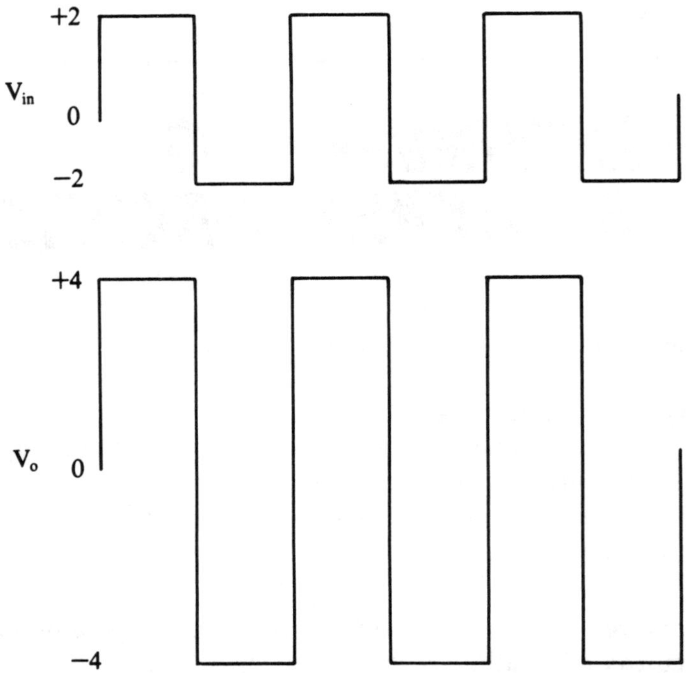

Figure 9-1. Input and output waveform of a noninverting amplifier.

In this circuit, the input signal is applied directly to +In, so the term $C = 1$ can therefore be ignored. The feedback factor, β, represents the transfer function of the feedback network. When that network is a resistor voltage-divider network, the value of β is a decimal fraction that represents the attenuation of the op-amp output voltage before it is applied to the op amp's inverting input. In the unity-gain follower circuit, the value of B is also one, so it too is ignored. The feedback amplifier equation therefore reduces to:

$$A_v = \frac{A_{vol}}{1 + A_{vol}} \tag{9-2}$$

Consider the implications of Eq. (9-2) for common operational amplifiers. With an open-loop voltage gain (A_{vol}) of 300,000 (not an unusual value), the closed-loop gain is 0.9999967. A gain of 0.9999967 is close enough to 1.0 to justify calling the circuit of Fig. 9-2 a unity-gain follower.

Applications of unity gain followers

What use is an amplifier that does not amplify? First of all, it is not strictly true that the circuit does not amplify. It has a unity voltage gain, but the

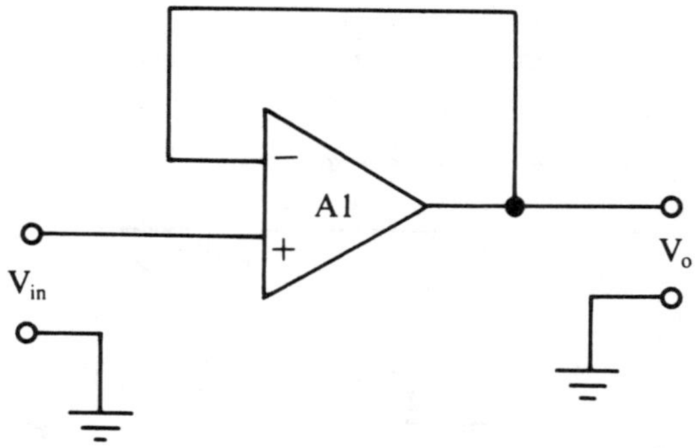

Figure 9-2. Noninverting unity-gain follower.

power gain is greater than unity. A unity-gain follower actually does have power gain (the unity-gain feature refers only to the voltage gain). If the input impedance is typically very much higher than the output impedance, yet $V_o = V_{in}$, then by V^2/R it stands to reason that the delivered power output is much greater than the input power. Thus, the circuit of Fig. 9-2 is unity gain for voltage signals and greater than unity gain for power. Therefore, it is a power amplifier.

Experiment 9-1

The unity-gain follower is a noninverting amplifier that has a gain of one. In this brief experiment, you can build such a circuit and test it. For temporary experiments, use an IC breadboard (see chapter 3) or some similar medium. Construct the circuit of Experiment 9-1 with either a 741 or a CA-3140 BiMOS op amp. If the CA-3140 is used, then capacitors C1 and C2 *must* be used.

We know that this circuit is a unity-gain follower because the output terminal (pin 6) is strapped directly to the inverting input (pin 2). The signal is applied to pin 3, the noninverting input. In this case, the "signal" is a dc-voltage level that is produced by a 10-kohm multi-turn potentiometer.

The output and input signal indicators can be 0 to 15 Vdc voltmeters (M1 and M2). Alternatively, only one meter is needed, and it can be switched back and forth between input and output. In any event, a digital voltmeter is preferred over analog meters, although either will work fine.

1. Adjust R1 to produce a voltage of approximately 1 Vdc at point A.
2. Measure the voltage a point A and record it on a piece of paper.
3. Without touching R1 any further, measure the voltage at point B (output). Record the results.

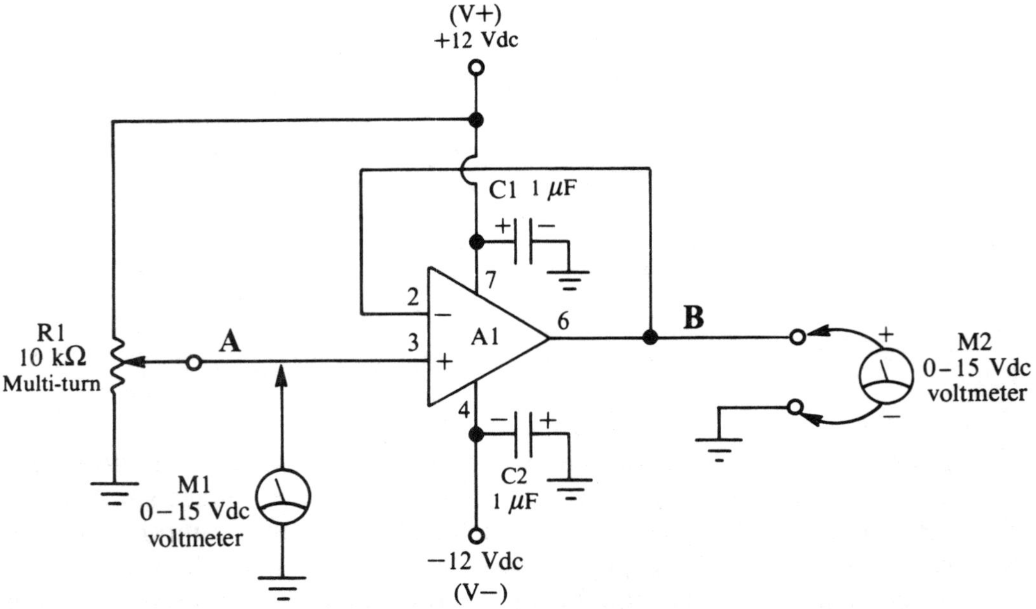

Experiment 9-1. Circuit for Experiment 9-1.

4. Calculate the gain of this circuit by dividing the voltage at point B by the voltage at point A.
5. The gain should be unity (1). Explain any discrepancies.
6. Repeat the experiment at various settings of potentiometer R1.

Discussion

Some deviation from a perfect value of 1.000 can be expected. If your instrument has sufficient resolution, the gain difference between A_v and A_{vol} that is inherent in Eqs. (9-1) and (9-2) will become apparent. However, unless you have a very good meter, these discrepancies are invisible. The most likely explanation is that measurement errors have occurred between the voltages at points A and B.

This experiment can be rerun for ac signals. Replace potentiometer R1 with an audio signal source. Initially apply a signal of 100 to 1000 Hz at an amplitude of 1 to 3 volts peak-to-peak. The input and output indicators can be either audio ac voltmeters or an oscilloscope. A dual-channel 'scope will allow both input and output voltages to be simultaneously monitored on the same instrument.

Three principal uses of the unity-gain noninverting follower exist: buffering, power amplifying, and impedance transforming.

A buffer amplifier is placed between a circuit and its load to improve the isolation between the two. An example is the use of a buffer amplifier between an oscillator or waveform-generator circuit and its load. The buffer is especially useful where the load normally exhibits a varying impedance that could "pull" the oscillator frequency. Such unintentional frequency modulation of the oscillator is very annoying because it makes some oscillator circuits unable to function and causes others function poorly.

Another common use for buffer amplifiers is to isolate an output connection from the main circuitry of an instrument. An example might be an instrumentation circuit that uses multiple outputs, perhaps one to the analog-to-digital (A/D) converter at the input of a digital computer and another to an analog oscilloscope or strip-chart paper recorder. A buffered analog output to the oscilloscope prevents short circuits in the display wiring from affecting the signal to the computer, and vice versa.

A special case of buffering is represented by using the unity-gain follower as a power driver. A long cable run might attenuate low-power signals, especially where a high output impedance drives a high cable capacitance. To overcome this problem, sometimes a low-impedance power source is used to drive a long cable.

The impedance transformation capability is obtained because an op amp has a very high input impedance and a very low output impedance. Figure 9-3A is a generic equivalent of a voltage source driving a resistive load, R_2. Resistance R_1 represents the internal impedance of the signal source impedance. The signal voltage V_1 is reduced at the output V_o by whatever voltage is dropped across internal source resistance R_1. The output voltage is found from:

$$V_o = \frac{V R_2}{R_1 + R_2} \tag{9-3}$$

If the ratio of R_1/R_2 is, say, 10:1, then a 1-Vdc potential is reduced to 0.091 Vdc across R_2. Ninety percent of the signal amplitude is lost. With a unity-gain noninverting amplifier, as in Fig. 9-3B, the situation is entirely different. If the amplifier input impedance is very much larger than the source resistance and the amplifier output impedance is very much lower than the load impedance, very little loss will occur and V will closely approximate V_o.

Noninverting followers with gain

Figure 9-4A shows the circuit for the noninverting follower with gain. In this circuit, the signal (V_{in}) is applied to the noninverting input and the feedback network (R_f/R_{in}) is almost the same as it was in the inverting-follower circuit. The difference is that one end of R_{in} is grounded.

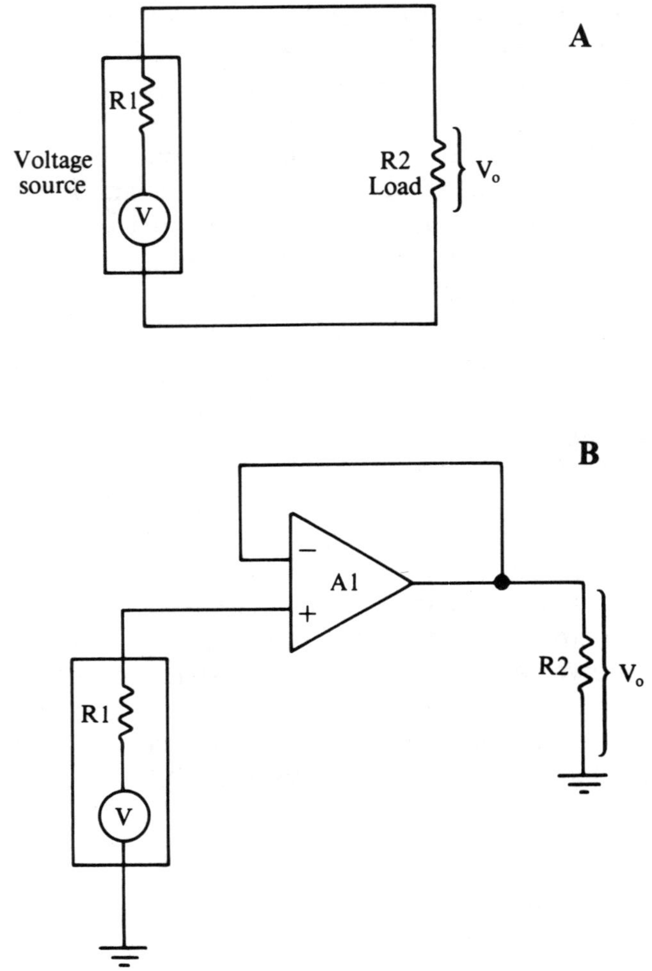

Figure 9-3. A) Equivalent circuit, B) equivalent circuit for a noninverting follower.

Evaluate this circuit using the same general method as was used in the inverting follower case. You know from Kirchhoff's current law and the fact that the op-amp inputs neither sink nor source current that currents I_1 and I_2 are equal. Thus, the Kirchhoff expression for these currents at the summing junction (point A) can be written as:

$$I_1 = I_2 \qquad (9\text{-}4)$$

From the properties of the ideal op amp, any voltage applied to the noninverting input (V_{in}) also appears at the inverting input. Therefore:

$$V_1 = V_{in} \qquad (9\text{-}5)$$

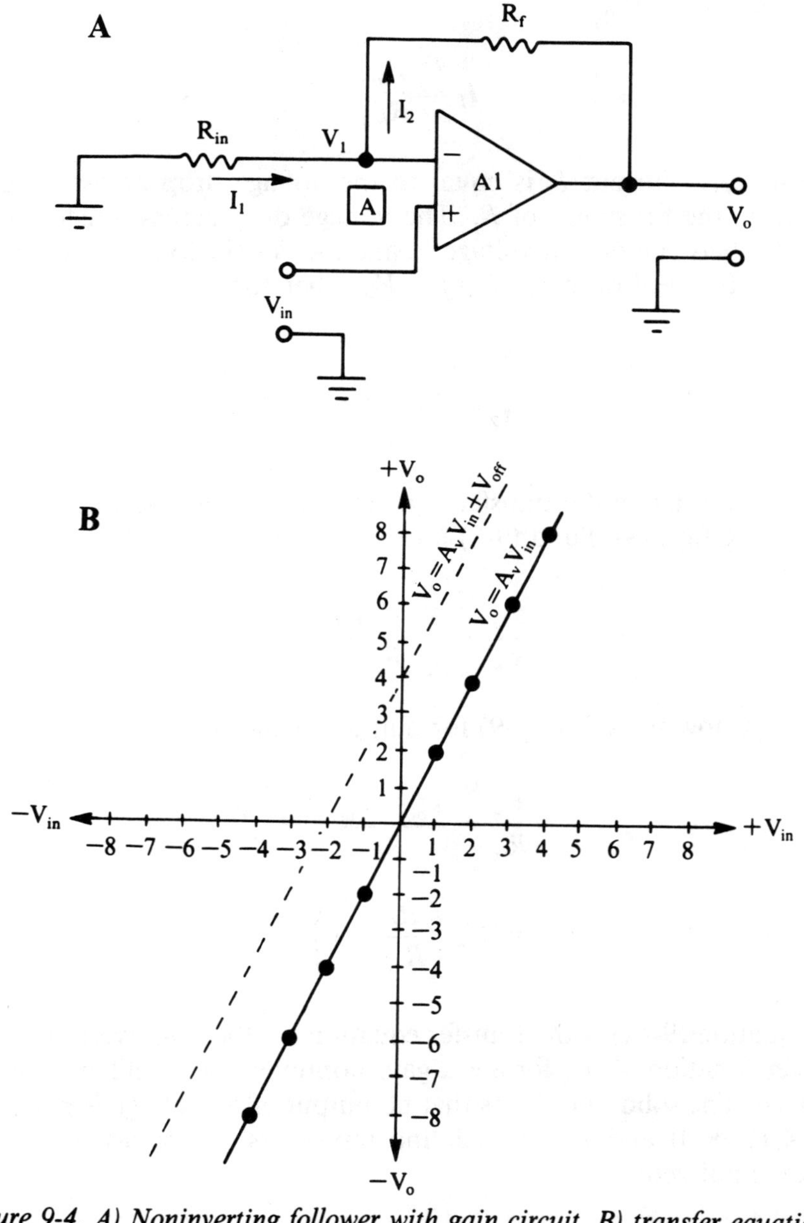

Figure 9-4. A) Noninverting follower with gain circuit, B) transfer equation.

From Ohm's law, the value of current I_1 is:

$$I_1 = \frac{V_1}{R_{in}} \tag{9-6}$$

or, because $V_1 = V_{in}$,

$$I_1 = \frac{V_{in}}{R_{in}} \tag{9-7}$$

Similarly, current I_2 is equal to the voltage drop across resistor R_f, divided by the resistance of R_f. The voltage drop across resistor R_f is the difference between output voltage V_o and the voltage found at the inverting input, V_1. By ideal property 7, $V_1 = V_{in}$. Therefore:

$$I_2 = \frac{V_o - V_{in}}{R_f} \tag{9-8}$$

You can derive the transfer equation of the noninverting follower by substituting Eqs. (9-7) and (9-8) into Eq. (9-4):

$$\frac{V_{in}}{R_{in}} = \frac{V_o - V_{in}}{R_f} \tag{9-9}$$

You must now solve Eq. (9-9) for output voltage V_o:

$$\frac{V_{in}}{R_{in}} = \frac{V_o - V_{in}}{R_f} \tag{9-10}$$

$$V_o = V_{in}\left(\frac{R_f}{R_{in}} + 1\right) \tag{9-11}$$

Equation (9-11) is the transfer equation for the noninverting follower. Transfer function V_o/V_{in} for a -2-gain noninverting amplifier is shown in Fig. 9-4B. The solid line shows that no output offset voltage is present (i.e., $V_o = A_v V_{in} + 0$) and the dotted line represents a case where the offset voltage is not zero.

Advantages of noninverting followers

The noninverting follower offers several advantages. During the unity-gain configuration section, I mentioned that buffering, power amplifying, and impedance transforming were advantages. Also, in the gain noninverting amplifier configuration, voltage gain can be provided with no phase reversal.

The input impedance of the noninverting followers shown thus far has been very high, being essentially the input impedance of the op amp itself. In the ideal device, this impedance is considered to be infinite, but in practical devices it can range from 500,000 ohms to more than 10^{12} ohms. Thus, the noninverting follower is useful for amplifying signals from any high-impedance source, regardless of whether impedance transformation is a circuit requirement.

Selecting resistor values

When the required gain is known (as it usually is in practical situations), you can select a trial value for R_{in}, and then solve the gain equation to find R_f. This new version of the equation is:

$$R_f = R_{in} (A_v - 1) \tag{9-12}$$

Determine from evaluating the application whether the trial result obtained from this operation is acceptable. If the result is not acceptable, work the problem again with a new trial value.

What is "acceptable?" If the value of R_f is exactly equal to a standard resistor value, then all is well. But, if the value is not standard, a decision has to be made. Will the nearest standard values result in an acceptable gain error (which is determined from the application)?

The ac response of noninverting followers

The noninverting amplifier circuits discussed in the preceding sections are dc amplifiers. Nonetheless, as with the inverting amplifiers in chapter 8, the noninverting amplifier will also respond to ac signals up to the upper frequency-response limit of the circuit.

Frequency-response tailoring

It is possible to custom tailor the upper-end frequency response of an op amp with a capacitor to shunt the feedback resistor. The same method was used for the inverting follower in chapter 8. This section will expand the subject and feature the tailoring of the upper and lower −3 dB frequency response.

Upper −3 dB point The capacitor across the feedback resistor (Fig. 9-5) sets the frequency at which the upper-end frequency response falls −3 dB

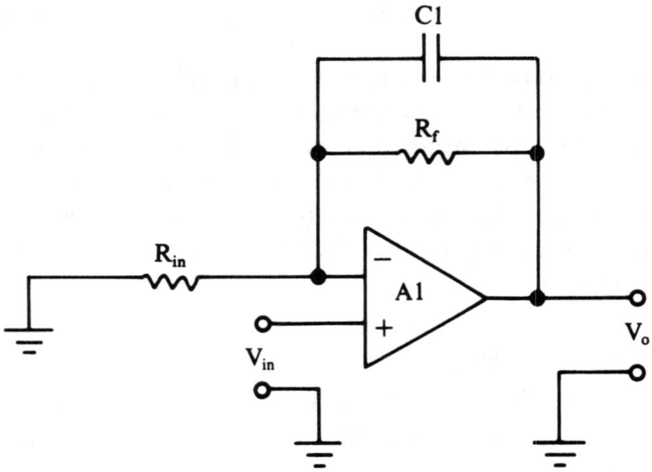

Figure 9-5. High-frequency roll-off by shunting feedback resistor with a capacitor.

below the low-end in-band gain. The gain at frequencies higher than this −3 dB frequency falls at a rate of −6 dB/octave (an octave is a 2:1 frequency ratio) or −20 dB/decade (a decade is a 10:1 frequency ratio). Note: −6 dB/octave and −20 dB/decade have the same roll-off slope. The value of capacitor C1 is found by:

$$C1_{\mu F} = \frac{1{,}000{,}000}{2\pi R_f F_c} \tag{9-13}$$

Where:

$C1_{\mu F}$ is the capacitance of *C1* (in *microfarads*)
R_f is the resistance (in *ohms*)
F_c is the upper −3-dB frequency (in *hertz*)

Experiment 9-2

Construct the circuit of Experiment 9-2 with a CA-3140 BiMOS op amp. If a CA-1340 is not available, use a 741 op amp and limit the frequencies of operation to 2000 Hz. For capacitors C3A through C3D, initially try the following values: C3A = 100 pF, C3B = 470 pF, C3C = 0.005 μF, C3D = 0.22 μF.

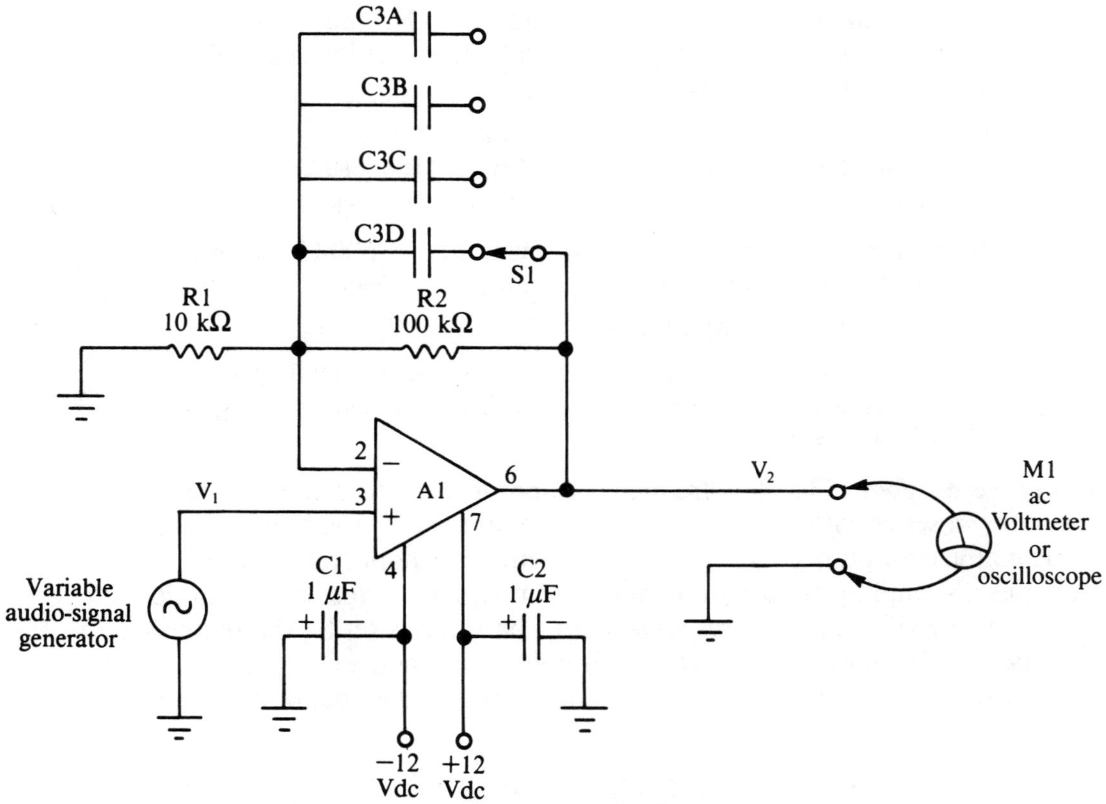

Experiment 9-2. Circuit for Experiment 9-2.

Create a table that will show the results. A typical table will look like this:

Trial	Frequency (Hz)	V_1	V_2	A_v (V_2/V_1)
1	10			
2	20			
3	30			
4	40			
5	50			
6	60			
7	70			
8	80			
9	90			

1. Set the frequency of the signal generator to 10 Hz and the amplitude to approximately 0.5 volts peak-to-peak. Keep the input signal voltage constant during this experiment.
2. Measure the input voltage (V_1) and output voltage (V_2) with either an ac voltmeter or an oscilloscope and record the result on the table.
3. Calculate the gain (V_2/V_1) and enter the value on the table.
4. Repeat for frequencies at 10-Hz intervals up to 5,000 Hz or until no further reading is obtainable on the output indicator.
5. Plot the results on graph paper.
6. Determine the −6 dB point by finding the frequency at which the gain is 6 dB lower than the mid-band or low-band gain (note: gain in dB = 20 LOG gain).

Lower −3 dB point The low-frequency response is controlled by placing a capacitor in series with the resistor, which makes the noninverting follower an ac-coupled amplifier. Figure 9-6 is the circuit for a noninverting follower that uses ac-coupling at both input and output circuits. Capacitor C2 limits the upper −3-dB frequency-response point. Its value is set by the method discussed in chapter 8. The lower −3 dB point is set by the combination of R1 and input capacitor C1. This frequency is set by the equation:

$$F = \frac{1{,}000{,}000}{2\pi R_1 C_1} \tag{9-14}$$

Where:

C_1 is the capacitance (in *microfarads*)
R_1 is the resistance (in *ohms*)
F is the lower −3-dB point (in *hertz*)

In some cases, you will want to ac-couple the output circuit (although it is optional in most cases). Capacitor C3 ac-couples the output, thus preventing any dc component that is present on the op-amp output from affecting the following stages. Resistor R_L is used to keep capacitor C3 from being charged by the offset voltage from op amp A1. The value of capacitor C3 is set to retain the lower −3-dB point, using the resistance of the stage following as the "R" in the equations.

Resistor R1 also sets the input impedance of the amplifier. Previous circuits had a very high input impedance because that parameter was determined only by the (extremely high) op-amp input impedance. In Fig. 9-6, however, the input impedance at the source is equal to R1.

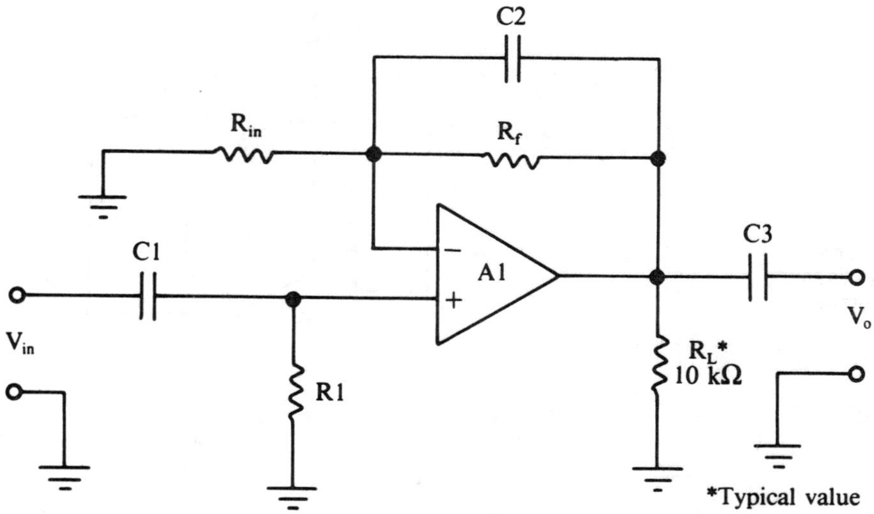

Figure 9-6. Ac-coupled noninverting follower with gain.

The form of Eq. (9-14) is backwards from the point of view of practical circuit-design problems. In most practical cases, the required frequency-response limit is known from the application. Also, the minimum value of R1 can be found from the application (derived from source impedance), and often it is set as high as possible (e.g., 10 megohms) for practicality. Thus, the equation should be solved for *C*, as shown below:

$$C_1 = \frac{1{,}000{,}000}{2\pi R_1 F} \tag{9-15}$$

Experiment 9-3

Build the circuit of Experiment 9-3 with a CA-3140 operational amplifier. Repeat Experiment 9-2 using an ac voltage of 3 to 5 volts peak-to-peak.

Single dc power-supply circuits

The technique of Fig. 9-6 works well for dual-polarity dc power-supply circuits. In single-polarity dc power-supply circuits, however, the method fails because of the large dc offset voltage that is present on the output. For these applications, use a circuit, as shown in Fig. 9-7.

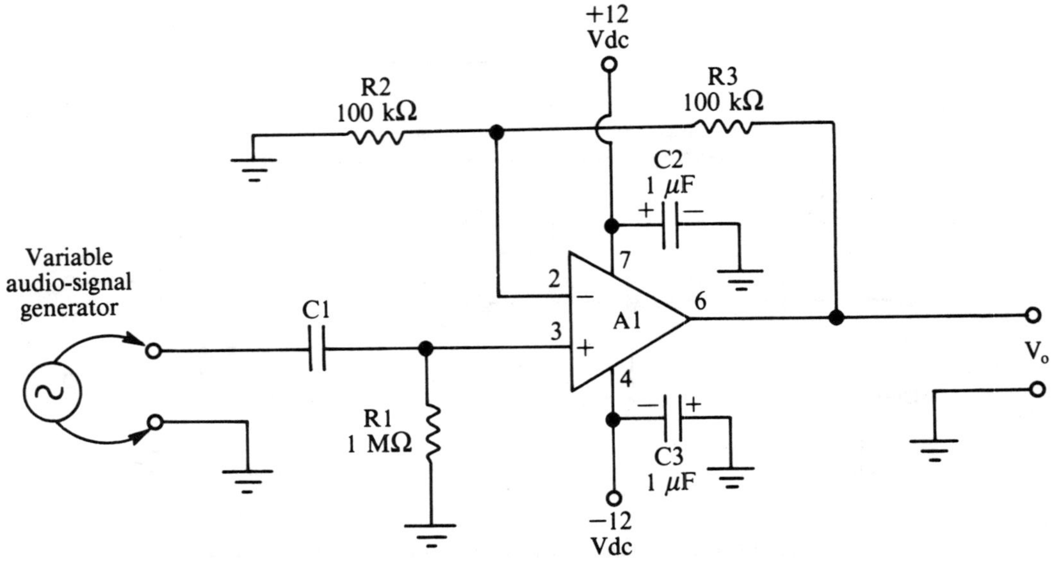

Experiment 9-3. Circuit for Experiment 9-3.

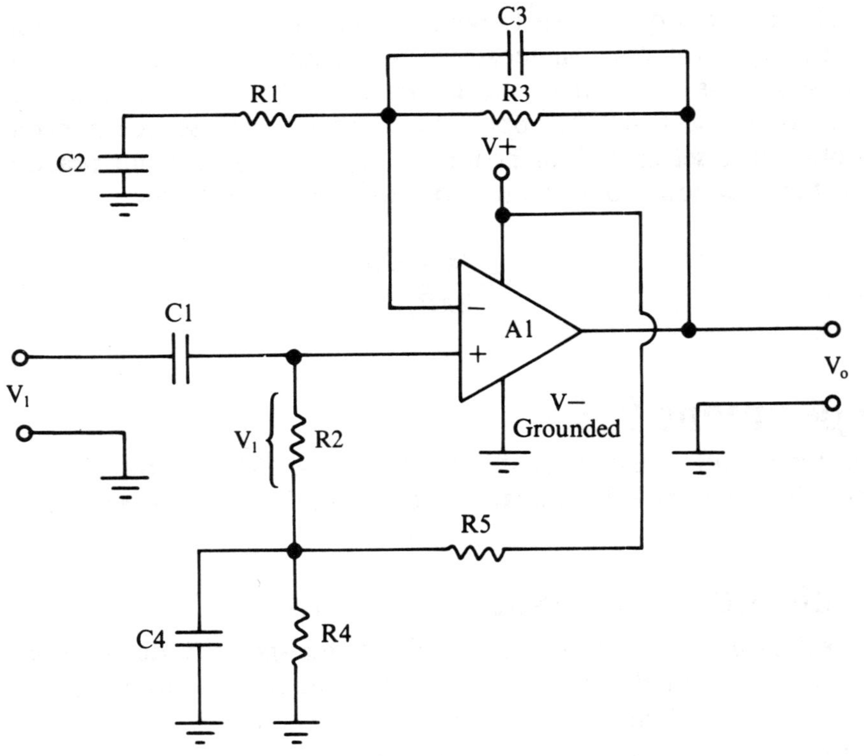

Figure 9-7. Noninverting ac-coupled amplifier that is operated from a single-polarity dc power supply.

The circuit in Fig. 9-7 is operated from a single V+ dc power supply (the V− terminal of the op amp is grounded). In order to compensate for the grounded V− supply, the noninverting input is biased to a potential of:

$$V_1 = (V+)\left(\frac{R_4}{R_4 + R_5}\right) \tag{9-16}$$

If $R_4 = R_5$, then V_1 will be $(V+)/2$. Because the noninverting input typically sinks very little current, the voltage at both ends of R_1 is the same (i.e., V_1). The circuit of Fig. 9-7 does not pass dc and some low ac frequencies because of the capacitor coupling. Also, because capacitor C3 shunts feedback resistor R3, the higher frequencies roll off. The high-frequency roll-off −3-dB point is found from:

$$F = \frac{1{,}000{,}000}{2\pi R_3 C_3} \tag{9-17}$$

Where:

F is the −3-dB frequency (in *hertz*)
R_3 is the resistance (in *ohms*)
C_3 is the capacitance (in *microfarads*)

We can restate Eq. (9-17) to take into account that the value of R3 is usually known from setting the gain, and the nature of the application sets the minimum value of frequency F. We can rewrite Eq. (9-17) in a form that yields the value of C3 from:

$$C_3 = \frac{1{,}000{,}000}{2\pi R_3 F} \tag{9-18}$$

The lower −3-dB frequency is set by any or all of several RC combinations within the circuit.

Resistor R1 is the input resistor, which serves the same purpose as the similar resistor in the previous circuit. At midband, the input impedance is set by resistor R1, although at the extreme low end of the frequency range, the reactance of C4 becomes a significant feature. In general, X_{C1} should be less than or equal to $R_1/10$ at the lowest frequency of operation.

Capacitor C1 is in series with the input signal path and it serves to block dc and certain very low-frequency signals. The value of C1 should be:

$$C_1 = \frac{1{,}000{,}000}{2\pi F R_1} \tag{9-19}$$

Where:

C_1 is the capacitance (in *microfarads*)
F is the frequency (in *hertz*)
R_1 is the resistance (in *ohms*)

Capacitor C5 keeps the dc output offset from affecting the succeeding stage. The 10-kohm output load resistor, R6, keeps C5 from being charged by the dc offset voltage. The value of C5 should be greater than:

$$C_5 > \frac{10^6}{2\pi FR_4} \qquad (9\text{-}20)$$

Where:

C_5 is the capacitance (in *microfarads*)
F is the low-end −3-dB frequency (in *hertz*)
R_L is the load resistance (in *ohms*)

Experiment 9-4

Construct the circuit of Experiment 9-4 with either a 741 or a CA-3140 op amp. Use either an oscilloscope or a digital ac voltmeter for the output indicator (M1).

1. Adjust the input voltage for 2 volts peak-to-peak.
2. Set potentiometer R5 to minimum resistance.
3. Measure the output voltage across R8.
4. Measure the output voltage at point A.
5. Compare the results of steps 3 and 4.
6. Repeat the experiment with various settings for R5.

Note: If an oscilloscope is available, then perform this experiment on the 'scope using dc coupling.

Project 9-1
Microphone preamplifier

The circuit of Project 9-1 is a simple communications or public-address preamplifier. The circuit consists of a 741 operational amplifier that is housed inside a shielded metal box. The circuit is a noninverting follower with a gain of approximately 100. The input impedance within the 300-to-3000-Hz nominal passband is 1 megohm.

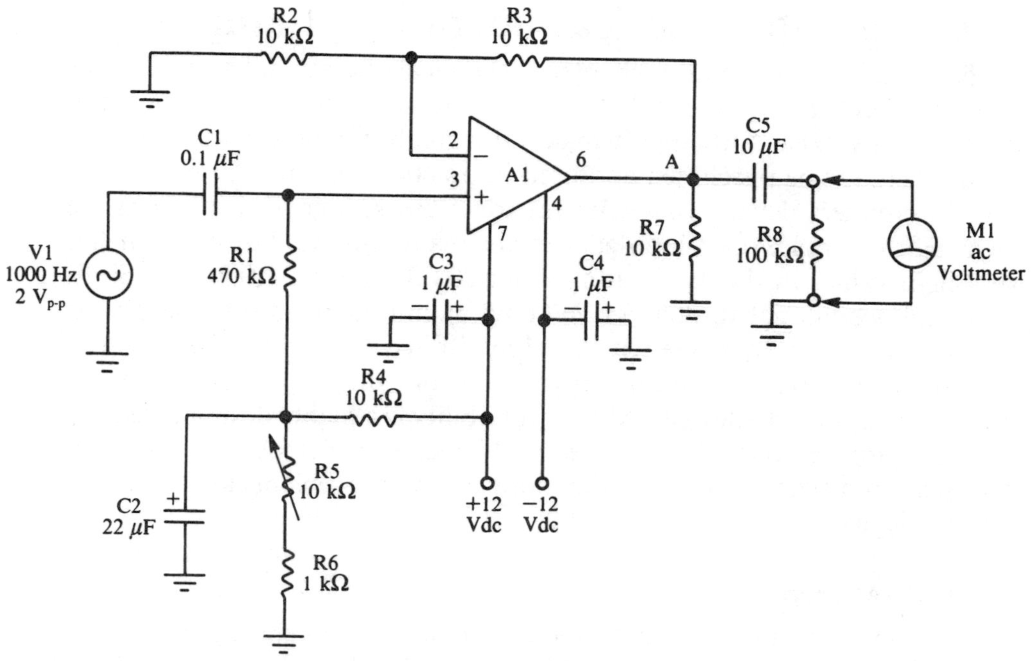

Experiment 9-4. Circuit for Experiment 9-4.

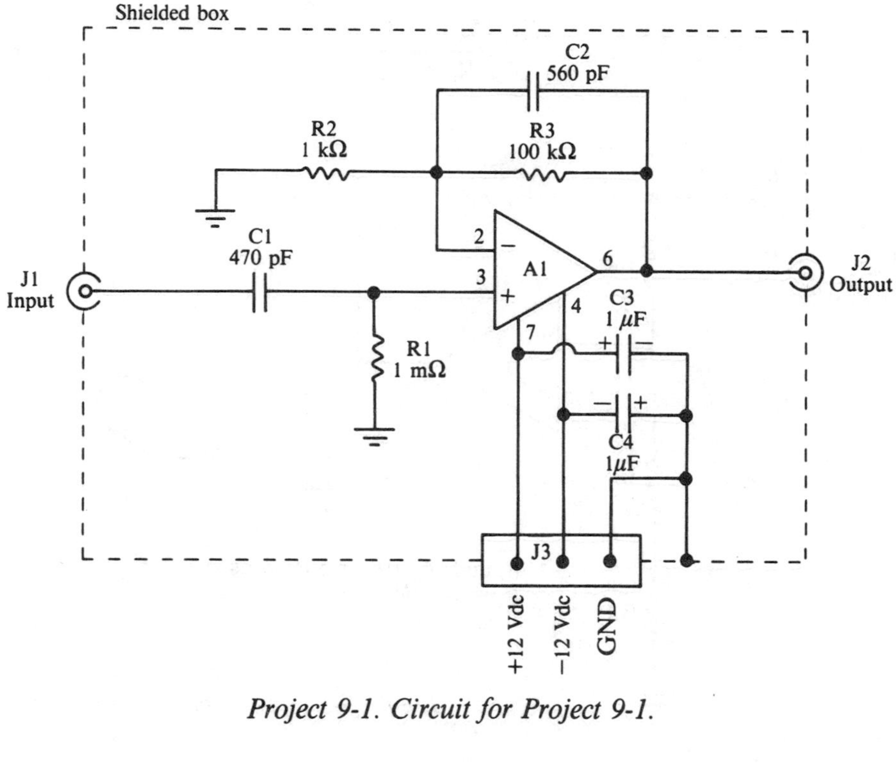

Project 9-1. Circuit for Project 9-1.

Dealing with practical operational amplifiers

Earlier in this chapter and in chapter 8, the ideal operational amplifier was featured. Such a hypothetical device is a good model-making tool, but doesn't really exist. Although it makes our math circuit analysis easier, it cannot actually be purchased and used in practical circuits. All real operational amplifiers depart somewhat (in some areas, a great deal) from the ideal. For example, open-loop gain is not really infinite, but rather it has very high values in the range from about 20,000 to over 1,000,000. Similarly, real operational amplifiers don't have an infinite bandwidth, and most are even intentionally made with severely limited bandwidth. Such amplifiers are said to be "unconditionally stable" or "frequency compensated" (for example, the 741 device). Although stability is a highly desirable feature for many applications, it is obtained at the expense of frequency response. This section deals with some of the more common problems and solutions for real devices.

Input offset current

Input offset current is measured in a test circuit, such as shown in Fig. 9-8. Input offset current can be specified by the relationship between two different output offset voltages that are taken under different input conditions:

$$I_{io} = \frac{R_{in}\,(V_{o1} - V_{o2})}{R_2\,(R_f + R_{in})} \tag{9-21}$$

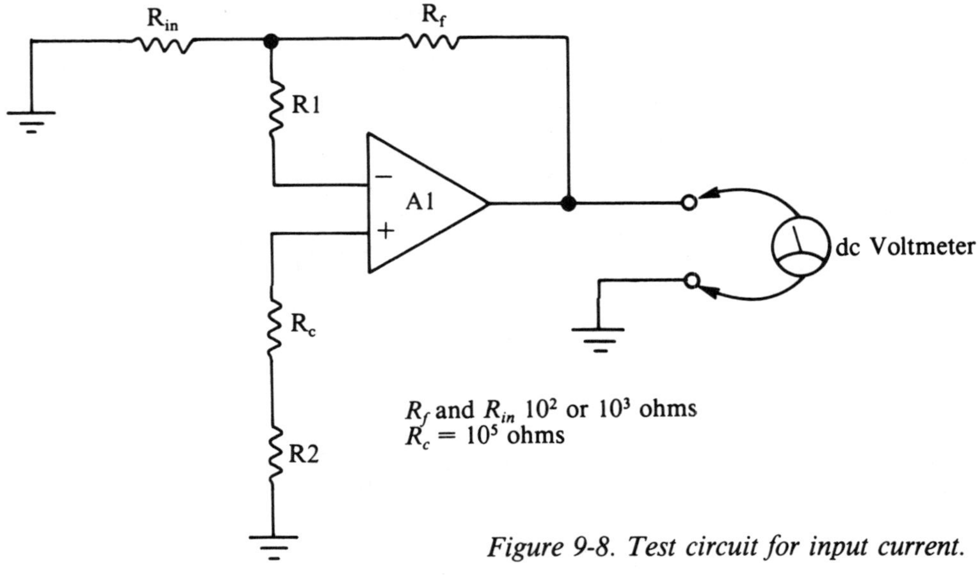

Figure 9-8. Test circuit for input current.

The first output voltage, V_{o1}, is measured with R1 and R2 connected in the circuit. Voltage V_{o2} is then measured with the resistors short-circuited, but all other conditions remain the same. The resulting output voltage can, along with V_{o1}, be used in Eq. (9-21) to determine the input offset current.

Input offset voltage

The input offset voltage is the voltage that is required to force the output voltage (V_o) to zero when the input voltage is also zero. The operational amplifier is connected in an inverting-amplifier configuration, such as Fig. 9-9. To make the measurement, the input terminal is connected to ground. The input offset voltage is found by measuring the output voltage (when $V_{in} = 0$) and then using the voltage divider equation:

$$V_{io} = \frac{V_o R_{in}}{R_f + R_{in}} \qquad (9\text{-}22)$$

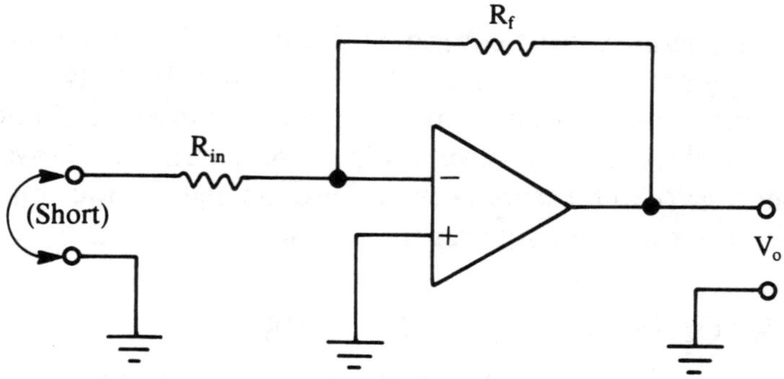

Figure 9-9. Test circuit for output offset.

Greater accuracy is achieved if the gain of the amplifier is 100, 1000 (or even higher), provided that such gains can be accommodated without saturating the amplifier.

Input bias current

This test requires a pair of closely matched resistors to be connected between the op-amp inputs (−In and +In) and ground (Fig. 9-10). Power is applied to the operational amplifier and the voltage is measured at each input. The value of resistors R1 and R2 must be high enough to create a measurable voltage drop at the level of the anticipated current. Although the actual value of these resistors is not too critical, the match between them ($R_1 = R_2$) is critical to the success of the measurement. The definition of

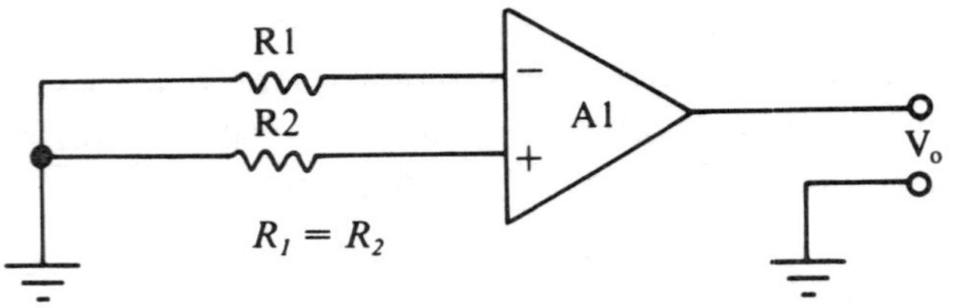

Figure 9-10. Output current offset from input-bias currents.

readable voltage drop depends upon the instrumentation that is available to do the job. For example, assume that an actual input bias current of 5 microamperes (μA) is flowing and the resistors are each 10 kohms. In this case, the measured voltage will be (by Ohm's law):

$$V = IR$$
$$V = (5 \times 10^{-6} \text{ amperes}) (10{,}000 \text{ ohms})$$
$$V = 5 \times 10^{-2} \text{ volts} = 50 \text{ mV}$$

If your voltage-measuring equipment is not capable of measuring these levels, then higher resistor values would be required. If the two inputs had ideally equal input-bias currents, only one measurement would be needed. Because real devices usually have unequal bias currents, however, it is sometimes necessary to measure both and use the higher input-bias current. Alternatively, the root sum squares (rss) value can be used.

Power supply sensitivity (pss)

Power-supply sensitivity Φ is the worst-case change-of-input offset voltage for a 1.0-Vdc change of one dc power-supply voltage (either V− or V+) with the other supply potential held constant. The same test configuration is used to measure this parameter as was used to measure the input offset voltage (Fig. 9-9). First, the two power-supply voltages are set to equal levels and the input offset voltage is measured. One of the power-supply voltages is then changed by precisely 1.00 Vdc and the input offset voltage is again measured. The power-supply sensitivity (Φ) is given by:

$$\Phi = \frac{\Delta V_{io}}{\Delta V_o} \qquad (9\text{-}23)$$

The actual power-supply sensitivity is the worst case when this measurement is made under four conditions: **1.** $V+$ increased 1 Vdc, **2.** $V+$

decreased 1 Vdc, **3.** *V*− increased 1 Vdc, and **4.** *V*− decreased 1 Vdc. The worst case of these four measurements is taken as the true power-supply sensitivity.

Slew rate

Slew rate is a measure of the op amp's ability to shift between the two possible opposite output voltage extremes while supplying full output power to the load. This parameter is usually specified in terms of volts per unit of time (e.g., 30 V/μS).

A saturating square wave is usually used to measure the slew rate of an operational amplifier. The square wave must have a rise time that substantially exceeds the expected slew rate of the operational amplifier. The rise-time value is found by examining the leading edge of the output waveform on an oscilloscope while the input is overdriven by the square wave. The time measured is that which is required for the output to slew from 10 to 90 percent of the final value. The slew rate can be affected by gain, so the value at unity gain will not match either the slew rate under open-loop or very high gain closed-loop conditions. Once the switching time is known, the slew rate (S_r) is closely approximated by:

$$S_r = \frac{(V+) + ABS(V-)}{T_s} \tag{9-24}$$

Where:

S_r is the slew rate (in *volts per microsecond*)
$V+$ is the positive supply voltage
$ABS(V-)$ is the absolute value of the negative supply voltage
T_s is the switching time.

Since most manufacturers specify the slew rate for the open-loop configuration in their data sheets, you can use this relationship to approximate the switching times of specific op amp "digital" circuits. It is possible to improve the closed-loop slew rate at any given gain figure through the use of appropriate lag compensation techniques (Fig. 9-11). Keep the values of R_f and R_{in} low when trying to improve slew rates; values under 10 kohms will be best. The compensation capacitor will have a value of:

$$C = \left(\frac{R_f + R_{in}}{4\pi R_f R_{in}}\right)\left(\frac{F_{oi}}{10^m}\right) \tag{9-25}$$

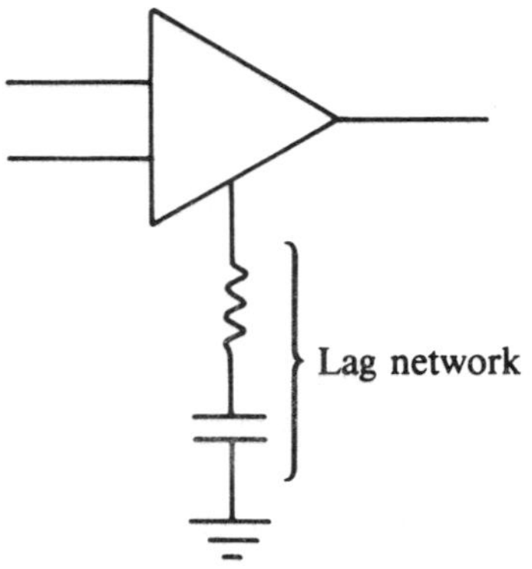

Figure 9-11. Lag compensation.

Where:

F_{oi} is the -3-dB half-power unit
R_f is the feedback resistance
R_{in} is the input resistance
m is the quantity $[A_{vol(dB)} - A_{v(dB)}]/20$

The resistor value is found from:

$$R = \frac{1}{2\pi F_{oi} C} \tag{9-26}$$

Usually the slew rate is measured in the noninverting unity-gain voltage-follower configuration because that circuit generally has the poorest slew rate in most op-amp devices. As in the previous test, the worst-case figure is used as the standard.

The unity-gain follower is driven by a square wave of sufficient amplitude to drive the device well beyond the full saturation point. This criterion is necessary to eliminate the rounded curves that exist at points just below full saturation. The output waveform can then be examined with a wide-band oscilloscope, which has a time base fast enough to allow for a meaningful examination. The trace (Fig. 9-12) will be a straight line with a certain slope. The standard practice is to measure rise time as the time of transition from 10 percent of full amplitude to 90 percent of full amplitude. Adjust the time-base triggering of the oscilloscope so that the slope covers several horizontal scale divisions. The slew rate (S_r) is then found from the slope of the trace on the oscilloscope:

$$S_r = \frac{V_o}{T} \tag{9-27}$$

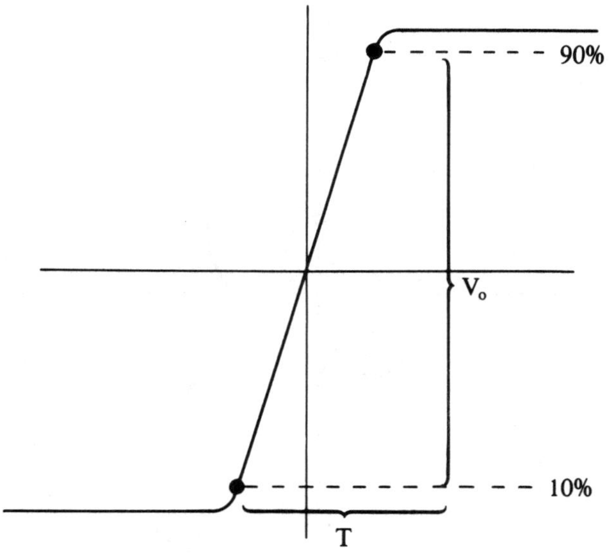

Figure 9-12. Slew-rate calculation.

Phase shift

The phase shift of an operational amplifier can be measured with a sine wave and an oscilloscope. In one version of this test, an X-Y oscilloscope (or a dual-channel model with an X-Y capability) is used. The input signal is applied to the vertical channel of the CRO and the operational-amplifier output is applied to the horizontal channel. The gains for the two channels are set to produce equal beam deflections.

The points marked Y1 through Y4 (Fig. 9-13A) are measured and the phase shift is calculated from:

$$\Phi = \arcsin\left(\frac{Y_1 - Y_2}{Y_4 - Y_3}\right) \tag{9-28}$$

An alternative approach uses a dual-trace CRO, in which the input signal is applied to one channel and the output signal is applied to the other channel. The noninverting unity-gain op-amp configuration is used. The CRO channel gains are adjusted to be identical and the traces are superimposed (Fig. 9-13B) on each other. The phase shift is found from:

$$\Phi = \frac{360B}{A} \tag{9-29}$$

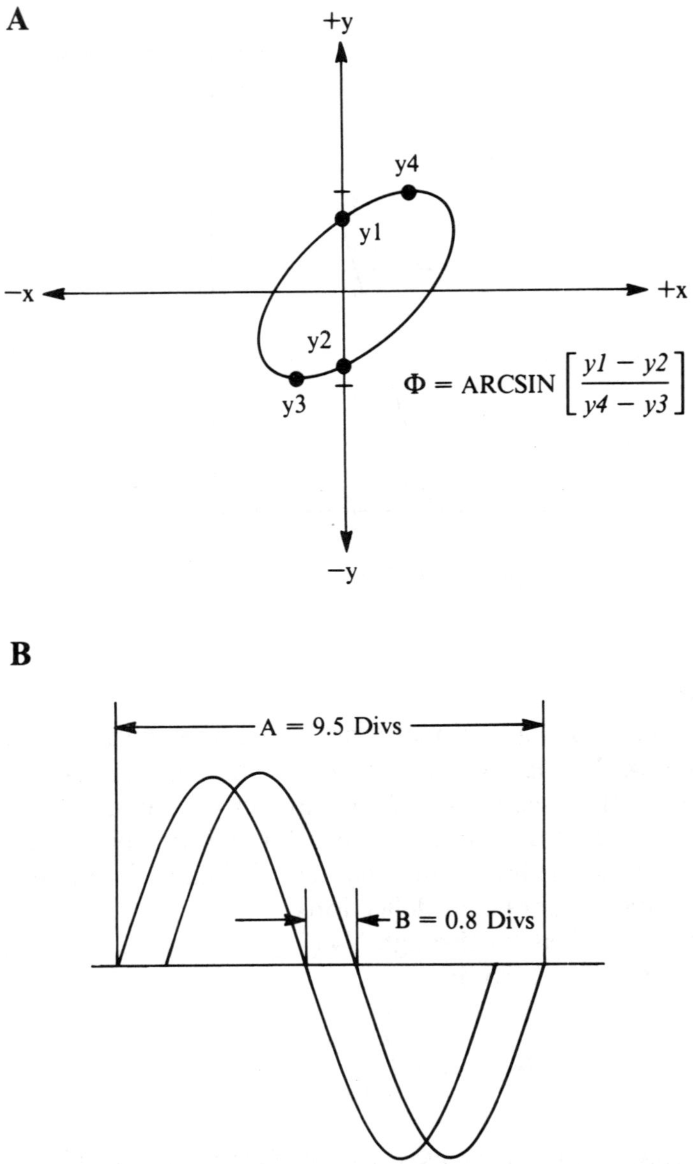

Figure 9-13. A) phase-error measurement using Lissajous pattern, B) using sine waves on Y-time oscilloscope.

This test only works properly if both channels of the CRO are identical, and thus have the same internal phase shift. In addition, the CRO must have a high internal chopping frequency in the dual-beam mode. The instrument limitations constrain the usefulness of this test.

Common-mode rejection ratio (CMRR)

The *common-mode rejection ratio* (*CMRR*) is the ratio of the differential gain to the common-mode gain:

$$CMRR = \frac{A_{vd}}{A_{vcm}} \tag{9-30}$$

Where:

$CMRR$ is the common-mode rejection ratio
A_{vd} is the differential voltage gain
A_{vcm} is the common-mode voltage gain

The common-mode voltage gain is not always a linear function of common-mode input voltage (V_{cm}), so it is usually specified at the maximum allowable value for V_{cm}. The amplifier is connected into a circuit, such as Fig. 9-14, and $CMRR$ is determined from:

$$CMRR = \frac{(R_f + R_{in})\ V_{in}}{R_{in} V_o} \tag{9-31}$$

Provided that $R_f > R_{in}$, i.e., 100:1

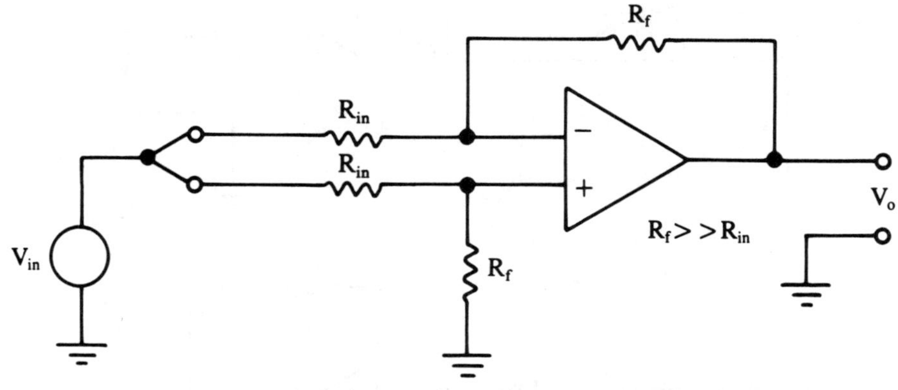

Figure 9-14. Common-mode signal applied to a dc differential amplifier.

Dc errors and their solutions

This section examines dc errors in operational amplifiers. These errors are in the form of current and voltage levels that force the output voltage to differ from the theoretical value in any given case. In many cases, the error will be specified in terms of an actual output voltage that is not zero at a time when it should be (e.g., when $V_{in} = 0$). Also, some circuit tactics will be proposed to minimize or eliminate certain errors.

Output offset compensation

The dc error factors result in output offset voltage V_{oo}, which exists between the output terminal and ground when V_o should be zero. This phenomena helps explain discrepancies between the voltages that actually exist and those that the equations say should exist. Two causes of output are: input offset voltage and input bias current.

Input offset voltage V_{io} (Fig. 9-15A) is the differential input voltage that is required to force the output to zero when no other signal is present ($V_{in} = 0$). A reasonably good model for the input offset-voltage phenomenon (Fig. 9-15B) is a voltage source that has one end connected to ground and the other end connected to the noninverting input. Although voltage-source polarity is shown, the actual polarity in any given situation can be either positive or negative to ground, depending on the device being tested. Values for input offset are typically from one to several millivolts. The popular 741 operational amplifier is specified to have a 1-to-5-mV input offset voltage (2 mV is typical).

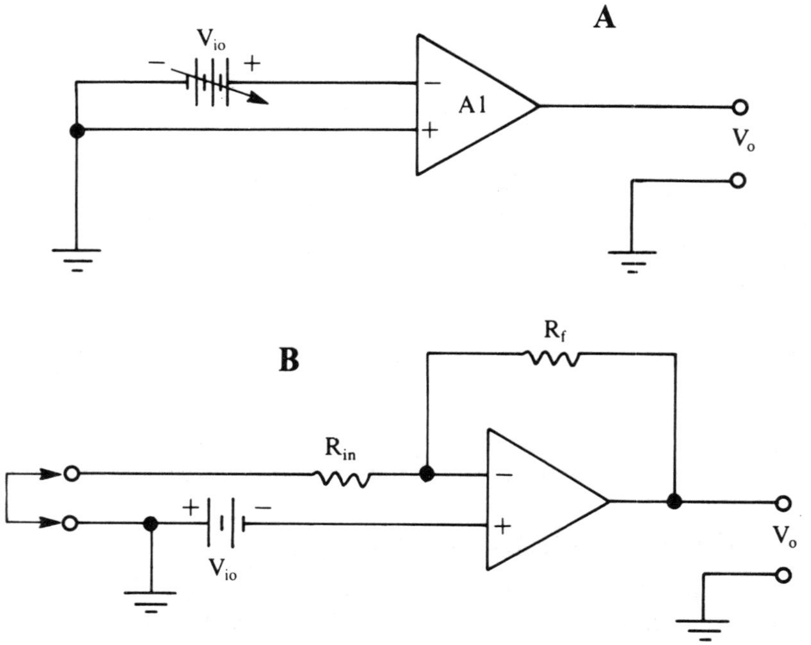

Figure 9-15. A) circuit for measuring input offset voltage, B) alternate method.

The value of the output offset voltage (V_{oo}) caused by an input offset voltage is:

$$V_{oo} = R_f V_{io} \tag{9-32}$$

If the circuit gain is low and V_{in} remains at relatively high values, the input offset voltage might be of little practical consequence. Primarily, the high values of A_v or the low values of V_{in} are encountered where the input offset voltage becomes a problem (these conditions are often encountered together).

The second major cause of output offset voltage is spurious input current. This category can be further subdivided into two more classes: normal input bias and input offset current.

Figure 9-16A shows a typical op-amp differential input stage. Whenever bipolar (npn or pnp) transistors are used in this stage, some small input-bias current will be required for normal operation. This is one of those unavoidable conditions that is inherent in the nature of the transistor, rather than any deficiency in the internal op-amp circuitry. The problem that input bias currents cause becomes acute when high values of input and feedback resistances are used (Fig. 9-16B). When these resistors are in the circuit, the bias current causes a voltage drop across the resistances (even when $V_{in} = 0$), causing an output voltage equal to:

$$V_{oo} = I_b R_f \tag{9-33}$$

Figure 9-16C shows the use of a compensation resistor, R_c, to reduce the offset potential as a result of input bias currents. This same resistor also improves thermal drift. The value of the resistor is equal to the parallel combination of the other two resistors:

$$R_c = \frac{R_f R_{in}}{R_f + R_{in}} \tag{9-34}$$

Approximately the same bias current flows in both inputs, so the compensation resistor will produce the same voltage drop at the noninverting input as appears at the inverting input. Because the op-amp inputs are differential, the net output voltage is zero.

The method of Fig. 9-16C is used where the source of the dc offset potential in the output signal is caused by the input bias currents. If the input signal contains an undesired dc component, the dc component will also create an amplifier output error. Depending on the specific situation,

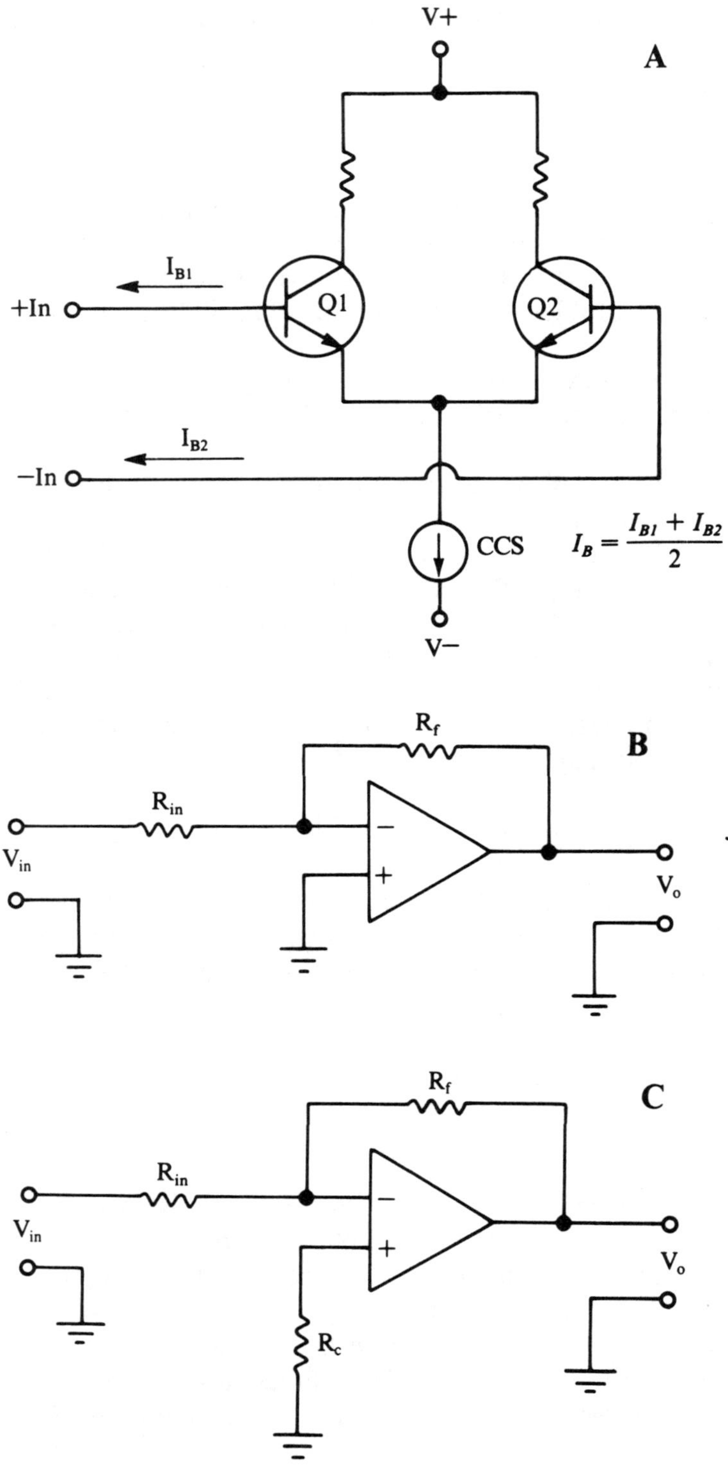

Figure 9-16. A) Dc differential amplifier input stage circuit, B) inverting-follower circuit, C) using a compensation resistor for eliminating offset voltage.

you might need a considerably greater range of control than is offered by Fig. 9-16C. For this purpose, turn to one of several external offset null circuit techniques.

Figure 9-17 shows two methods for nulling output offsets, regardless of whether the source is bias currents, other op-amp defects, or input-signal dc components. Figure 9-17A shows the use of the offset-null terminals that are found on some op amps. A potentiometer is placed between the terminals while the wiper is connected to the V− power supply. This potentiometer is adjusted to produce the required null. The input terminals are shorted together at ground potential and the potentiometer is adjusted to produce zero-Vdc output. In the LM-101/201/301 family of devices, the wiper terminal of the potentiometer is connected to common, rather than V− (see Fig. 9-17B).

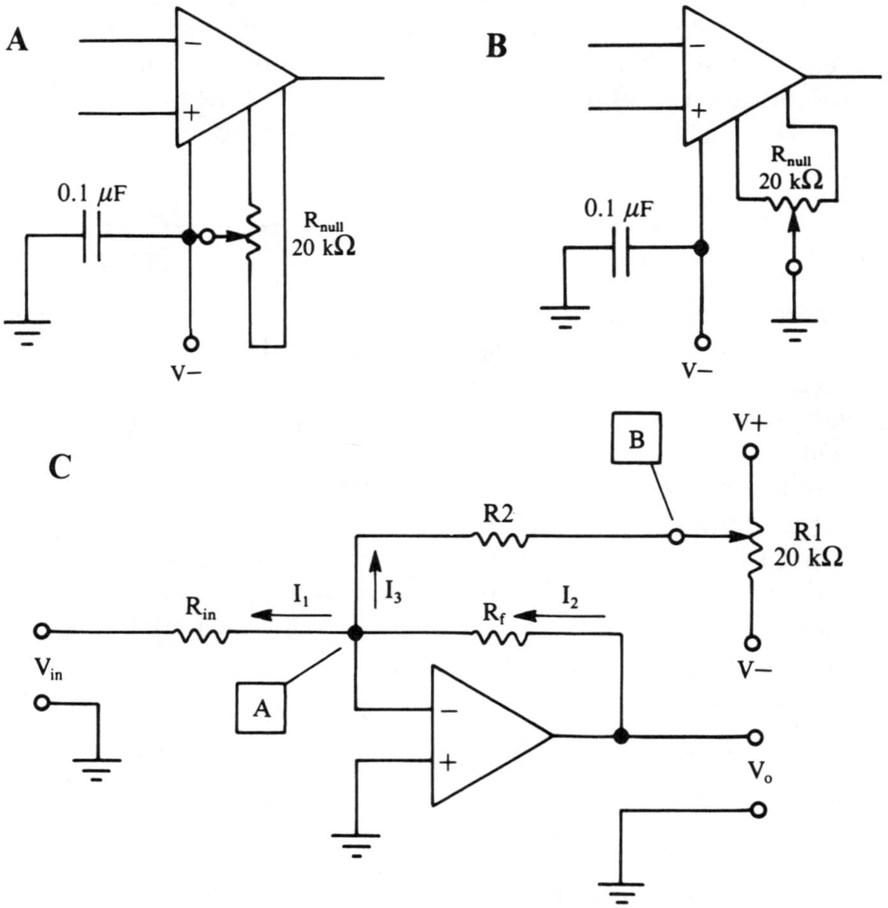

Figure 9-17. A) Using null terminals to null offset error, B) alternate method, C) universal offset null circuit.

Figure 9-17C shows a nulling circuit that can be used on any operational amplifier (inverting or noninverting), except the unity gain noninverting follower. A counter current (I_3) is injected into the summing junction (point A) of a magnitude and polarity to cancel the output offset voltage. The voltage at point B is set to produce a null offset at the output. The output voltage component as a result of this voltage is:

$$V'_o = -V_b \left(\frac{R_f}{R_2}\right) \tag{9-35}$$

If finer control over the output offset is required, the potentiometer can be replaced with another resistor voltage divider, which consists of a low-value potentiometer (e.g., 1 kohm) in series with two fixed resistors—one connected to each end of the potentiometer. Thus, the potentiometer resistance is only a fraction of the total voltage-divider resistance. In some cases, narrower limits are set by using zener diodes or three-terminal IC voltage regulators at each end of the potentiometer to reduce the range of V− and V+.

Summary

Noninverting follower circuits provide an output in phase with the input; the minimum gain is one. The simplest of these is the unity-gain amplifier, in which the input and the output have the same amplitude. The unity-gain amplifier provides buffering action and power gain. Varying the size of input and feedback resistors makes larger gains possible. Frequency response of noninverting followers can be controlled by the use of capacitors in both input and feedback circuits. Some op amps can use a single dc supply. In order for a single dc supply to be used for ac inputs, a voltage divider network is used to bias the noninverting input to a dc level. Differences between ideal and practical op amps include input offset current, input offset voltage, and input bias current.

Questions

1. What is a noninverting follower?
2. What is a unity gain amplifier? What are some of its applications?
3. How are resistor values for a noninverting follower determined?

4. What technique is used to set lower and upper breakpoint frequencies for ac-coupled noninverting amplifiers with gain?
5. How can a single dc supply provide biasing so that an ac signal can be amplified?
6. Explain how each of the following is determined: input offset current, input offset voltage, and input bias current.
7. What is slew rate? How does it relate to switching time?
8. How can the phase shift of an op amp be measured?
9. What is common mode rejection ratio?
10. Describe how to compensate for an output offset voltage.

Problems

Note: Assume that power supply voltages are ±12 volts unless indicated otherwise.

1. For the circuit in Fig. 9-4A, R_f = 100 kohms and R_{in} = 20 kohms. If the input voltage is +1.5 Vdc, what is the output voltage?
2. Repeat Problem 1 for R_f = 25 kohms and R_{in} = 10 kohms and an input voltage of −2 Vdc.
3. For the circuit in Fig. 9-5, R_f = 39 kohms, R_{in} = 13 kohms, and C_1 = 0.22 μF. **a.** What is the maximum amplitude of input signal that can be used before clipping occurs? **b.** What is the breakpoint frequency?
4. For the circuit in Fig. 9-5, R_{in} = 20 kohms and C1 = 10 μF. The input signal has an amplitude of 1 volt. **a.** What value of R_f should be used to have maximum gain without clipping? **b.** What is the breakpoint frequency for this value of R_f?
5. For the circuit in Fig. 9-6, R_{in} = R_1 = 10 kohms. If the required gain is 50, lower breakpoint frequency 400 Hz, and upper breakpoint frequency 800 Hz, determine R_{in}, C_1, and C_2.
6. For the circuit in Fig. 9-6, R_f = R_1 = 20 kohms. If the required gain is 20, lower breakpoint frequency 200 Hz, and upper breakpoint frequency 2 kHz, determine R_f, C_1, and C_2.
7. An op amp has a differential voltage gain of 10 and a common mode gain of 0.15. Calculate the common mode rejection ratio.
8. An op amp has a differential voltage gain of 50 and a common mode gain of 0.025. Calculate the common mode rejection ratio.

10 DC DIFFERENTIAL AMPLIFIERS

AFTER COMPLETING THIS chapter you will be able to:

- ☐ Explain the transistor differential amplifier
- ☐ Predict the result of having two input signals to the op amp
- ☐ Know the ways in which common-mode rejection can be improved

Most operational amplifiers have differential inputs. That is, the two inputs on the op amp (−In and +In) each provide the same voltage gain, but at the opposite polarity. The *inverting input* (*−In*) of the operational amplifier provides an output signal that is 180 degrees out of phase with the input signal. In other words, a positive-going input signal will produce a negative-going output signal, and vice-versa. The *noninverting input* (*+In*) provides an output signal that is in-phase with the input signal. For this type of input, a positive-going input signal will produce a positive-going output signal. These properties will now be used to understand a class of amplifiers in which both inputs are used. But before going further with differential amplifiers, let's revisit the basic differential-input stage of an operational amplifier.

Differential input stage

Figure 10-1 shows a hypothetical input stage for a differential-input op amp. Transistors Q1 and Q2 are a matched pair that share common V− and V+ dc power supplies through separate collector load resistors. The collector of transistor Q2 serves as the output terminal for the differential amplifier. Thus, collector voltage V_o is the output voltage of the stage. The exact value of the output voltage is the difference between the supply voltage ($V+$) and the voltage drop across resistor R2 (i.e., V_{R2}).

The emitter terminals of the two transistors are connected together, and are fed from a constant-current source (CCS). For analysis, you can

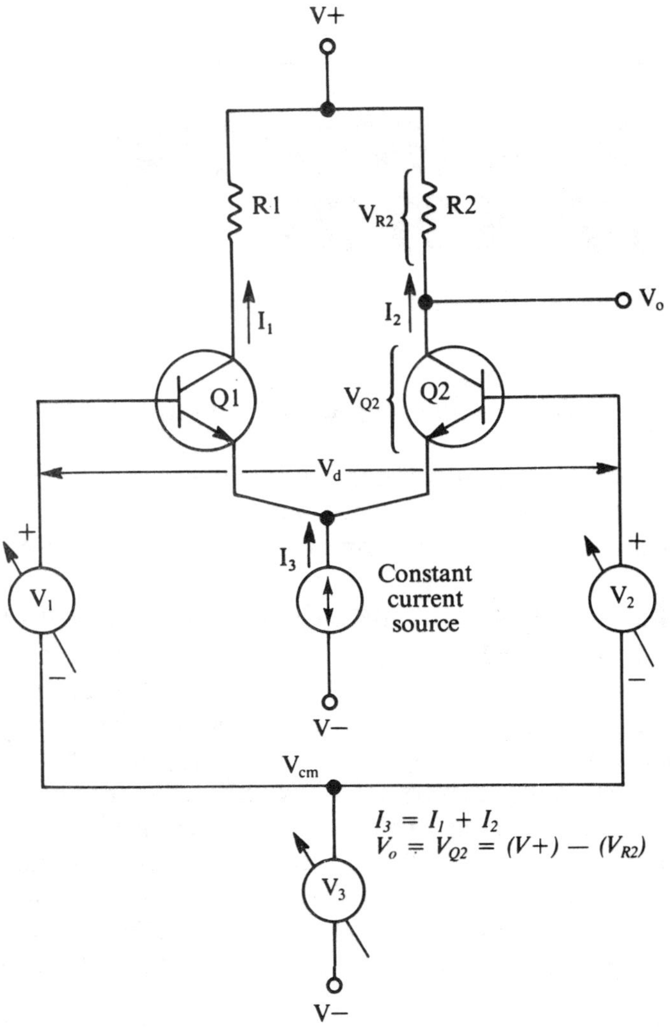

Figure 10-1. Dc differential-amplifier circuit.

assume that the collector and emitter currents (I_1 and I_2) are equal. Even though these currents are not actually equal, they are very close to each other. This convention is nonetheless useful for this purpose. Because of Kirchhoff's Current Law (KCL), we know that:

$$I_3 = I_1 + I_2 \tag{10-1}$$

Because current I_3 is a constant current, a change in either I_1 or I_2 will change the other. For example, an increase in current I_1 must force a decrease in current I_2 in order to satisfy Eq. (10-1).

Now, consider how the stage works. First, the case where $V_1 = V_2$. In this case, both Q1 and Q2 are biased equally, so the collector currents (I_1 and I_2) are equal. Under the normal situation, this case will cause V_{R2} to be equal to V_o, and also to $(V+)/2$.

Next, consider the case where V_1 is greater than V_2. Transistor Q1 is biased-on harder than Q2, so I_1 will increase. Because $I_1 + I_2$ is always equal to I_3 (which is held constant), an increase in I_1 will force a decrease in I_2. A decrease of I_2 will also reduce V_{R2} and increase V_o. Thus, an increase in V_1 increases V_o, so the base of transistor Q1 is the noninverting input of the differential amplifier.

Now consider the case where V_2 is greater than V_1. In this case, transistor Q2 is biased-on harder than Q1 so I_2 will increase and I_1 will decrease. An increase in I_2 forces a larger voltage drop (V_{R2}) across resistor R2, so V_o will decrease. Thus, an increase in V_2 forces a decrease in V_o, so the base of transistor Q2 is the inverting input of the differential amplifier.

Voltage V_3 is a common-mode signal, so transistors Q1 and Q2 are affected equally. For this kind of signal, output voltage V_o will not change.

The base bias currents that are required to keep transistors Q1 and Q2 operating become the input bias currents that make the practical op amp not ideal. In order to make the input impedance high, the device manufacturer makes these currents very low. Some manufacturers offer op amps that use MOSFET transistors (called *BiMOS op amps*) or JFET transistors (*BiFET op amps*) for the input stage. These types of transistors inherently have lower input bias currents than bipolar npn or pnp transistors. In fact, the bias currents approach mere leakage currents. Because of this low current level, BiMOS and BiFET op amps have extremely high input impedance levels ($Z_{in} = V_{in}/I_{in}$). One family of BiMOS devices, for example, offers $Z_{in} > 10^{12}$ ohms.

Input signal conventions

Figure 10-2 shows a generic differential amplifier with the standard signals applied. Signals V_1 and V_2 are single-ended potentials that are applied to the −In and +In inputs, respectively. The differential input signal, V_d, is the difference between the two single-ended signals: $V_d = (V_2 - V_1)$. Signal V_{cm} is a common-mode signal, that is, it is applied equally to both −In and +In inputs.

Common-mode signals

First, consider the common-mode signal, V_{cm}. A common-mode signal is a signal that is applied to both inputs at the same time. Such a signal might be either a voltage, such as V_{cm}, or a case where voltages V_1 and V_2 are equal to each other and are of the same polarity (i.e., $V_1 = V_2$). The implication of

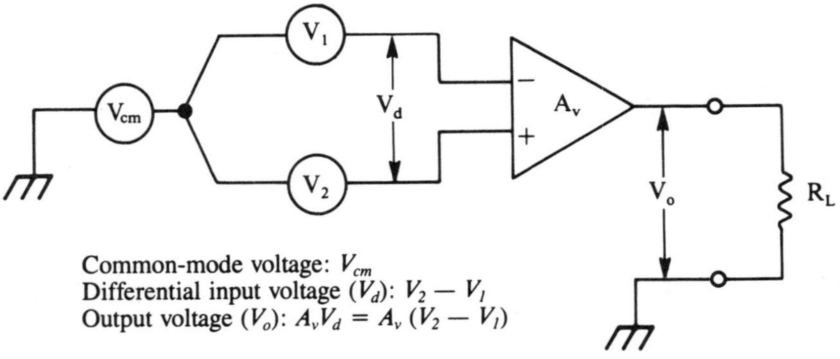

Figure 10-2. Signals applied to a dc differential amplifier.

the common-mode signal is that when it is applied equally to inverting and noninverting inputs, the output voltage is zero. Because the two inputs have equal but opposite polarity gains for common-mode signals, the net output signal in response to a common-mode signal is zero. This property is very useful, as you will see in later chapters.

An op amp with differential inputs cancels common-mode signals. An example of the usefulness of this property is in the performance of a differential amplifier, with respect to 60-Hz hum pick-up from local ac power lines. Almost all input-signal cables for practical amplifiers receive 60-Hz radiated energy and convert it to a voltage that is seen by the amplifier as a genuine input signal. In a differential amplifier, however, the 60-Hz field will affect both inverting and noninverting input lines equally, so the 60-Hz artifact signal will disappear in the output.

The practical operational amplifier will not perfectly reject common-mode signals. The common-mode rejection ratio (CMRR) shows the ability of any given op amp to reject such signals. The *CMRR* is usually specified in decibels (dB) and is:

$$CMRR = \frac{A_{vd}}{A_{cm}} \qquad (10\text{-}2)$$

or, in decibel form:

$$CMRR_{dB} = 20\text{LOG}\left(\frac{A_{vd}}{A_{cm}}\right) \qquad (10\text{-}3)$$

Where:

CMRR is the common-mode rejection ratio
A_{vd} is the voltage gain to differential signals
A_{cm} is the voltage gain to common-mode signals

In general, the higher the CMRR, the better the op amp. Typical low-cost devices have CMRR ratings of 60 dB or more, but better devices exhibit CMRR values up to 120 dB.

Differential signals

Signals V_1 and V_2 in Fig. 10-2 are single-ended. The total differential signal that is seen by the operational amplifier is the difference between the single-ended signals:

$$V_d = V_2 - V_1 \tag{10-4}$$

The output signal from the differential operational amplifier is the product of the differential voltage gain and the difference between the two input signals (hence the term *differential amplifier*). Thus, the transfer equation for the op amp is:

$$V_o = A_v(V_2 - V_1) \tag{10-5}$$

Differential amplifier gain equation

As is true with the other basic op-amp circuits, it is more instructive to develop the gain equation, than to simply list it. The math gets a little tedious and long-winded, but it is essentially quite easy.

The basic circuit for the dc differential amplifier is shown in Fig. 10-3A. This circuit uses only one operational amplifier, so it is the simplest possible configuration. Later, you will see additional circuits that are based on two or three op amps. In its most common form, the circuit of Fig. 10-3A is balanced so that $R_1 = R_2$ and $R_3 = R_4$.

Consider the redrawn differential-amplifier circuit in Fig. 10-3B. Assume that source resistances R_{S1} and R_{S2} are zero. Further assume that $R_1 = R_2 = R$, and that $R_3 = R_4 =$ kR, where k is a constant multiplier of R.

1. Set $V_2 =$ o and assume V_1 is nonzero. In this case V_o is found from:

$$V_{o1} = \frac{-kR}{R} V_1 \tag{10-6}$$

$$V_{o1} = -kV_1 \tag{10-7}$$

2. Now assume that $V_1 = 0$ and V_2 is nonzero:

$$V_a = \frac{V_2 KR}{KR + R} \tag{10-8}$$

$$V_a = \frac{V_2 K}{K + 1} \tag{10-9}$$

$$V_{o2} = V_a \left(\frac{KR}{R} + 1 \right) \tag{10-10}$$

$$V_{o2} = \left(\frac{KR}{R} + 1 \right) \left(\frac{V_2 k}{K + 1} \right) \tag{10-11}$$

$$V_{o2} = (K + 1) \left(\frac{V_2 k}{K + 1} \right) \tag{10-12}$$

$$V_{o2} = V_2 k \tag{10-13}$$

3. Now, by superimposing the two expressions for V_o:

$$V_o = V_{o2} + V_{o1} \tag{10-14}$$

$$V_o = (V_2 k) - (V_1 k) \tag{10-15}$$

$$V_o = k\,(V_2 - V_1) \tag{10-16}$$

According to Eq. (10-16), the output voltage is the product of the difference between single-ended input potentials V_1 and V_2 and a factor k. The differential input voltage (V_d) is:

$$V_d = V_2 - V_1 \tag{10-17}$$

And factor k represents the differential voltage gain, A_{vd}. Thus, the output voltage is:

$$V_o = V_d A_{vd} \tag{10-18}$$

You can also use a less parametric analysis by appealing to Fig. 10-3B. The following relationships are true (assuming $R_{s1} = R_{S2} = 0$):

$$V_a = \frac{(V_{cm} + V_2)\, R_4}{R_2 + R_4} \tag{10-19}$$

$$I_1 = -I_3 \tag{10-20}$$

$$I_1 = \frac{V_{cm} + V_1 + V_a'}{R_1} \tag{10-21}$$

$$I_3 = \frac{V_o - V_a'}{R_3} \tag{10-22}$$

From the properties of the ideal op amp, voltage $V_{a'} = V_a$, so;

$$I_1 = \frac{V_{cm} + V_1 - V_a}{R_1} \tag{10-23}$$

$$I_3 = \frac{V_o - V_a}{R_3} \tag{10-24}$$

Combine equations and solve for V_o:

$$V_o = V_{cm}\left(\frac{R_3\, R_4 + R_1\, R_4 - R_2\, R_3 - R_3\, R_4}{R_1(R_2 + R_4)}\right) - \left(\frac{R_3\, V_1}{R_1}\right) + \left(V_2\left(\frac{R_4}{R_2}\right)\left(\frac{1 + \left(\frac{R_3}{R_1}\right)}{1 + \left(\frac{R_4}{R_2}\right)}\right)\right) \tag{10-25}$$

Assuming that $R_1 = R_2$ and $R_3 = R_4$, Eq. (10-25) resolves to:

$$V_o = \left(\frac{R_3}{R_1}\right)(V_2 - V_1) \tag{10-26}$$

Equation (10-26) is similar to Eqs. (10-5) and (10-16). In this case, A_{vd} is R_3/R_1 and V_d is $(V_2 - V_1)$. The standard transfer equation for the single op-amp dc differential amplifier is:

$$V_o = V_d A_{vd} \tag{10-27}$$

$$V_o = V_d \left(\frac{R_3}{R_1}\right) \tag{10-28}$$

It is difficult to build a dc differential amplifier with a variable gain control. It is, for example, very difficult to find two ganged potentiometers (used to replace R3 and R4 in Fig. 10-3A) that track well enough to vary gain while maintaining the required balance. Figure 10-3C shows one attempt to solve the problem. In this case, a potentiometer is connected between the midpoint of the two feedback resistances. This circuit works, but the gain control is a nonlinear function of the potentiometer setting. Generally, a better practice is to use a *postamplifier stage* following the differential amplifier and control the gain in that stage. Alternatively, use one of the differential-amplifier circuits (shown later in this chapter), which use more than one operational amplifier.

Experiment 10-1

This experiment is designed to demonstrate the operation of the dc differential amplifier. Build the circuit of Experiment 10-1A on a breadboard (such as the Heathkit ET-3300) or on a piece of perfboard. Use either a CA-3140 (preferred) or a 741 IC.

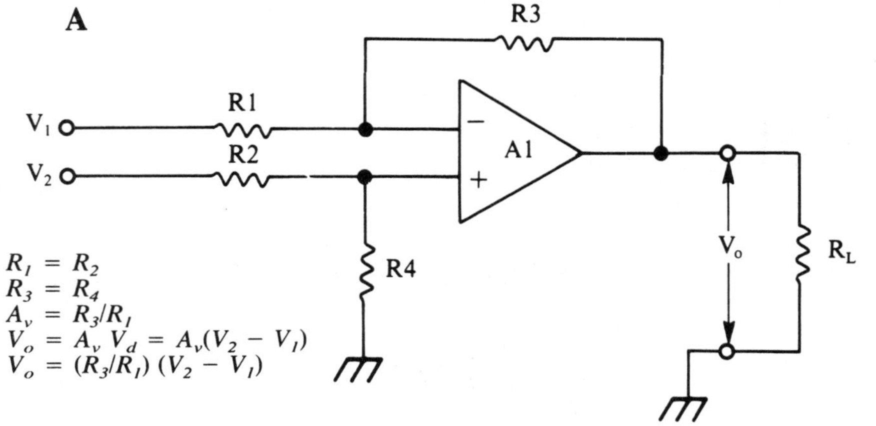

Figure 10-3. A) Circuit for a dc differential amplifier, B) signals and currents, C) gain-control differential amplifier.

B

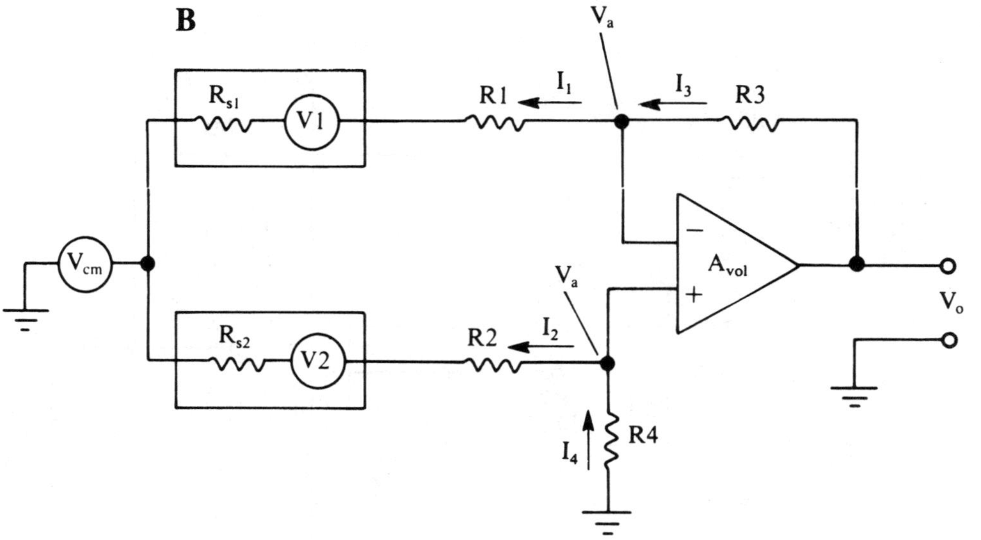

C

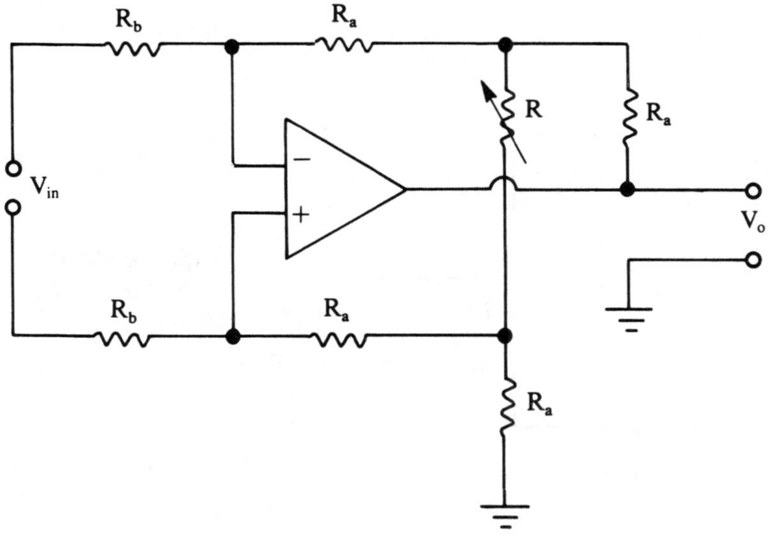

The input signal is a dc voltage from a pair of 1.5-volt batteries (alternatively, 1.3-volt mercury cells, such as HG1 or HG2, can be used). Points A and B (shown in squares) are the input points where input signal voltages are measured. Because only a few settings are used, however, you can calculate the input voltages once the dc battery voltages are known. Switches S1 and S2 select the input voltage parameters and meter M1 measures the dc output voltage (a digital voltmeter can be used at this point). Fill in the following table as you perform the experiment.

A

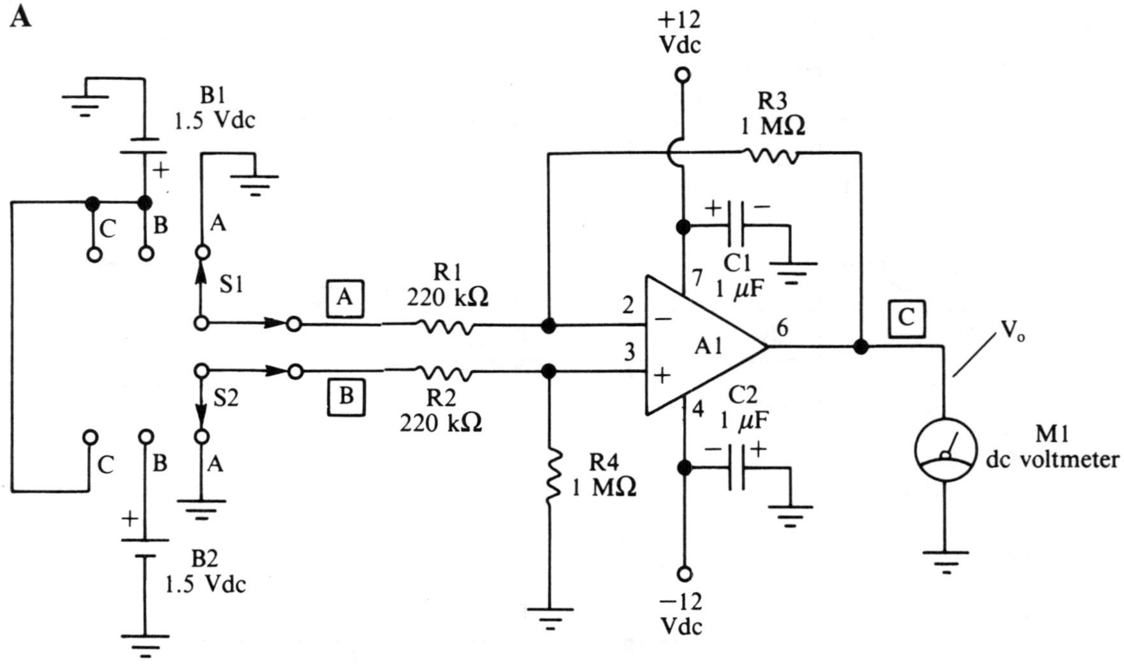

B

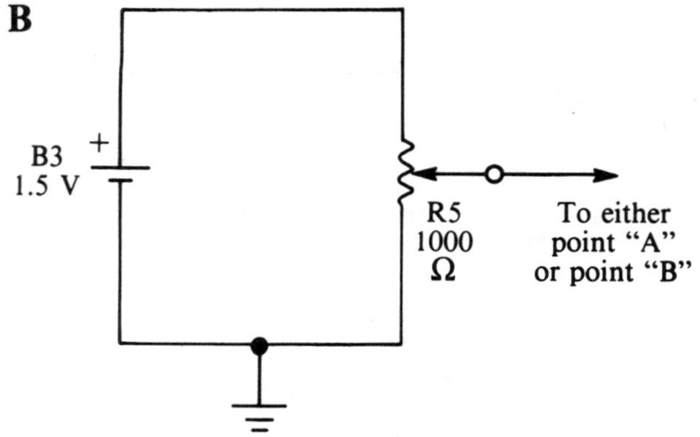

Experiment 10-1. A) Circuit for Experiment 10-1, B) dc signal source.

Trial	Input V_{in} $(V_B - V_A)$	Dc offset V_{off}	Measured V_o	Corrected Output V_{corr}
1.				
2.				
3.				
4.				
5.				

1. Calculate the dc differential voltage gain of the amplifier shown in Experiment 10-1. This is quantity A_v.
2. Set switches S1 and S2 to position A on each. This setting corresponds to the zero input voltage because the inputs are grounded.
3. Measure the output voltage on meter M1. This voltage represents the dc offset voltage (V_{off}) of the system, and it should be subtracted from future readings. The corrected output voltage in future cases is: $V_{corr} = V_o - V_{off}$.
4. Set switch S1 to position B and leave S2 at position A.
5. Measure the voltage from point A (the wiper of S1) to ground; this voltage is V_A. Similarly, measure the voltage from point B (the wiper of S2) to ground; this voltage is V_B.
6. Measure the dc output voltage.
7. Perform the calculation $V_{corr} = V_o - V_{off}$ (remember the signs of negative voltages, if any). How does this result compare with the theoretical that is obtained by multiplying the input voltage by the gain ($A_v V_{in}$)?
8. Repeat steps 3 through 7 using the following switch settings:

Trial	**S1**	**S2**	
2	A	B	
3	B	B	*Common-mode case 1*
4	C	C	*Common-mode case 2*
5	*Repeat Trial 3 with B1 reversed*		

9. Repeat the entire experiment with the circuit of Experiment 10-1B, but replace either battery B1 or B3 (i.e., connected to either point A or B in Experiment 10-1A). Use various potentiometer settings to supply different input voltages.

Common-mode rejection

Figure 10-4 shows two different situations. Figure 10-4A shows a single-ended amplifier and Fig. 10-4B shows a differential amplifier in a similar situation. In these circuits, a noise signal, V_n, is placed between the input ground and the output ground. This noise signal might be either ac or dc noise. Figure 10-4A is a case where the noise signal is applied to a single-ended input amplifier. The input signal seen by the amplifier is the algebraic sum of the two independent signals: $V_{in} + V_n$. Because of this fact, the amplifier output signal (V_o) will see a noise artifact that is equal to the product of the noise-signal amplitude and the amplifier gain, $V_o = -A_v V_n$.

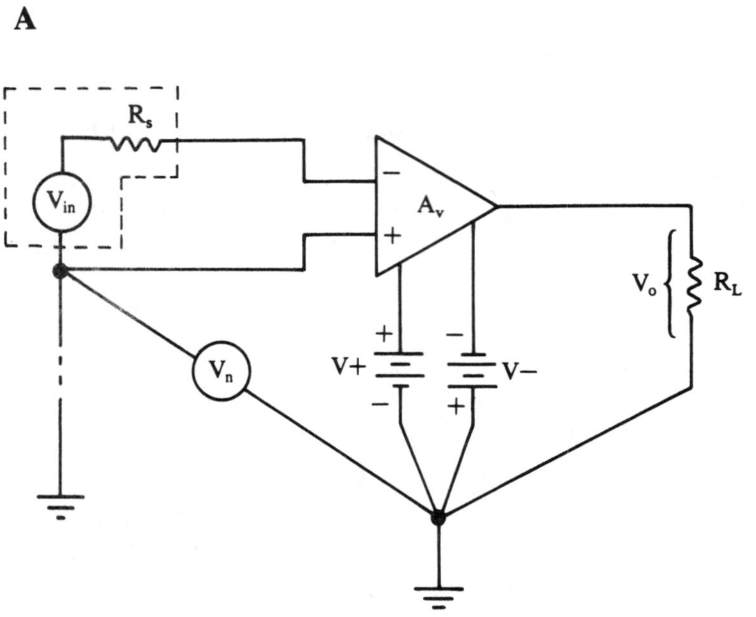

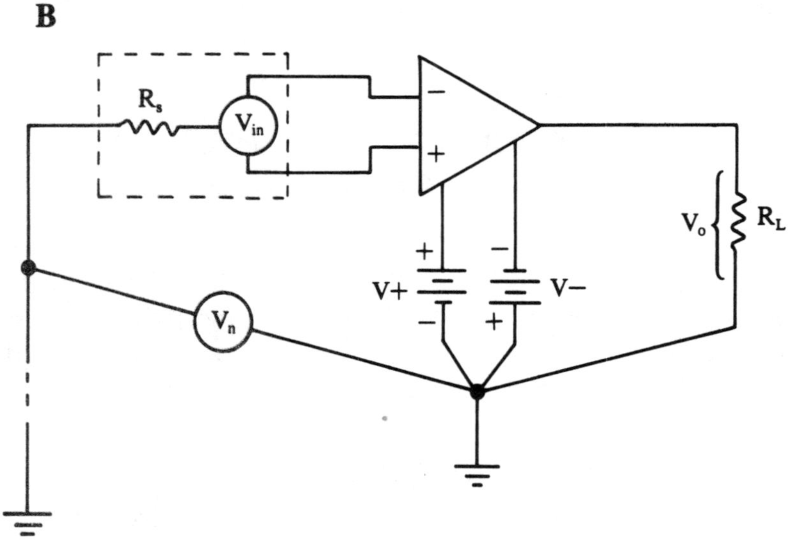

Figure 10-4. A) single-point grounding in an amplifier circuit, B) alternate circuit.

Now consider the situation of the differential amplifier that is depicted in Fig. 10-4B. The noise signal in this case is common mode, so it is essentially cancelled by the common-mode rejection ratio of the amplifier. Of course, in nonideal amplifiers, the input signal, V_{in}, is subject to the differential gain and the noise signal, V_n, is subject to the common-mode gain. If an amplifier has a CMRR of 90 dB, for example, the gain seen by the noise signal will be 90 dB below the differential gain.

The CMRR of the op amp dc differential circuit is principally dependent upon two factors: the natural CMRR of the op amp used as the active device and the balance of the resistors ($R_1 = R_2$ and $R_3 = R_4$). Unfortunately, the balance is typically difficult to obtain with common fixed resistors (not considering precision types). A circuit, such as Fig. 10-5, can be used to compensate for the mismatch. In this circuit, R_1 through R_3 are exactly the same as in previous circuits. The fourth resistor, however, is a potentiometer (see R4). The potentiometer will "adjust out" the CMRR errors that are caused by resistor and other circuit mismatches.

A version of the circuit that has greater resolution is shown in the inset to Fig. 10-5. In this version, the single potentiometer is replaced by a fixed resistor and a potentiometer in series. The sum resistance ($R_{4A} + R_{4B}$) is equal to approximately 20 percent more than the normal value of R_3. Ordinarily, the maximum value of the potentiometer is 10 to 20 percent of the overall resistance.

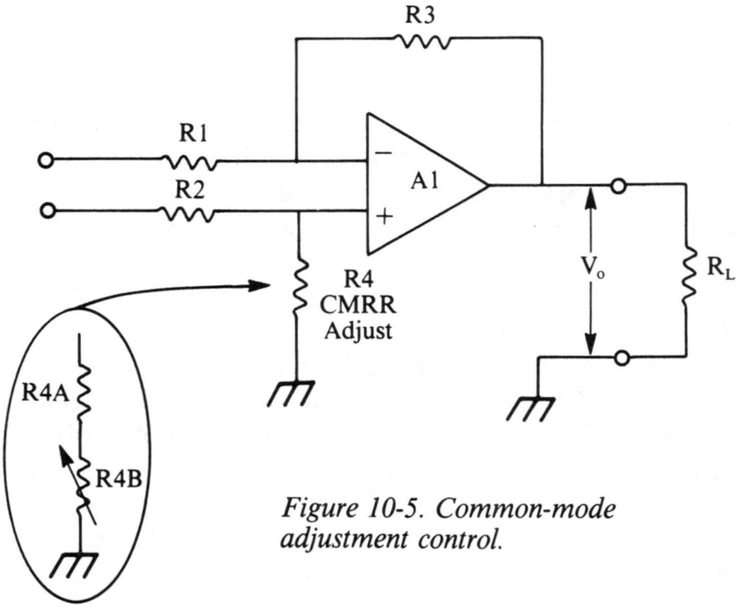

Figure 10-5. Common-mode adjustment control.

The adjustment procedure for either version is the same (see Fig. 10-6):

1. Connect a zero-center dc voltmeter to the output terminal (M1).
2. Short inputs A and B together, then connect them to either a signal voltage source or ground.

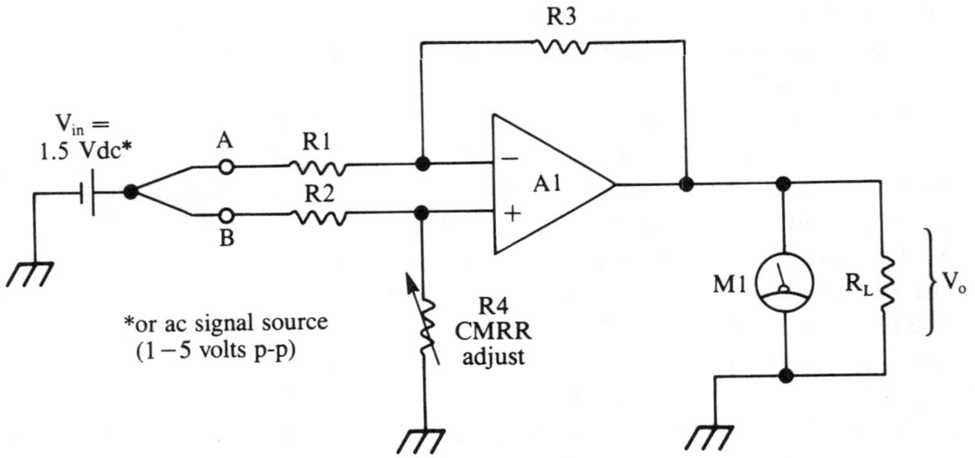

Figure 10-6. Circuit for adjusting CMRR.

3. Adjust potentiometer R4 (CMRR ADJ) for zero volts output.
4. If the output indicator (meter M1) has several ranges, then switch to a lower range and repeat step 3 until no further improvement is possible.

Alternatively, connect the output to an audio voltmeter or to an oscilloscope and connect the input to a 1-to-14-volt peak-to-peak ac signal that is within the frequency range of the particular amplifier. For audio amplifiers, a 400-to-1000-Hz 1-volt signal is typically used.

Experiment 10-2

This experiment is designed to teach the method for adjusting the dc differential amplifier for improved CMRR. Connect the circuit of Experiment 10-2 with a CA-3140 or 746 IC op am. Do not apply the ac input signal until it is specified in the directions.

1. Connect the circuit and check for errors in wiring.
2. Ground the inputs (points A and B) and set potentiometer R4 to about mid-range.
3. Turn on dc power to the experiment.

Note: The signal source should be an audio signal generator set to a frequency of 100 to 1000 Hz (400 Hz nominal) with an output voltage of

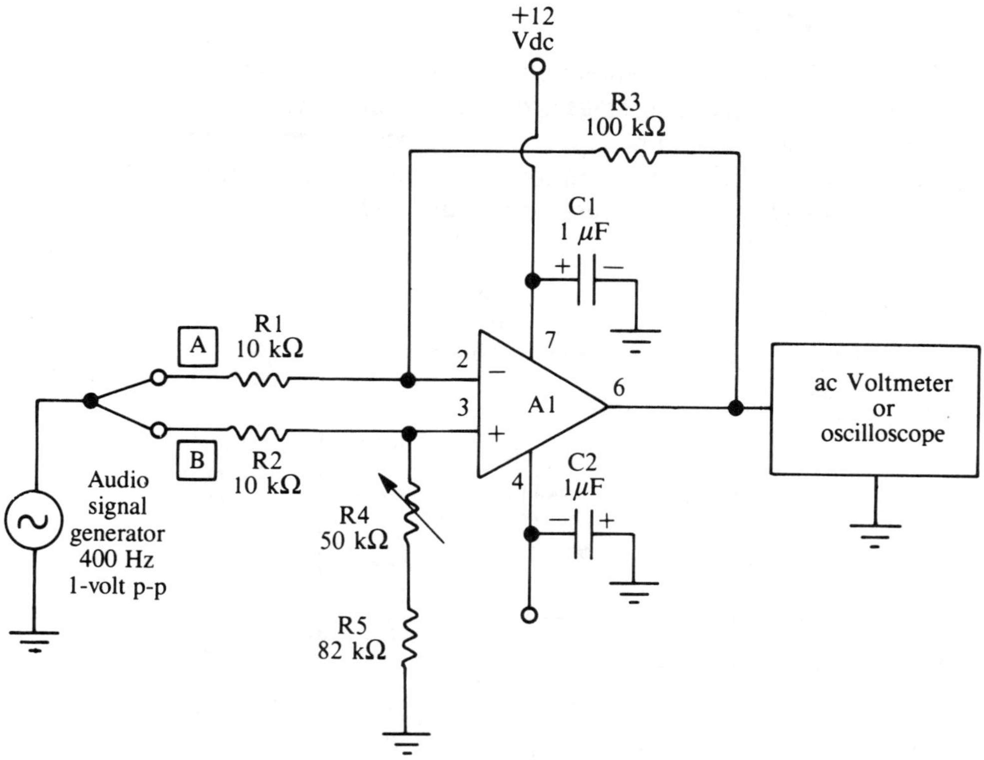

Experiment 10-2. Circuit for Experiment 10-2.

approximately 1 volt peak-to-peak. The input signal should remain constant during this test.

4. Unground the inputs and connect both A and B to the output of the signal generator.
5. Adjust the vertical input controls of the oscilloscope until a readable trace is seen. It should be a sine wave at the same frequency as the input signal.
6. Adjust R4 for minimum signal amplitude on the 'scope. Continually readjust the 'scope input controls for more sensitivity and repeat the nulling process until no further improvement is possible.
7. Measure the peak-to-peak ac output voltage.
8. Measure the peak-to-peak ac input voltage.
9. Compare V_{in} and V_o.

Project 10-1

Design a dc differential amplifier with a gain of 100. Assume that the source impedance of the preceding stage is about 100 ohms (refer to Project 10-1).

Because the source impedance is 100 ohms, we need to make the input resistors of the dc differential amplifier 10 times larger (or more). Thus, the input resistors (R_1 and R_2) must be at least 1000 ohms; $R_1 = R_2 = 1000$ ohms.

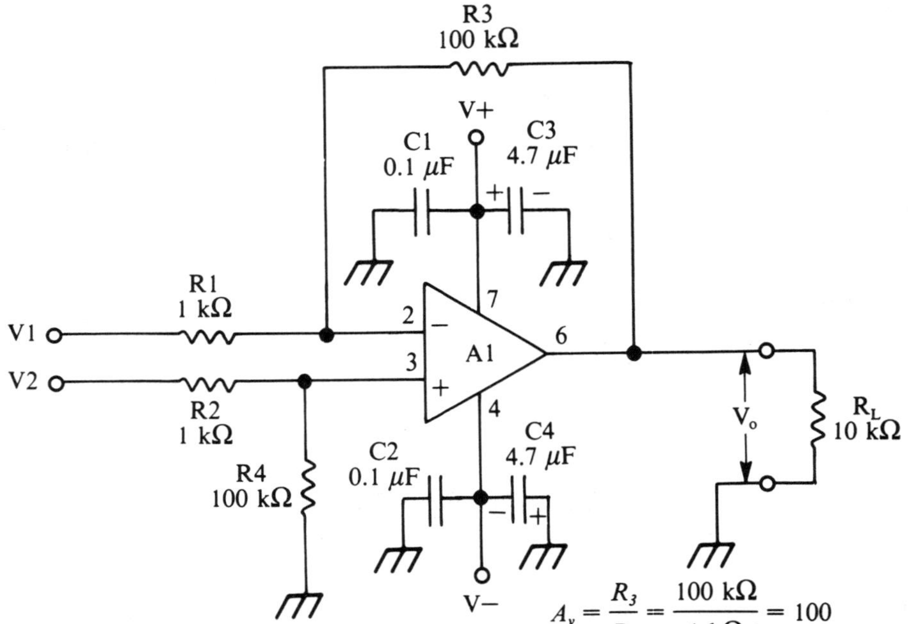

$$A_v = \frac{R_3}{R_1} = \frac{100\ \text{k}\Omega}{1\ \text{k}\Omega} = 100$$

Project 10-1. Circuit for Project 10-1.

Set the value of the two feedback resistors (keep in mind that $R_3 = R_4$). The value of these resistors is found from rewriting Eq. (10-29).

$$R_3 = A_v R_1 \qquad (10\text{-}29)$$
$$R_3 = (100)\,(1000) \qquad (10\text{-}30)$$
$$R_3 = 100{,}000 \text{ ohms} \qquad (10\text{-}31)$$

Project 10-1 shows the finished circuit of this amplifier. The values of the resistors are:

$R_1 = 1$ kohms
$R_2 = 1$ kohms
$R_3 = 100$ kohms
$R_4 = 100$ kohms

For best results, make R_1, R_2, R_3, and R_4 one-percent precision resistors. The pinouts shown in Project 10-1 are for the industry-standard 741 family of operational amplifiers. These same pinouts are found on many other different op-amp products as well.

The dc power-supply voltages are usually either ±12 Vdc, or ±15 Vdc. Lower power-supply potentials can be accommodated, however, a corresponding reduction in the output voltage swing is tolerable. A typical lower-grade op amp will produce a maximum output voltage that is approximately 3 volts lower than the supply voltage. For example, when V+ is 12 Vdc, the maximum positive output voltage permitted is $(12 - 3) = 9$ volts.

Project 10-1 shows the circuit of the simple dc differential amplifier that is based on the operational amplifier device. The gain of the circuit is set by the ratio of two resistors:

$$A_v = \frac{R_3}{R_1} \qquad (10\text{-}32)$$

or,

$$A_v = \frac{R_1}{R_2} \qquad (10\text{-}33)$$

Provided that $R_1 = R_2$ and $R_3 = R_4$.

The output voltage of the dc differential amplifier is given by:

$$V_o = \frac{R_3}{R_1}(V_2 - V_1) \qquad (10\text{-}34)$$

or,

$$V_o = A_V (V_2 - V_1) \qquad (10\text{-}35)$$

Where:

V_o is the output voltage
R_3 is the feedback resistance
R_1 is the input resistance
A_v is the differential gain
V_1 is the signal voltage applied to the inverting input
V_2 is the signal voltage applied to the noninverting input

Again, the constraint is that $R_1 = R_2$, and $R_3 = R_4$. These two balances must be maintained or the CMRR will deteriorate rapidly. In many common applications, the CMRR can be maintained within reason by specifying 10-percent tolerance resistors for R1 through R4. However, where superior CMRR is required, especially where the differential voltage gain is high, closer tolerance resistors (1 percent or better) are required.

Circuit stability

The decoupling bypass capacitors (in Project 10-1) are used to keep the circuit stable (Fig. 10-7), especially in cases where the same dc power supplies are used for several stages. The low-value 0.1-μF capacitors (C1 and C2) are used to decouple high-frequency signals. These capacitors should be placed as close as possible to the body of the operational amplifier.

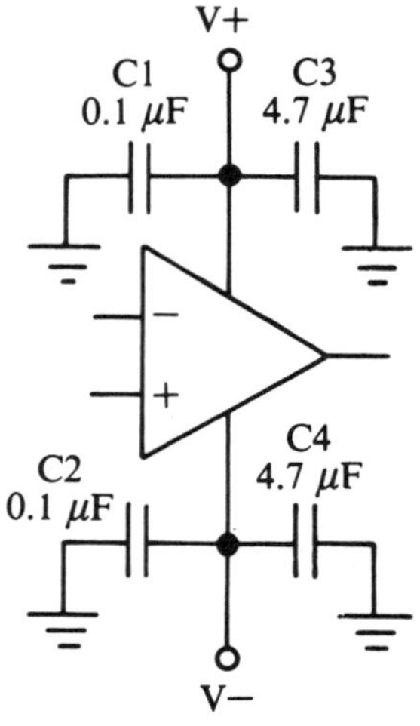

Figure 10-7. Decoupling capacitors for added stability.

The high-value capacitors (C3 and C4) are needed to decouple low-frequency signals. Two values of capacitors are needed on some op amps because the high-value capacitors that are needed for low-frequency decoupling are typically electrolytics (tantalum or aluminum), some of which are ineffective at high frequencies. Thus, you must provide smaller-value capacitors that have a low enough capacitive reactance to do the job, and are effective at high frequencies. The extra capacitor is used on uncompensated op amps that have a high gain-bandwidth product.

A different situation is shown in Fig. 10-8. Most differential amplifiers have relatively low input impedances, which is a function of factors, such as input bias currents and so forth. The amplifier in Fig. 10-8 uses a high input impedance by virtue of the high values of resistors R1 and R2. In order to attain this input impedance, however, you need to specify an op amp for A1

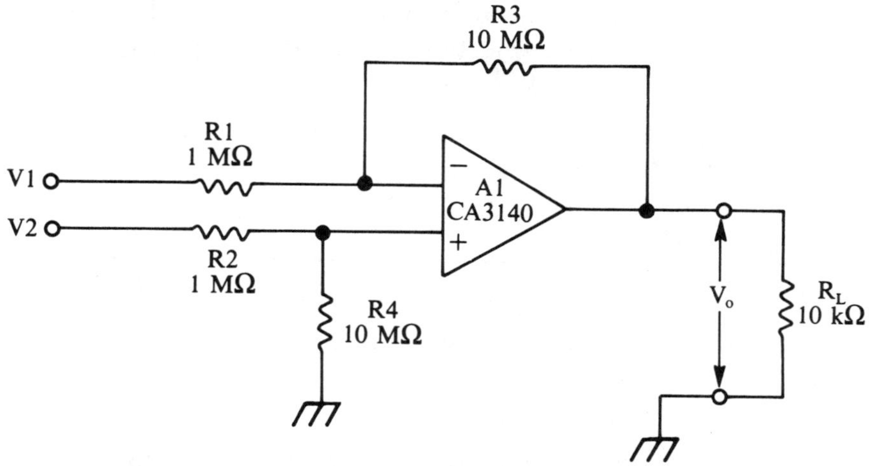

Figure 10-8. Gain-of-10 dc differential amplifier.

that has a very low input bias current (i.e., a very high natural input impedance). The BiMOS devices, which use MOSFET input transistors, and the BiFET, which uses JFET input transistors, are good selections for dc differential amplifier circuits when very high input impedances are needed.

Summary

The differential amplifier takes two different inputs and amplifies the difference between them. This can be done by using two matched transistors, but it is most commonly done by using an op amp circuit. In this circuit, both inputs have two resistors. The inverting input contains an input and feedback resistor, and the noninverting input contains two resistors acting as a voltage divider for the op amp. In order to maximize common-mode rejection, one of the resistors in the noninverting network is made variable.

Questions

1. What is a differential amplifier?
2. For the transistor differential amplifier circuit in Fig. 10-1, explain why V_1 is the noninverting input and V_2 is the inverting input.

3. What is common-mode rejection? Why is a differential amplifier better for common-mode rejection than an inverting or noninverting amplifier?
4. What is the purpose of using capacitors in the positive and negative power supplies of an op amp?

Problems

1. An op amp has a differential voltage gain of 20 and a common mode gain of 0.1. **a.** Calculate the common-mode rejection ratio. **b.** Convert this value to decibels.
2. An op amp has a differential voltage gain of 50 and a common mode gain of 0.02. **a.** Calculate the common-mode rejection ratio. **b.** Convert this value to decibels.
3. For the circuit in Fig. 10-9, $R_1 = R_2 = 50$ kohms, and $R_3 = R_4 =$ 250 kohms. Calculate the output voltage for the 1.5-Vdc input.

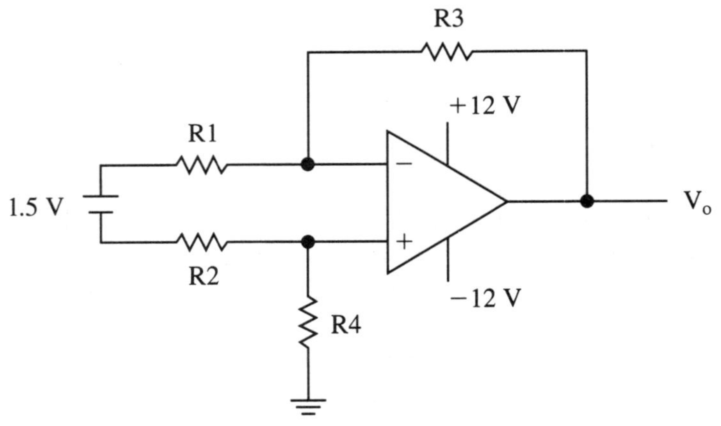

Figure 10-9.

4. For the circuit in Fig. 10-9, $R_1 = R_2 = 25$ kohms, and $R_3 = R_4 =$ 100 kohms. What is the output voltage for the 1.5-Vdc input?
5. For the circuit in Fig. 10-10, $R_1 = R_2 = 20$ kohms, and $R_3 = R_4 =$ 100 kohms. If $V_1 = 1$ volt and $V_2 = 1.5$ volts, calculate **a.** the voltage at point A **b.** the output voltage

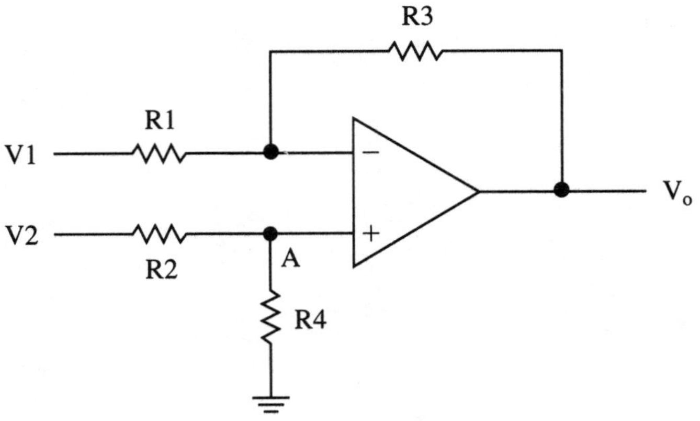

Figure 10-10.

6. Repeat Problem 5 for $R_1 = R_2 = 10$ kohms, $R_3 = R_4 = 100$ kohms, $V_1 = 1.2$ volts, and $V_2 = 1.1$ volts.

11

ADDITIONAL DIFFERENTIAL AMPLIFIER CIRCUITS

AFTER COMPLETING THIS chapter, you will be able to:

- ☐ Understand the characteristics of the instrumentation amplifier
- ☐ Describe some sample instrumentation amplifier circuits
- ☐ Discuss commercial instrumentation amplifier ICs

In past chapters, you learned several different op-amp circuit configurations. The single-ended circuits (i.e., those with voltage signal sources that are unbalanced, with respect to ground) were the inverting and noninverting followers. A balanced-input circuit, the dc differential amplifier, was introduced in chapter 10.

These simple dc differential amplifier circuits are useful for low- to medium-gain applications and for applications where a low-to-moderate input impedance is permissible (e.g., 300 to 1,000,000 ohms). Where a higher voltage gain is required, it is prudent to use a more complex circuit, called the *operational-amplifier instrumentation amplifier*, *OAIA* or simply *IA*. It is possible to obtain the IA in a single small IC package.

Instrumentation amplifiers

The simple dc differential amplifier, which was discussed earlier, suffers from several important drawbacks. First, the input impedance is limited (Z_{in} is approximately equal to the sum of the two input resistors).

Second, available gain from the simple single-device dc differential amplifier of chapter 10 is limited. If high gain is attempted, either the input bias current causes large output offset voltages or the input impedance becomes too low. This chapter demonstrates a solution to these problems in the form of the instrumentation amplifier. All of these amplifiers are differential amplifiers, but they offer superior performance over the simple dc differential amplifiers of the last chapter. The instrumentation amplifier can offer high input impedance, higher gain, and better common-mode rejection than the single-device dc differential amplifier.

Simple IA circuit

The simplest form of instrumentation amplifier circuit is shown in Fig. 11-1. In this circuit, the input impedance is improved by connecting inputs of a simple dc differential amplifier (A3) to two input amplifiers (A1 and A2), which are each of the unity-gain noninverting-follower configuration (used here as buffer amplifiers). The input amplifiers offer an extremely large input impedance (a result of the noninverting configuration) while driving the input resistors of the actual amplifying stage (A3). The overall gain of this circuit is the same as for any simple dc differential amplifier:

$$A_v = \frac{R_3}{R_1} \tag{11-1}$$

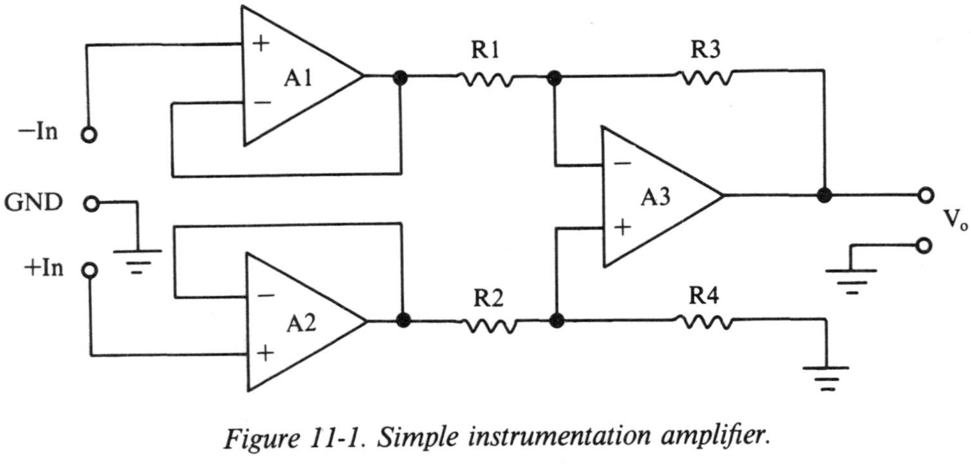

Figure 11-1. Simple instrumentation amplifier.

Where:

A_v is the voltage gain

Assuming that $R_1 = R_2$ and $R_3 = R_4$.

A1 and A2 should be identical operational amplifiers. In fact, it is advisable to use a dual op amp for both A1 and A2 (i.e., two op amps in a common IC package). The common thermal environment of the dual op amp will reduce thermal-drift problems. The very high input impedance of superbeta (i.e., Darlington), BiMOS, and BiFET op amps make them ideal for use as input amplifiers in this type of circuit.

One of the biggest problems with the circuit of Fig. 11-1 is that it wastes two good operational amplifiers. The most common instrumentation amplifier circuit uses the input amplifiers to provide voltage gain in addition to a higher input impedance. Such amplifier circuits are discussed in the next section.

Standard instrumentation amplifiers

The standard instrumentation amplifier is shown in Fig. 11-2. Like the simple circuit discussed above, this circuit uses three operational amplifiers. The biggest difference is that the input amplifiers (A1 and A2) are now used in the noninverting-follower-with-gain configuration. Like the circuit of Fig. 11-1, the input amplifiers are ideally BiMOS, BiFET, or superbeta input types in order to obtain the maximum possible input impedance. Again, for best thermal performance use a dual, triple, or quad op amp for this application. The signal voltages, shown in Fig. 11-2, follow the standard pattern: voltages V_1 and V_2 form the differential input signal ($V_2 - V_1$) and voltage V_{cm} represents the common-mode signal, because it affects both inputs equally.

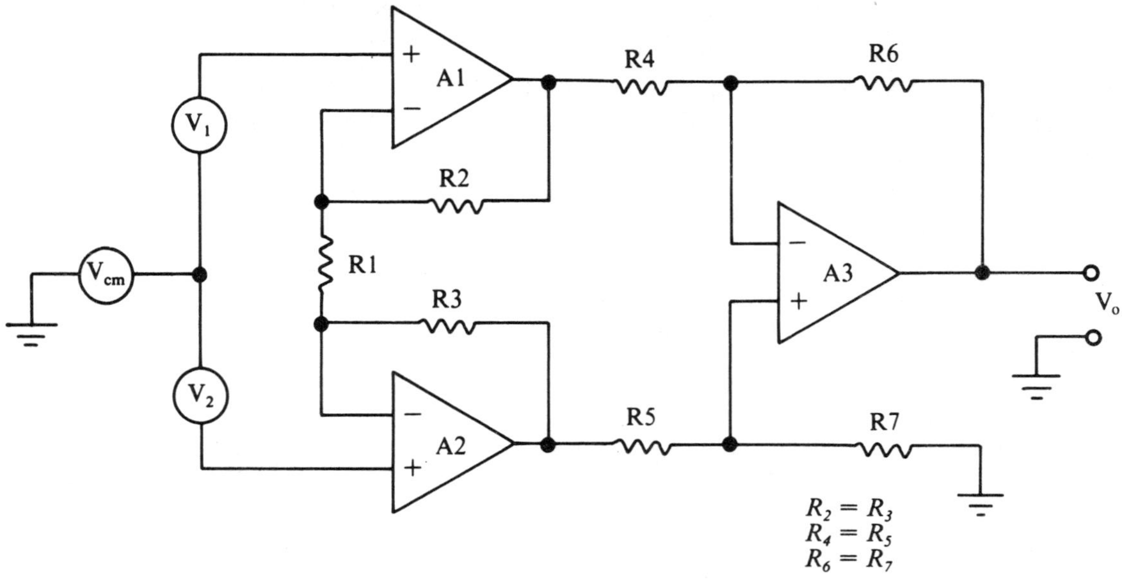

Figure 11-2. Standard instrumentation amplifier.

Gain equation for the IA circuit

The instrumentation amplifier consists of a pair of noninverting followers that drive a simple dc differential amplifier. Not surprisingly, the gain equation for the IA is, therefore, a combination of the two:

$$A_v = \left(\frac{2R_2}{R_1} + 1\right)\left(\frac{R_6}{R_4}\right) \tag{11-2}$$

Provided that $R_2 = R_3$, $R_4 = R_5$ and $R_6 = R_7$.

Notice in the gain equation that resistance R_1 is in the denominator. Further, notice that it does not have to be equal to any other resistance. It is therefore convenient to use R_1 as a gain control, as long as R_1 does not become zero.

Gain control for the IA

It is difficult to provide a gain control for a simple dc differential amplifier without adding an extra amplifier stage (for example, an inverting follower postamplifier with a gain of 0 to -1). For the instrumentation amplifier, however, resistor R1 can be used as a gain control provided that the resistance does not drop near zero ohms.

Figure 11-3 shows a revised circuit with resistor R1 replaced by a series circuit that consists of fixed resistor R1A and potentiometer R1B. This circuit prevents the gain from rising to above the level set by R1A. Don't use a potentiometer alone in this circuit because it can have a disastrous effect on the gain. Note in Eq. (11-2) that the term R_1 appears in the denominator. If the R_1 drops close to zero, the gain becomes very high (in fact, supposedly to infinity if $R_1 = 0$). The gain of the circuit in Fig. 11-3 varies from a minimum of 167 (when R_{1B} is 2000 ohms) to a maximum of 1025 (When R_{1B} is zero). The gain expression for Fig. 11-3 is:

$$A_v = \left(\frac{2\,R_2}{R_{1a} + R_{1b}} + 1\right)\left(\frac{R_6}{R_4}\right) \tag{11-3}$$

or, rewriting Eq. (11-3) to take into account for the fact that R_{1A} is fixed,

$$A_v + \left(\frac{2\,R_2}{390 + R_{1b}} + 1\right)\left(\frac{R_6}{R_4}\right) \tag{11-4}$$

Where:

R_{1b} varies from 0 to 2000 ohms.

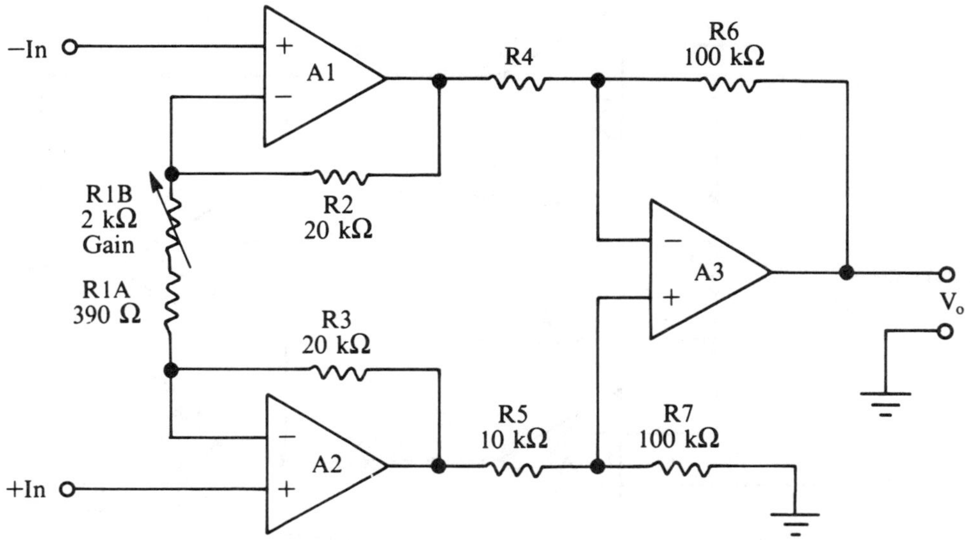

Figure 11-3. Variable-gain instrumentation amplifier.

Common-mode rejection ratio adjustment

The instrumentation amplifier is no different from any other practical dc differential amplifier in that the common-mode signals will be in imperfect balance. The operational amplifiers are not ideally matched, so a gain imbalance will exist. This gain imbalance is further deteriorated by the mismatch of the resistors. The result is that the instrumentation amplifier will respond at least to some extent to common-mode signals. As in the simple dc differential amplifier you can provide a common-mode rejection-ratio adjustment by making resistor R7 variable (see Fig. 11-4).

One configuration of Fig. 11-4 uses a single potentiometer (R7), which has a value that is 10 to 20 percent larger than the required resistance of R_6. For example, if R_{14} is 100 kohms, then R_7 should be 110 to 120 kohms. Unfortunately, these values are somewhat difficult to obtain, so pick a standard value for R_7 (e.g., 100 kohms), then select a value for R_6 that is somewhat lower (e.g., 82 kohms or 91 kohms).

The second configuration of Fig. 11-4 uses a fixed resistor in series with a potentiometer. The general rule is to make R_{7A} approximately 80 percent of the total required value, and R_{7B} 40 percent of the required value. As was true in the other configuration, the sum of R_{7A} and R_{7B} is approximately 110 to 120 percent of resistance R_6. Resistance values of this sort permit the total resistance to vary from less than to greater than the nominally required value.

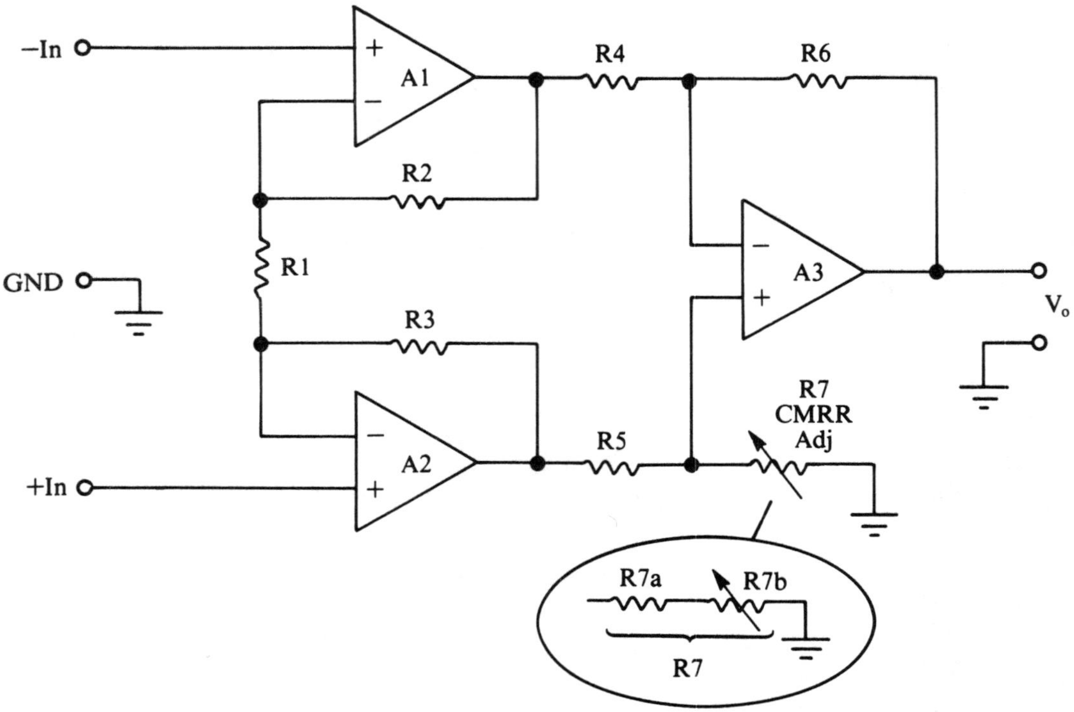

Figure 11-4. Instrumentation amplifier with CMR adjust.

Experiment 11-1

In this experiment, you repeat Experiment 10-2 with an instrumentation-amplifier circuit. Construct the circuit of Fig. 11-4 with the following values for the resistors:

R1:	1000 ohms
R2, R3:	47 kohms
R4, R5:	51 kohms
R6:	100 kohms
R7:	100-kohm 10-turn linear-taper potentiometer

Use CA-3140 operational amplifiers for A1, A2, and A3. The pinouts are the industry standard for the 741:

Pin	Function
2	Inverting input
3	Noninverting input

4	−12 volts
6	Output
7	+12 volts

Bypass pins 4 and 7 with 1μF tantalum capacitors (observe polarity). With this circuit, repeat the CMRR experiment in chapter 10.

Project 11-1 Galvanometer/amplifier

One little box that is useful on almost everyone's workbench is a variable-gain amplifier. Such an amplifier should offer a medium gain, a high gain, and a variable gain over a wide range. The amplifier should also have a very high input impedance (in order to keep from loading the circuit being monitored), and operate well into the audio region. A wider frequency-response range might be nice, but no currently available low-cost devices have this. Another desirable feature is that the amplifier should have both an analog output to external devices (such as oscilloscopes or paper recorders), as well as an on-board output indication. The amplifier should be able to be used as an inverting amplifier, a noninverting amplifier, or a differential amplifier. Above all, the amplifier should also be inexpensive.

It is also nice to have another "little box," a *zero-center dc voltmeter.* Many circuits, such as the Wheatstone bridge, require the ability to find a *null condition.* Zero-center analog-meter movements are often called *galvanometers.* A large number of instrumentation and radio-alignment tasks require a zero-center dc meter that offers a high-impedance load to the circuit being measured (low-impedance meters load down the circuit and either distort the readings or make the circuit inoperative).

This project is a combination of the two "little boxes," i.e., an amplifier and galvanometer in the same package—the *galviamp* (see Fig. 11-5). The circuit is described in detail; construction hints, the initial adjustment procedure, and the operating tips are included.

The circuit

The basic circuit for the galviamp (Fig. 11-6) is known as an instrumentation amplifier (IA). An IA uses three op amp ICs to form a differential amplifier with a very high input impedance. Four op amps used are in this project, not just three. The fourth op amp is used as a driver for a pair of back-to-back *light-emitting diodes* (*LEDs*) that help find the null condition. These four op amps are divided between two dual op amps (i.e., ICs that contain two independent op amps each). IC1 must be a CA-3240 BiMOS

Figure 11-5. Photo of galviamp front panel.

operational amplifier, but the other (IC2) can be a lower-grade part, such as the 1458.

The differential-input amplifier is formed from IC1A and IC1B. Each op amp is connected in the noninverting follower configuration. This circuit is used because the input impedance of the overall amplifier is essentially the input impedance of the device (which is not true if the inverting-follower circuit is used). For a BiMOS op amp, the input impedance is on the order of 10^{12} ohms!

Op-amp IC2A is a dc differential-amplifier circuit that combines the outputs of IC1A and IC1B ($V_C = V_B - V_A$). This output voltage is applied to both a meter (M1) and an output jack (J4), which can be used to send the analog output signal to an oscilloscope, external voltmeter, or other display device.

The voltage gain A_v) of the galviamp is a function of the resistance between points A and B in Fig. 11-6. The resistance (in ohms) to connect between A-B is found from $R = 112{,}000/(A_v - 1)$. In the circuit shown in Fig. 11-6, the resistance A-B is selected by a three-position double-pole (3PDP) rotary switch (S1A and S1B). This switch selects a 220-ohm resistance in position A, a 2,200-ohm resistance in position B, and a variable resistance of 150 to 1,150 ohms in position C. These resistances translate to fixed gains of about 500 and 50 for positions A and B, and a variable gain of 96 to 745 in position C. In position C, the gain is set by adjusting potentiometer R4.

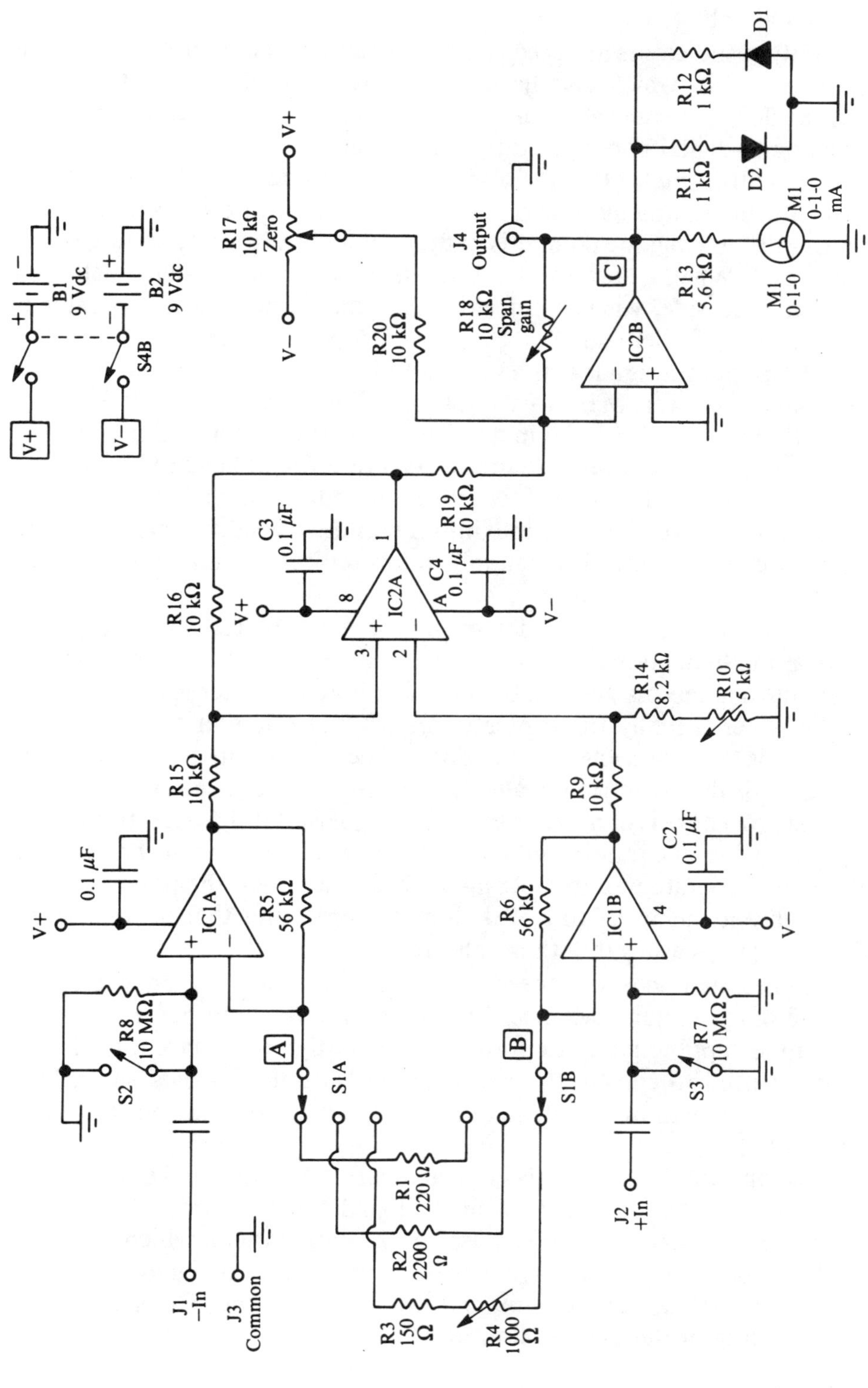

Figure 11-6. Galviamp circuit

Output indicators

Three output indicators are used on the galviamp. First, there is the analog output jack (J4), which can be used to drive any of several display or recording devices (external voltmeter, oscilloscope, paper-chart recorder, or the analog-to-digital converter input of a computer).

The built-in meter (M1) is also an output indicator. This device is a zero-center milliammeter with a range of 1-0-1 mA. In other words, when no current is flowing, the pointer is in the center of the meter face (see meter in Fig. 11-5). When a current of +1 mA is flowing, the pointer is deflected to the far right side; when a current of −1 mA is flowing, the pointer is deflected to the far left side.

Because output signal V_c is a voltage and the meter is a current meter, it is necessary to convert M1 into a voltmeter. This job is easily done: connect resistor R13 in series with the meter. The value shown for R13 is probably suitable for most common milliammeters, but it might have to be changed for certain models. The value of R13 is set to produce a 1-mA current in M1 when output voltage V_c is at its maximum permissible value. Because 9-Vdc batteries are used for the power supply in this project, the maximum output voltage will be about 8.5 volts.

Several methods exist to determine the value of R13. In one purely empirical method, connect a potentiometer in series with M1, and connect the two of them across a 9-volt battery. The potentiometer is connected so that the wiper (usually the center terminal) and one end are used—the other end is not connected to anything. The potentiometer resistance is reduced until the current indicated on M1 is 1 mA, and no more.

This procedure is dangerous to the health of M1! If the potentiometer is set too low when the series combination is connected across the battery, the meter will evaporate in a puff of smoke. Make sure that the potentiometer has a high enough value (try 50 kohms to start), and that it is set to its highest value (measure it with an ohmmeter!).

Another method is to proceed from a little calculation. First, determine the value of the meter resistance. You might find this number inscribed in nearly imperceptible type in a lower corner on the meter face. Otherwise, you might need to consult the manufacturer's catalog or data sheet. In a pinch, you can measure it with an ohmmeter, but again, it can be dangerous to the meter!

If the ohmmeter is an old-fashioned analog type, then make sure that it doesn't use an internal battery of more than 1.5 volts. Some other model *vacuum-tube voltmeters* (*VTVMs*) used a 22.5-volt battery, which are guaranteed to wipe out the meter being tested. Better yet, use a digital ohmmeter (part of a digital multimeter) on the low-power setting. Do *not* use the "diode" setting of the digital ohmmeter!

In either event, start at the highest ohms scale on the meter and work down in scale until a reading is obtained (that doesn't produce an overrange deflection) on the meter (as is indicated by the meter pointer slamming to one peg or the other). It is safest to perform this little test by connecting one lead of the ohmmeter to the milliammeter (M1). Then, quickly and lightly tap the other lead to the other M1 terminal while watching the pointer of M1. If the M1 pointer doesn't deflect rapidly, it's probably safe to continue. If you know the value of the meter resistance (R_m) by whichever means, then the value of R_{13} (in ohms) can be calculated from: $R_{13} = [1000 V_{c(max)} - R_m]$.

Once resistance R_{13} is determined, select an actual resistor that is the next higher standard value. Alternatively, make a series combination of a potentiometer and a fixed resistor. The potentiometer should be adjusted so that the combination is at the calculated required resistance value.

The third output indicators for the galviamp are LEDs D1 and D2. These devices are special pn-junction diodes that will emit a light when they are forward-biased. Because D1 and D2 are connected back-to-back, they become forward-biased with opposite polarities. In other words, when a positive voltage is applied to point C in the circuit of Fig. 11-6, LED D2 will turn on and emit light and D1 is turned off. But when the voltage at point C is negative, D2 is off and D1 is turned on to emit light. The resistors (R11 and R12), in series with diodes (D1 and D2), are needed to limit the current to a value that will not destroy the diodes. Most LEDs are tolerant of maximum currents of 15 to 20 mA, although some have ratings higher than these values.

The diodes to use for D1 and D2 are red LEDs; you might find that the high-intensity types (available at Radio Shack) work better. In any event, LEDs are not uniform in output light levels; as a result, it is advisable to select two that seem to match from a collection of LEDs. Connect an LED in series with a 470-ohm resistor, then connect the combination across a 9-volt battery. Also, connect a second 470-ohm resistor to the battery so that several LEDs can temporarily be connected into the circuit to compare with the first LED. Try several and use the pair that seem to match the best. A pot luck selection might yield good results, but LEDs that are typically sold in hobbyist electronics outlets are generally not well tested for light output-level consistency!

The purpose of the LEDs is to indicate an imbalanced output voltage. When the output voltage is zero, neither LED will glow. However, if the output voltage is not zero, the LED at the appropriate polarity will glow at a brightness that is proportional to the voltage (it is not linear, so it cannot be used as a good guide to output level, except that bright is "a lot" and dim is "not a bunch.")

The LEDs are driven from the output of the fourth op-amp IC2B. This op amp is connected in the unity-gain noninverting-follower configuration (i.e., the output, pin 7, is connected to the inverting input, pin 6). The voltage gain is +1, so the voltage at point C is equal to V_c at the output of IC2A.

Dc power-supply circuit

The dc power for the galviamp is ±9 Vdc supplied from a pair of 9-volt dc transistor-radio batteries. The op amps require two power supplies because their outputs might swing either positive or negative, depending on the input signals and their polarities. One of the power supplies (V+) is positive, with respect to ground or common, while the other (V−) is negative, with respect to ground or common. The 9-volt transistor radio batteries were selected safety reasons.

If you want the project to be ac operated, it is permissible to use regulated electronic dc power supplies that rate up to ±15 Vdc (although R_{11}, R_{12}, and R_{13} will have to be proportionally increased to the voltage change over ±9 volts). However, remember that certain inherent safety issues apply to ac power supplies—you must be knowledgeable of these circuits before attempting their use. Unfortunately, the topic is beyond the scope of this section so a warning must be sufficient.

The power supplies are switched on and off by a double-pole single-throw (DPST) toggle switch (S4A/S4B). A double-pole double-throw (DPDT) switch can also be used if the third pair of terminals are ignored. The DPDT switch is a little easier to find in local parts stores than the DPST type.

Notice that 0.1 μF capacitors are connected between the power-supply terminals of each op amp and ground. These capacitors are needed to prevent the op amp from bursting into self-oscillation—an undesirable problem! These capacitors must be mounted as close to the power-supply pins (4 and 8) of the op amp as possible and the leads of the capacitor must be kept as short as possible.

Other controls

Two additional controls are on the galviamp: zero (R17) and CMRR (R10). The zero control is essentially an offset-null control. In other words, the zero control is adjusted to force output voltage V_c to zero when the input voltage is zero. This job is accomplished with both inputs shorted to the common (COM) terminal. The function of the zero control is to suppress any dc offset that might exist, either internally to the amplifier or in the input signal, so it is made a front-panel adjustment. If only internal offsets are a concern, then R17 can be a screwdriver-adjusted trimmer pot. It would be adjusted once when the amplifier is built, and only readjusted when the sum of all drift elements has forced the offset too high (maybe once per year).

The CMRR control (R10) is used to balance the amplifier to make it truly differential. A differential amplifier does not respond to common-mode signals (i.e., signals that are applied equally to both inputs). CMRR control R10 is adjusted with 1-to-3-volt peak-to-peak ac signal (in the 100 to 2000 Hz range) that is applied simultaneously to J1 (−In) and J2 (+In). This operation requires J1 and J2 to be shorted together for the test.

Input circuitry

The galviamp uses a differential input. That is, two inputs exist, and the amplifier responds to the difference between the signal voltages that are applied to these points. A differential input is actually two inputs, each of which are referenced to ground. The +In input (see Fig. 11-6) is noninverting, so it produces an output that is in-phase with the output. In other words, a positive signal at +In will produce a positive-polarity output signal at J4/M1; similarly, a negative input on +In places a negative signal at J4/M1. The −In input is inverting (which is what the "−" sign means in "−In"), so it produces an output signal that is 180 degrees out of phase with the input signal. So, a positive signal at −In produces a negative signal at J4/M1; a negative signal at −In produces a positive signal at J4/M1.

Four modes exist for the switches (S2/S3) at the inputs:

1. **S2 and S3 both closed** Both inputs are shorted to ground. This mode is used for testing, and for ensuring the zero-volts baseline is, indeed, zero volts (use zero-adjust control).
2. **S2 open, S3 closed** This mode grounds the +In input, so the galviamp operates as a single-ended (one input) inverting amplifier.
3. **S2 closed, S3 open** this mode grounds the −In input, so the circuit operates as a noninverting amplifier.
4. **S2 and S3 both open** This is the differential mode. The galviamp operates as a true differential amplifier.

Construction

Construction of the galviamp is not terribly critical because the circuit is usually well behaved. The amplifier electronics were constructed on a general-purpose PC board (*Radio Shack* 276-168A), as shown in Fig. 11-7A. This particular PC board is very inexpensive and it has a copper foil pattern printed to match the dual-inline packages (DIP) that the IC op amps (IC1

A

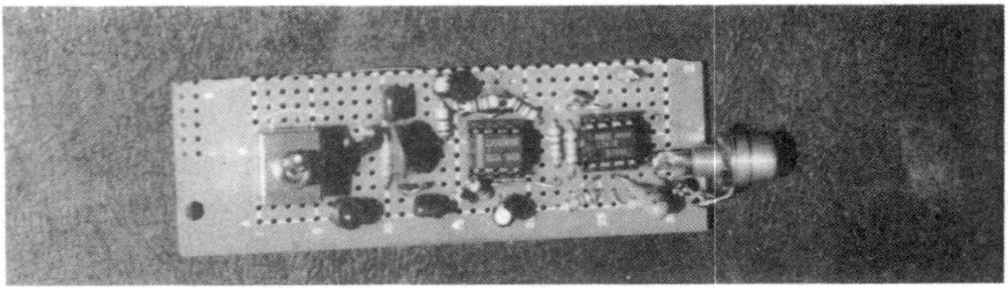

B

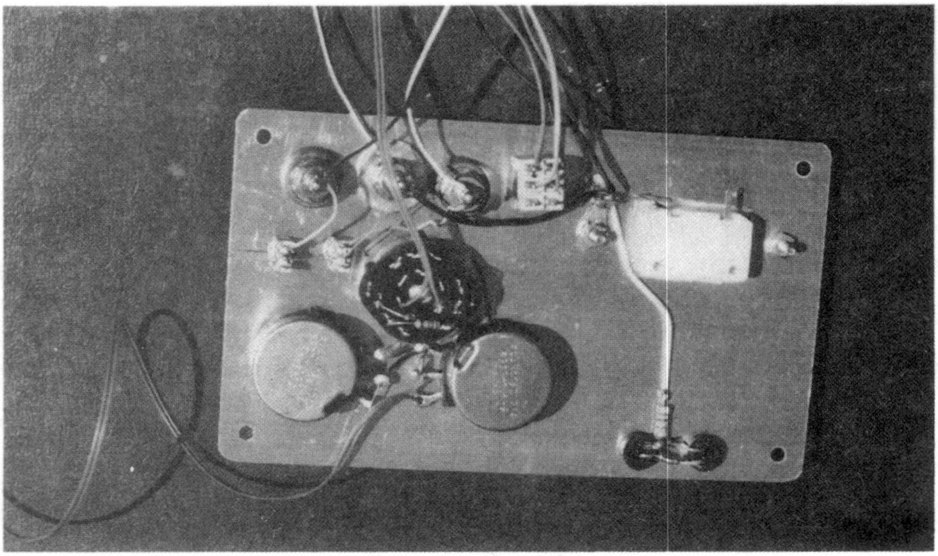

Figure 11-7. A) Circuit board, B) rear of front panel.

and IC2) need. It also has a ground line on the outer perimeter and a pair of power busses. I found it convenient to short the two power busses and the ground bus together and use them as a massive single ground bus.

The cabinet for the galviamp (Fig. 11-5) is the Radio Shack 270-627 experimenter's box, which measures 6.25 × 3.75 × 1 15/16 inches (15.9 × 9.6 × 5 cm). It is made of plastic, but it has an aluminum front panel. If shielding is desired for the cabinet (as when the galviamp is to be used in the presence of strong RF, power line, or other fields) an all-aluminum cabinet can be substituted for the experimenter's box that I used. Any well-stocked electronics parts store will carry a line of these boxes.

The front-panel wiring is shown in Fig. 11-7B. Ordinary #22 or #24 insulated hook-up wire can be used for all of the wiring on this panel. However, I have salvaged the parallel-conductor ribbon cable from many computer projects. I had a section of 25-conductor ribbon cable, in which all wires were color coded. That feature allowed me to keep connections to the same component together and to establish my own little color-coding scheme to keep track of the wires. This is not necessary, but it facilitates work on the bench when all wires are the same color, the eyes water and head aches!).

Interconnections can be made with #24 solid hook-up wire, or the very light gauge wire that is used for wire wrapping (although that wire is a tad expensive!). Don't forget to save the snipped-off ends of resistor and capacitor leads in a safe place. They can be salvaged for use as jumpers or for short runs on the wiring board.

Adjustment

Adjusting the galviamp is relatively simple, especially if you have an oscilloscope or ac voltmeter available to use as an ac-output indicator.

1. Turn the galviamp on (S4), and note the indication on M1. If M1 slams to one peg, *quickly* turn off S4 and look for a wiring error.
2. If M1 is either at zero, or some moderate positive or negative deflection, proceed with adjustment after about 5 minutes of warm-up (to control drift).
3. Short J1, J2, and J3 together by closing switches S2 and S3.
4. Set R10 approximately to the middle of its range and S1 to position B.
5. Adjust R17 (zero control) for a zero mA indication on M1.
6. Set S1 to position A and repeat step 5.
7. Connect either an ac voltmeter or an oscilloscope to analog output jack J4. Open switches S2 and S3.
8. Connect a short piece of wire from J2 to J3. Connect an audio signal generator (use 100 to 2000 Hz if an oscilloscope or ac voltmeter is used at J4 or 100 to 400 Hz if the ac function of an ordinary multimeter is used) to J2/J3, with the ground or common of the signal generator to J3. Adjust the signal-generator output to about 1 to 3 volts p-p.
9. Set S1 to position B.
10. Adjust R10 for minimum ac signal output.
11. Set switch S1 to position A and repeat step 10 until no further improvement is possible.
12. Disconnect the audio signal generator and remove the short from between J2 and J3.

The galviamp is now ready for use. As a matter of practice, keep S2 and S3 closed until the amplifier is used. Otherwise, the open inputs on IC1A and IC1B will force the amplifier output to rise to full-scale, even when no signal is present.

Typical applications

The galviamp can be used as a single-ended amplifier with either inverting or noninverting characteristics, or it can be used as a differential amplifier. Its high input impedance and relatively wide (audio frequencies) bandwidth make the galviamp useful for a wide variety of workbench and experimenter applications. This concluding section, however, examines an application in scientific instrumentation. Here, the galviamp is used as an amplifier for a Wheatstone bridge.

The Wheatstone bridge

Figure 11-8 shows how a simple resistive Wheatstone bridge can be developed from the simple resistor voltage-divider network (Fig. 11-8A). Output voltage V_o is a fraction of applied voltage V. The value of V_o is found from:

$$V_o = \frac{V R_2}{R_1 + R_2} \tag{11-5}$$

In the context of the Wheatstone bridge, Fig. 11-8A is sometimes called a *half-bridge circuit*. If R2 is a *thermistor* (temperature-sensitive resistor) or a *photoresistor* (light-sensitive resistor), the output voltage will be proportional to either the temperature or light level, respectively. A disadvantage of this circuit is that a perpetual dc offset exists, unless R_2 is zero ohms. If you place a temperature probe at R2, therefore, a zero-degree temperature does not yield a zero-volts output state. This problem can be solved by the Wheatstone-bridge circuit of Fig. 11-8B.

If you connect two half bridges in parallel, as in Fig. 11-8B, the classic Wheatstone-bridge circuit is created; R1/R2 forms one half bridge and R3/R4 forms the other. The differential output voltage V_{od} is the difference between the output voltages of the two half bridges (V_2 and V_4). The output voltage is $V_2 - V_4$:

$$V_{od} = V\left(\frac{R_2}{R_1 + R_2} - \frac{R_4}{R_3 + R_4}\right) \tag{11-6}$$

When the two half bridges are in balance, the output voltage is zero. In this state, the bridge is in *null condition*. If resistances R_1 through R_4 are

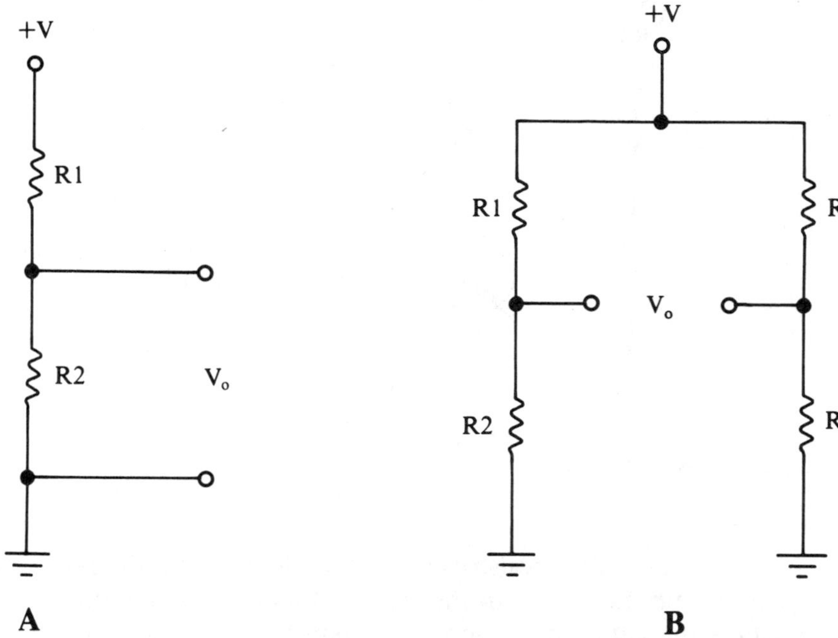

Figure 11-8. A) Resistor voltage divider, B) Wheatstone bridge.

correctly selected, the null condition can exist when the physical parameter (e.g., temperature) is also zero. Exact values of the applied parameter can be read as deviations of V_{od} from zero.

The Wheatstone bridge can also be used to compare measurements. Suppose that R2 is a potentiometer (variable resistance) and R4 is a temperature sensor. Calibrate a dial that is connected to the shaft of R2 in units of temperature from the characteristics of the thermistor sensor at R4. Thus, allow the thermistor to stabilize, then adjust R2 for a null condition on M1. When this is achieved, read the temperature (or whatever parameter is calibrated) from the dial that is attached to R2.

When designing Wheatstone bridge circuits for an experiment, remember that the resistances of R_1 through R_4 do not have to be equal. The ratio of the two half bridges must be equal for null to occur, but the actual resistance values can be different:

$$\frac{R_1}{R_2} = \frac{R_3}{R_4} \tag{11-7}$$

Figure 11-9 shows a slightly modified Wheatstone-bridge circuit that permits either bridge balancing under a stated value of the stimulus parameter (e.g., zero degrees Celsius), or to set the bridge to a specific offset point.

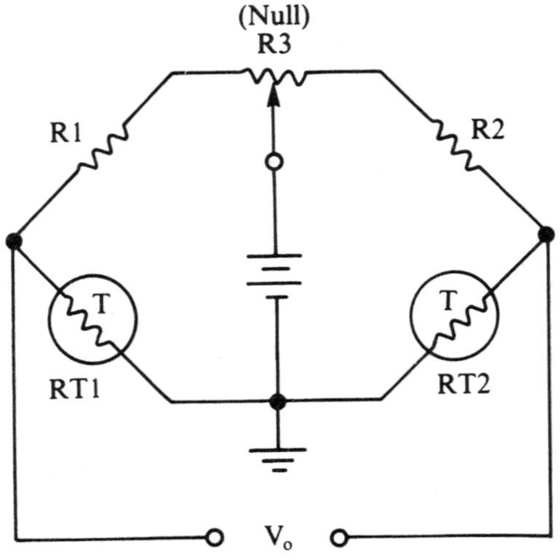

Figure 11-9. Wheatstone bridge with null control.

In this case, the values of the resistors should be close so that the potentiometer (R5) will be approximately in the center of its range when the bridge is balanced. Make the value of R5 approximately 50 percent of the values of the bridge-element resistances.

Figure 11-10 shows how to connect the Wheatstone bridge to the galviamp. The output voltage V_{od} is connected across the differential inputs (−In to +In) of the galviamp. Output voltage V_o is equal to the product of the differential gain of the galviamp and bridge-output voltage V_{od}.

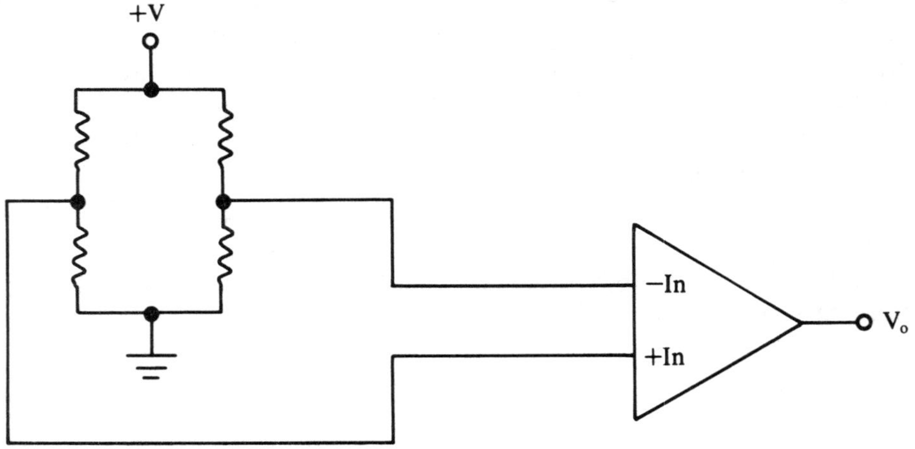

Figure 11-10. Connecting a Wheatstone bridge to a dc differential amplifier.

Figure 11-11 shows an experiment that uses photoresistor cells, PC1 and PC2. This little circuit is called a *colorimeter*, and it is the basis for a large family of medical, scientific, and engineering instruments. Radio Shack usually has a small selection of photoresistors, including an assortment or two. Find two matching photoresistors for PC1 and PC2, then use an ohmeter to measure their nominal resistance at normal room-illumination levels. Select fixed resistors R1 and R2 so that they are approximately the same resistance (±50%). R_3 should range from $0.3R_1$ to $1.5R_1$ to make it easy to balance the bridge. If the resistance at room light is less than about 500 ohms, the battery life will be short and the photoresistors might self-heat and cause errors.

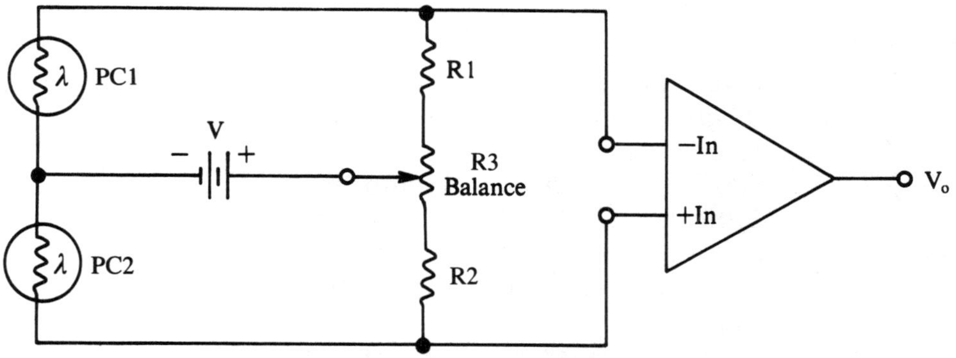

Figure 11-11. A photocolorimeter Wheatstone bridge.

Construct this little photo-bridge circuit on a piece of perfboard with PC1 and PC2 about 2 to 3 inches apart. If the photoresistors that are selected for PC1 and PC2 have clear edges (which would allow light to enter), cover the perimeter with black tape or black paint. Make sure that only light from the front of the lens over the active region enters.

Connect the output of the circuit (V_{od}) to the input of the galviamp, then expose the photoresistors to room light so that they are equally illuminated (with no shadows). Adjust R3 for a null condition on the output meter. With interposing colored filters between the light source and one (and only one) photoresistor, it is possible to get a rough idea of the percentage of the white light that this color represents. However, remember that the response of the photoresistor is not uniform across the spectrum, so the measure is rough.

An application for this type of circuit was provided by a friend of mine who was performing physics experiments on light reflected from the bottom of a salt-water bay on the East Coast. An instrument containing a circuit like Fig. 11-11 was towed about 10 feet beneath a boat in salt water that was 20 to 30 feet deep.

A clear filter was placed over one photoresistor; it received the entire spectrum as filtered only by the sea water. A revolving mechanism placed several different wavelength light filters in front of the other photoresistor. The experiment was repeated each trial with the assembly aimed toward the surface and toward the bottom. The scientist was attempting to ferret out information about the colors that are reflected and absorbed in the sea water. It was theorized that the colors of light were differentially reflected from the bottom.

Just how my scientist friend telemetered the color-percentage data to the surface is . . . well, it's a story for another time.

Parts list

B1,B2	9-volt transistor-radio batteries
C1,C2,C3,C4	0.1-μF 15-wvdc (or higher) mylar radial-lead capacitors
D1,D2	Red LEDs
IC1	CA-3240 dual BiMOS op amp
IC2	Either CA-3240 or 1458 dual op amp
J1,J2,J3	Five-way binding posts (J1/J2 red, J3 black)
J4	RCA chassis-mounted phono jack (BNC may be substituted)
M1	1-0-1 zero-center milliammeter
R4	1000-ohm linear-taper potentiometer*
R10,R17	10-kohm linear-taper potentiometer*
S1	2-pole 3-position (2P3P) rotary switch, or alternatively, a 3P3P switch (which might be easier to buy, but do not use third section)
S2,S3	SPST or SPDT miniature toggle switch
S4	DPST or DPDT miniature toggle switch
PC board	Radio Shack 276-168A equivalent
Box	Radio Shack 270-627 or equivalent
Misc.	2 battery clips to fit B1 and B2, 8-pin miniDIP IC circuit sockets for IC1 and IC2

*All resistors are ¼-watt carbon or metal film, except R11 or R12 which are ½ watt.

Ac instrumentation amplifiers

What is the principal difference between dc amplifiers and ac amplifiers? A dc amplifier will amplify both ac and dc signals up to the frequency limit of the particular circuit that is used. The ac amplifier, on the other hand, will not pass or amplify dc signals. In fact, ac amplifiers will not pass ac signals of frequencies from close to dc to some lower −3-dB bandpass limit. The gain in the region between near-dc and the full-gain frequencies within the passband rises at a rate determined by the design, usually +6 dB/octave (an *octave* is a 2:1 frequency change). The standard low-end point in the frequency-response curve is defined as the frequency at which the gain drops −3 dB from the full gain.

Figure 11-12A shows a modified version of the instrumentation amplifier that is designed as an ac amplifier. Modify the input circuitry of A1 and A2 by placing a capacitor in series with each op-amp's noninverting input. Resistors R8 and R9 keep the input bias currents of A1 and A2 from charging capacitors C1 and C2. In some modern low-input-current op amps, these resistors are optional because of the extremely low levels of bias current that are normally present.

The −3-dB frequency of the amplifier (Fig. 11-12A) is a function of the input capacitors and resistors (assuming that $R_8 = R_9 = R$ and $C_1 = C_2 = C$):

$$F = \frac{1{,}000{,}000}{2\pi R C_{\mu F}} \tag{11-8}$$

Where:

F is the −3-dB frequency (in *hertz*)
R is the resistance (in *ohms*)
$C_{\mu F}$ is the capacitance (in *microfarads*)

For frequency response, Eq. (11-8) is not the most useful form. In most practical cases, you will know the required frequency response from evaluating the application.

Furthermore, you will know the value of the input resistors (*R9* and *R11*) because they are selected for high-input impedance, and are by convention, either > 10× or > 100× the source impedance (depending upon the application). Typically, these resistors are selected to be 10 megohms. You must, therefore, select the capacitor values from Eq. (11-9) below:

$$C_{\mu F} = \frac{1{,}000{,}000}{2\pi R F} \tag{11-9}$$

Where:

$C_{\mu F}$ is the capacitance of C1 and C2, (in *microfarads*)
R is the resistance of R9 and R10 (in *ohms*)
F is the −3-dB frequency (in *hertz*)

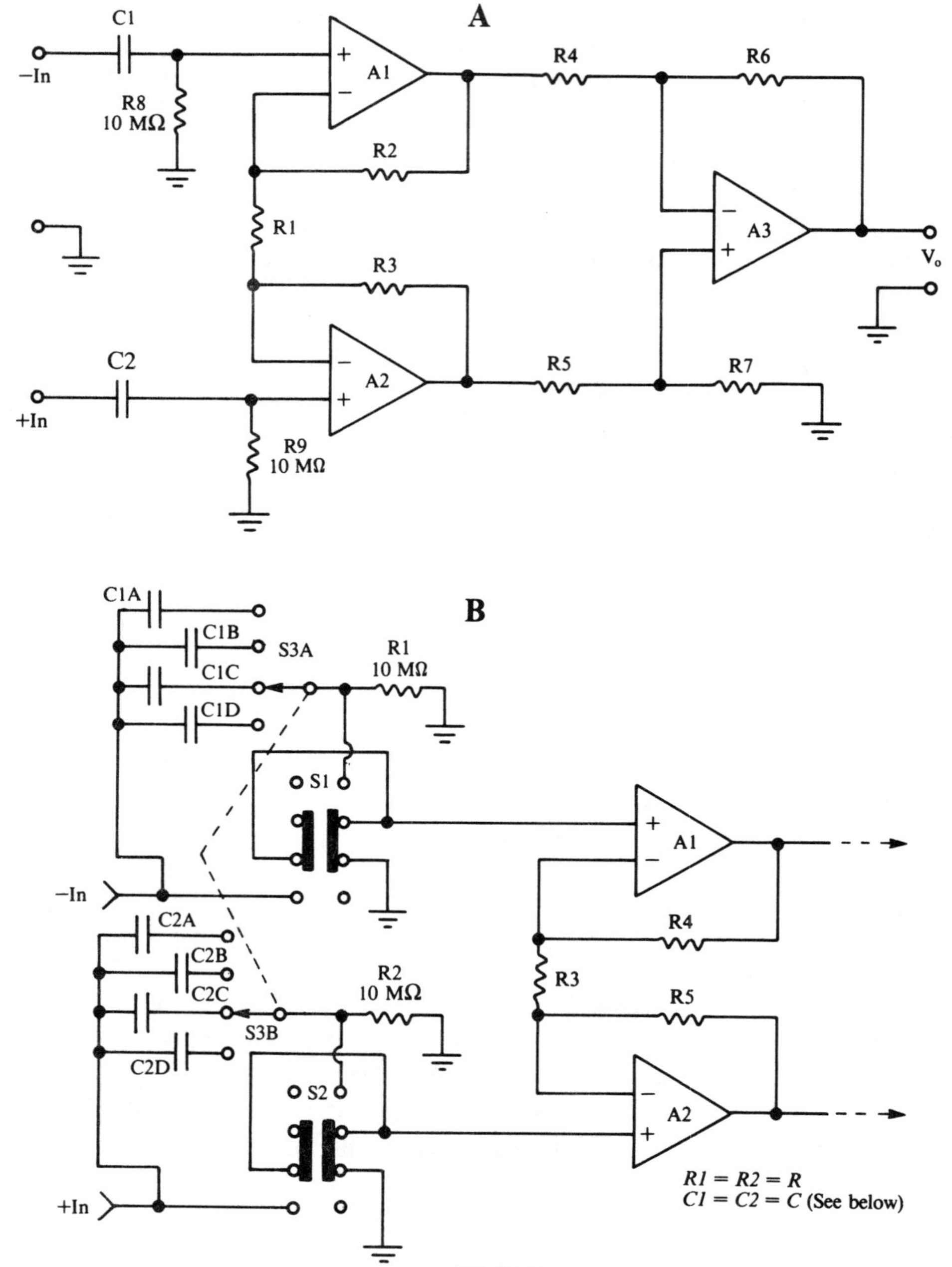

Pos	S1 & S2
1	ac
2	GND
3	dc

Freq	C = ?
0.05	0.33 μF
5	0.0033 μF
30	530 pF
100	160 pF

Figure 11-12. A) Ac-coupled instrumentation amplifier, B) universal ac-coupled instrumentation amplifier.

The ac instrumentation amplifier can be adapted to all the other modifications of the basic circuit that was discussed earlier in this chapter. We can, for example, use a gain control (replace R1 with a fixed resistor and a potentiometer), or add a CMRR-ADJ control. In fact, these adaptations are probably necessary in most practical ac IA circuits.

In many IA applications, it is desirable to provide selectable ac or dc coupling, as well as to ground the input of the amplifiers. This latter feature is especially desirable in circuits where an oscilloscope, strip-chart paper recorder, digital-data logger, or computer is used to receive the data. By grounding the input of the amplifier (without also grounding the source, which could be dangerous), it is possible to set (or at least determine) the $V_d = 0$ baseline. Figure 11-12B shows a modified input circuit that uses a switch to select ac-GND-dc coupling. In addition, a second switch is provided to set the low-end −3-dB frequency-response point, according to Eq. (11-9).

Project 11-2 ECG amplifier

The human and animal produces a small electrical signal that can be recorded through skin-surface electrodes and displayed on an oscilloscope or a paper strip-chart recorder. This electrical signal is called an *electrocardiograph* (*ECG*) signal. The peak value of the ECG signal at the skin-surface electrodes is on the order of one millivolt (1 mV). In order to produce a 1-volt signal to apply to a recorder or an oscilloscope, the ECG amplifier must provide a gain of 1000. Furthermore, because skin has a relatively high electrical resistance (1 to 20 kohms with 1-cm Ag/AgCl electrodes), the ECG amplifier must also have a very high input impedance.

The ECG amplifier must also be an ac amplifier. The reason for this requirement is that metallic electrodes (such as Ag/AgCl) that are applied to the electrolytic skin produce a dc halfcell potential. This potential is usually between 1 and 2 volts, so it is more than 1000 times greater than the signal voltage. By making the amplifier respond only to ac, the artifact caused by the dc halfcell potential is eliminated.

The frequency selected for the −3-dB point of the ECG amplifier must be very low and close to dc, because the standard ECG waveform contains very low frequency components. The typical ECG signal has significant frequency components that range from 0.05 to 100 Hz, the industry-standard frequency response for diagnostic ECG amplifiers. However, some clinical monitoring ECG amplifiers use 0.05 to 40 Hz to eliminate muscle artifact as a result of patient movements.

The typical ECG amplifier has differential inputs because the most useful ECG signals are differential in nature and because they can suppress the 60-Hz hum that is picked up on the leads and the patient's body. In the most simple case, the right-arm (RA) and left-arm (LA) electrodes form the inputs to the amplifier, and the right leg (RL) is the common (see Fig. 11-13). The basic configuration of the amplifier in Fig. 11-13 is the ac-coupled instrumentation amplifier that was discussed earlier. The gain for this amplifier is set to slightly more than ×1000, so a 1-mV ECG peak signal will produce a 1-volt output. Because of the high gain, it is essential that the amplifier is well-balanced. This requirement suggests a dual amplifier for A1 and A2. One example might be the CA-3240 device—a dual BiMOS, which is essentially two CA-3140s in a single 8-pin miniDIP package. Also, in the interest of balance, resistors with 1 percent or less tolerance should be used for the equal pairs.

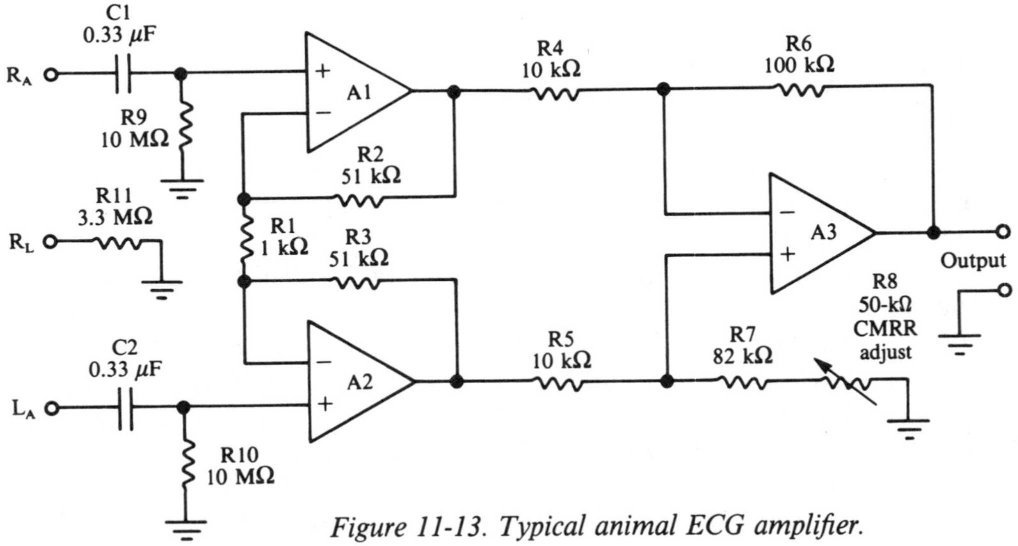

Figure 11-13. Typical animal ECG amplifier.

The lower-end −3-dB frequency-response point is set by the input resistors and capacitors. In this case, the particular combination of values shown produces a response of 0.048 Hz.

The CMRR-adjust control in Fig. 11-13 is usually a 10- to 20-turn trimmer potentiometer. It is adjusted in the following manner:

1. Short together the RA, LA, and RL inputs.
2. Connect a dc voltmeter to the output (either a digital voltmeter or an analog meter with a 1.5-Vdc scale (alternatively, a dc-coupled oscilloscope can be used, but be sure to identify the zero baseline).

3. Adjust the CMRR-adjust control (R7) for zero-volts output. Disconnect the RL input and connect a signal generator between RL and the still-connected RA/LA terminal.
4. Adjust the output of the signal generator for a sine-wave frequency in the range 10 to 40 Hz and a peak-to-peak potential of 1 to 3 volts.
5. Using an ac scale on the voltmeter (or oscilloscope), again adjust the CMRR adjust (R7) for the smallest possible output signal. It might be necessary to re-adjust the voltmeter or oscilloscope input range control for greater sensitivity, in order to observe the best null.
6. Remove the RA/LA short. The ECG amplifier is ready to use.

A suitable post-amplifier for the ECG preamplifier is shown in Fig. 11-14. This amplifier is placed in the signal line between the output of Fig. 11-13 and the input of the oscilloscope or paper chart recorder that will be used to display the waveform. The gain of the post-amplifier is variable from 0 to +2; it will produce a 2-volt maximum output when a 1-mV ECG signal provides a 1-volt output from the preamplifier. Because of the high-level signals, this amplifier can use ordinary 741 operational amplifiers.

The frequency response of the amplifier is set to an upper −3-dB point of 100 Hz and the response drops at a −6-dB/octave rate above that frequency. This frequency-response point is determined by capacitor C3 and with resistor R12:

$$F_{Hz} = \frac{1{,}000{,}000}{2\pi RC}$$

$$F_{Hz} = \frac{1{,}000{,}000}{(2)\ (3.14)\ (10^5\ (0.015\mu F)}$$

Three controls are in the post-amplifier circuit: span, position, and dc balance. The *span control* is the 0-to-2 gain control—*span* reflects instrumentation users' language, rather than electronics language. The *position control* sets the position of the output waveform on the display device. Resistors R21 and R22 are selected to limit the travel of the beam or pen to full-scale. Set these resistors so that the maximum potential at the end terminals of R20 corresponds to a full-scale deflection of the display device.

The *dc balance control* is used to cancel the collective effects of offset potentials that are created by the various stages of amplification. Follow the CMRR-adjust procedure, then reconnect the short-circuit at RA, LA, and RL. Move the voltmeter to the output of Fig. 11-14. Adjust the position control for zero volts at point A. Adjust the dc-balance control for zero volts

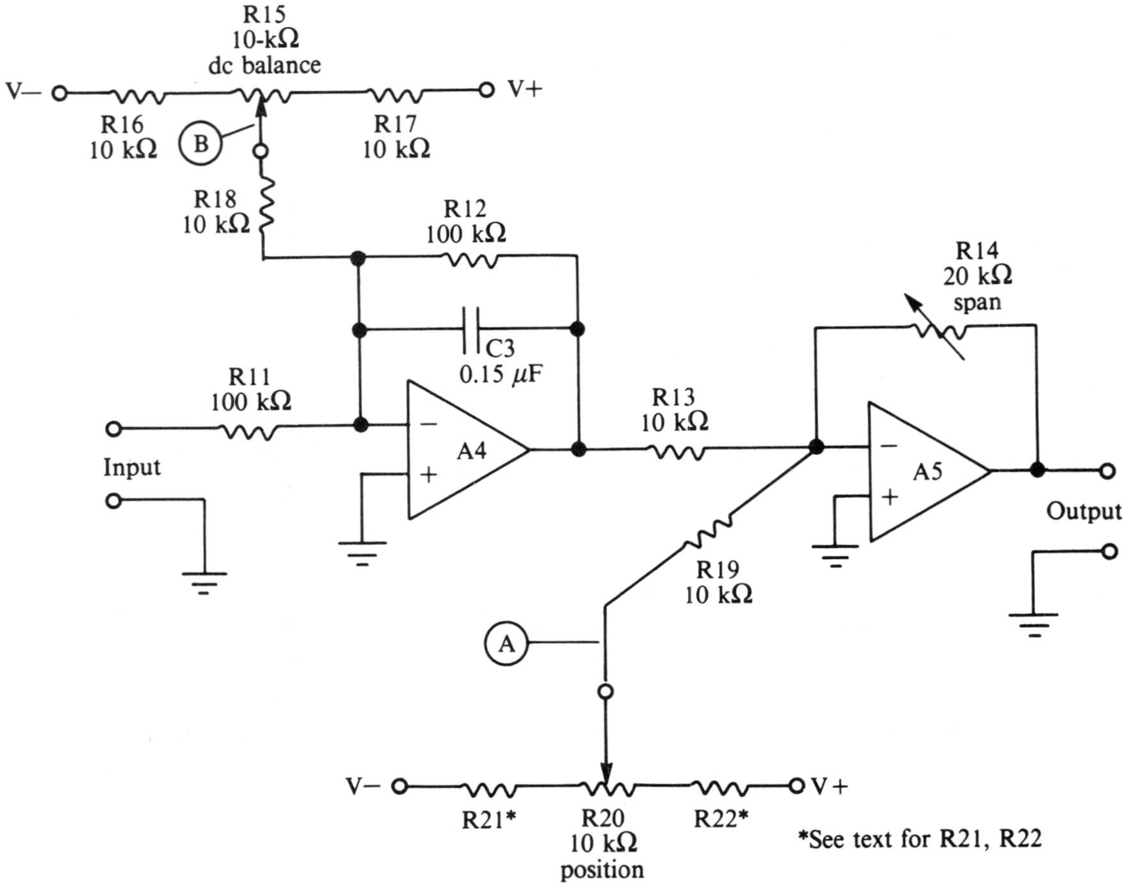

Figure 11-14. Post-amplifier circuit.

at point B. Adjust the span control (R14) through its entire range, from zero to maximum, while monitoring the output voltage. If the output voltage does not shift, no further adjustment is needed.

If the output voltage in the last step varied as the span control was varied, then adjust the dc balance until the span control does not produce an output voltage shift when you turn it across its full range. Repeat this step several times until no further improvement is possible. Remove the RA, LA, RL short; the amplifier is ready for use.

IC instrumentation amplifiers: some commercial examples

The operational amplifier truly revolutionized analog circuit design. For a long time, the only additional advances were that op amps became vastly improved (nearer to the ideal). Although these developments were exciting, the "new" devices were not truly new. The next big breakthrough came when the analog-device designers made an IC version of the instrumentation amplifier (Fig. 11-15), the integrated-circuit instrumentation amplifier (ICIA). Today, several manufacturers offer substantially improved ICIA devices.

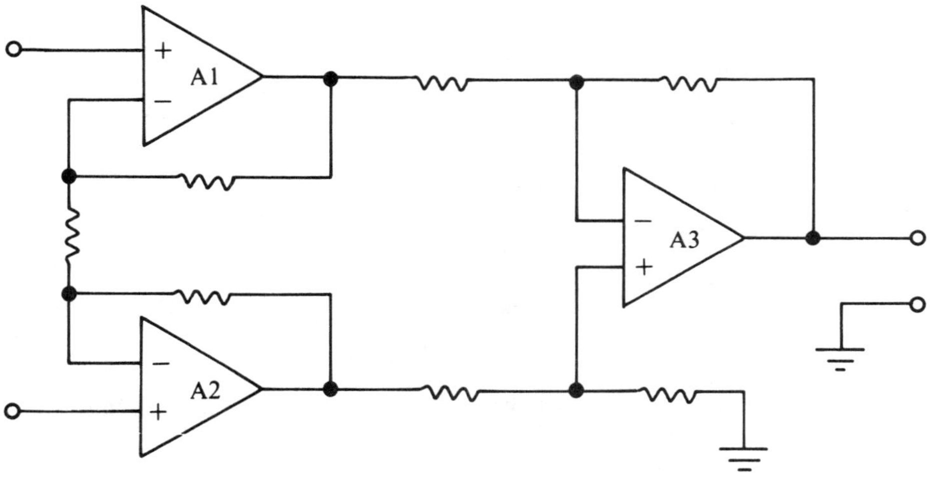

Figure 11-15. Generic instrumentation amplifier.

The Burr-Brown INA-101 (Fig. 11-16) is a popular ICIA device; a sample INA-101 circuit is shown in Fig. 11-17. This ICIA amplifier is simple to connect and use: it only has dc-power connections, differential-input connections, offset-adjust connections, a ground, and an output. The gain of the circuit is set by:

$$A_{vd} = \left(\frac{40\ kohms}{R_g} + 1\right) \qquad (11\text{-}10)$$

The INA-101 is a low-noise low-input bias-current IC version of the IA. Resistors R2 and R3 are internal to the INA-101 and are 20 kohms each, hence the *40 kohms* in Eq. (11-9).

Potentiometer R1 in Fig. 11-17 is used to null the offset voltages that appear at the output. An offset voltage is a voltage that exists on the output at a time when it should be zero (i.e., when $V_1 = V_2$, so that $V_1 - V_2 = 0$). The offset voltage might be internal to the amplifier, or alternatively, it might be a component of the input signal. Dc offsets in signals are common, especially in biopotentials amplifiers (such as ECG and EEG) and chemical transducers (such as pH, pO_2, and pCO_2.

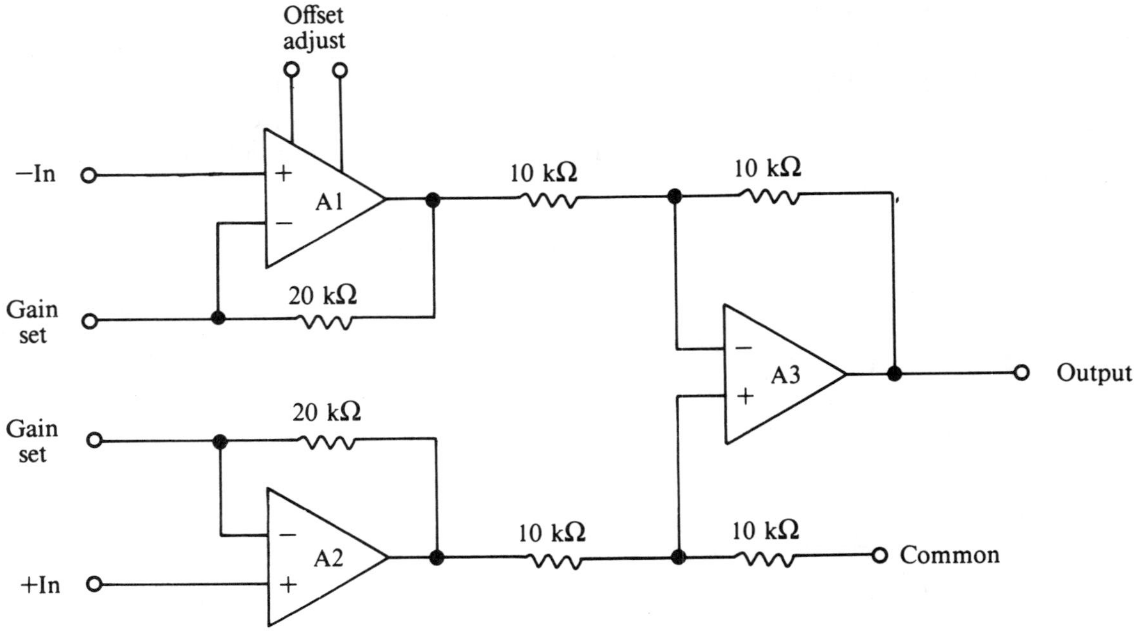

Figure 11-16. Selectable-gain instrumentation amplifier.

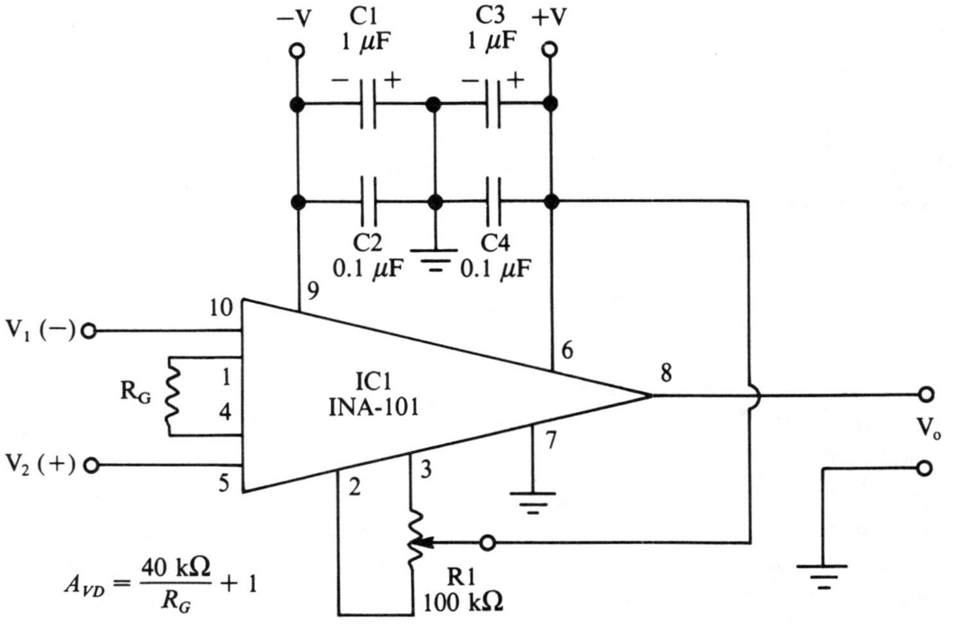

$$A_{VD} = \frac{40\ \text{k}\Omega}{R_G} + 1$$

R_G is in kΩ

Figure 11-17. Burr-Brown INA-101 instrumentation amplifier.

Another ICIA is the LM-363-xx device in Fig. 11-18, the miniDIP version is shown in Fig. 11-18A (an 8-pin metal can is also available), and a typical circuit is shown in Fig. 11-18B. The LM-363-xx device is a fixed-gain ICIA. The three versions of the LM-363-xx are enumerated according to gain:

Designation	Gain(A_v)
LM-363-10	×10
LM-363-100	×100
LM-363-500	×500

The LM-363-xx is useful in places where one of the standard gains is required, but only minimum space is available. Two examples spring to mind. The LM-363-xx can be used as a transducer preamplifier, especially in noisy-signal areas; the LM-363-xx can be built onto (or into) the transducer to build up its signal before sending it to the main instrument or signal-acquisition computer. Another possible use is in biopotentials amplifiers. Biopotentials are typically very small, especially when used in lab animals. The LM-363-xx can be mounted on the subject and a higher-level signal can be sent to the main instrument.

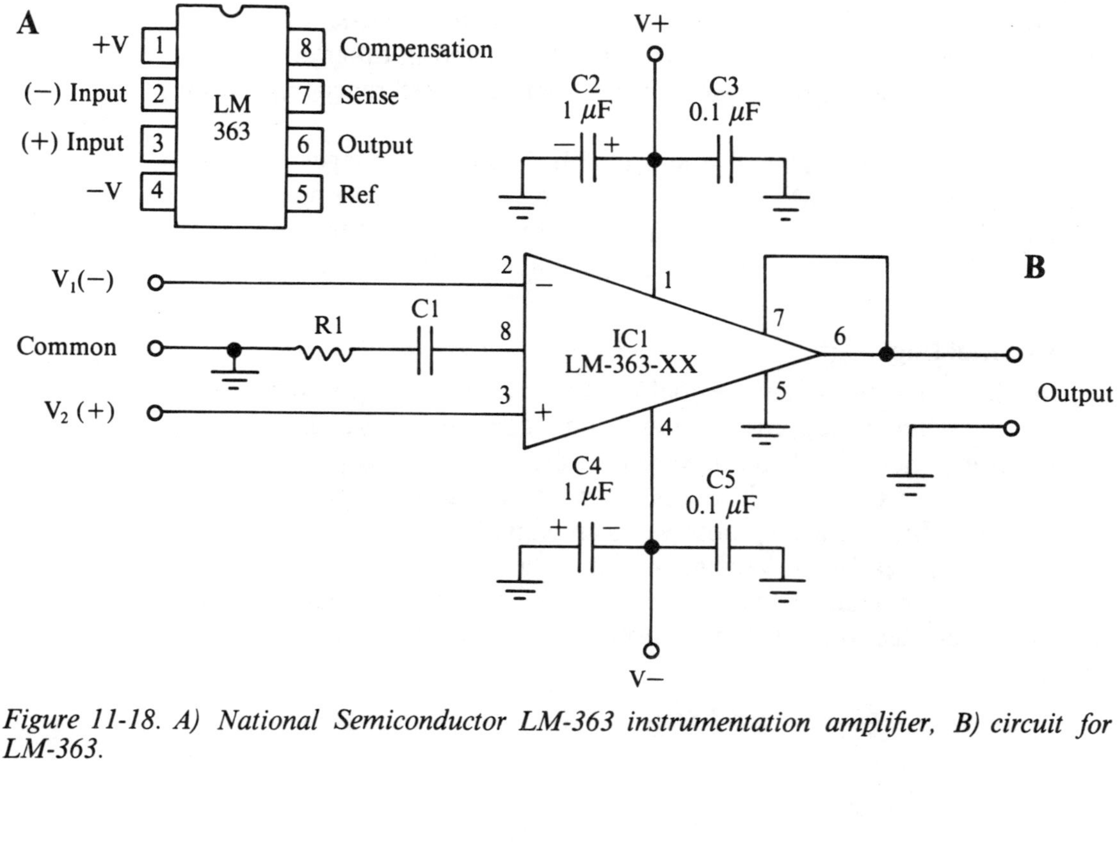

Figure 11-18. A) National Semiconductor LM-363 instrumentation amplifier, B) circuit for LM-363.

A selectable-gain version of the LM-363 is shown in Fig. 11-19A; the 16-pin DIP package is shown in Fig. 11-19A, and a typical circuit is shown in Fig. 11-19B. The type number of this device is LM-363-AD, which distinguishes it from the LM-363-xx devices. The gain can be ×10, ×100, or ×1000, depending on the programming of the gain-setting pins (2, 3, and 4). The programming protocol is.

Gain desired	Jumper pins
×10	(All open)
×100	3 and 4
×1000	2 and 4

S1 in Fig. 11-19B is the gain-select switch. This switch should be mounted close to the IC, but it is quite flexible in mechanical form. The switch could also be made from a combination of CMOS electronic switches (e.g., 4066). The dc power-supply terminals are treated in a manner similar to the other amplifiers. Again, the 0.1-μF capacitors must be mounted as close as possible to the body of the LM-363-AD. Pins 8 and 9 are guard-shield outputs. These pins make the LM-363-AD more useful for many instrumentation problems than other models. By outputting a signal sample back to the shield of the input lines, you can increase the CMRR. This feature is used a lot in bipotential amplifiers and in other applications where a low-level signal must pass through a strong interference (high-noise) environment.

The LM-363 devices will operate with dc supply voltages of ±5- to ±18-volts dc, with a CMRR of 130 dB. The 7-nV/Hz noise figure makes the device useful for low-noise applications (a 0.5-nV model is also available).

Guard shielding

One of the properties of differential amplifiers, including instrumentation amplifiers, is that it suppresses interfering signals from the environment. The common-mode rejection process is at the root of this capability. When an amplifier is used in a situation where it is connected to an external signal-source through wires, those wires are subjected to strong local 60-Hz ac fields from nearby power-line wiring. Fortunately, in the differential amplifier the field affects both lines equally, so the induced interfering signal is cancelled out by the common-mode rejection property of the amplifier.

Programming	
Gain	Jumper pins
X10	(All open)
X100	3, 4
X1000	2, 4

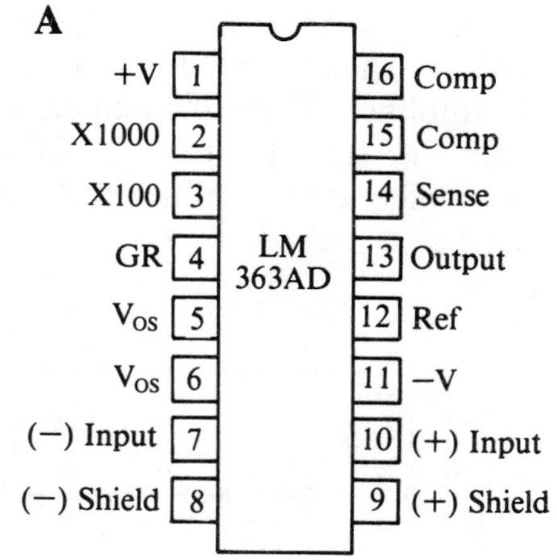

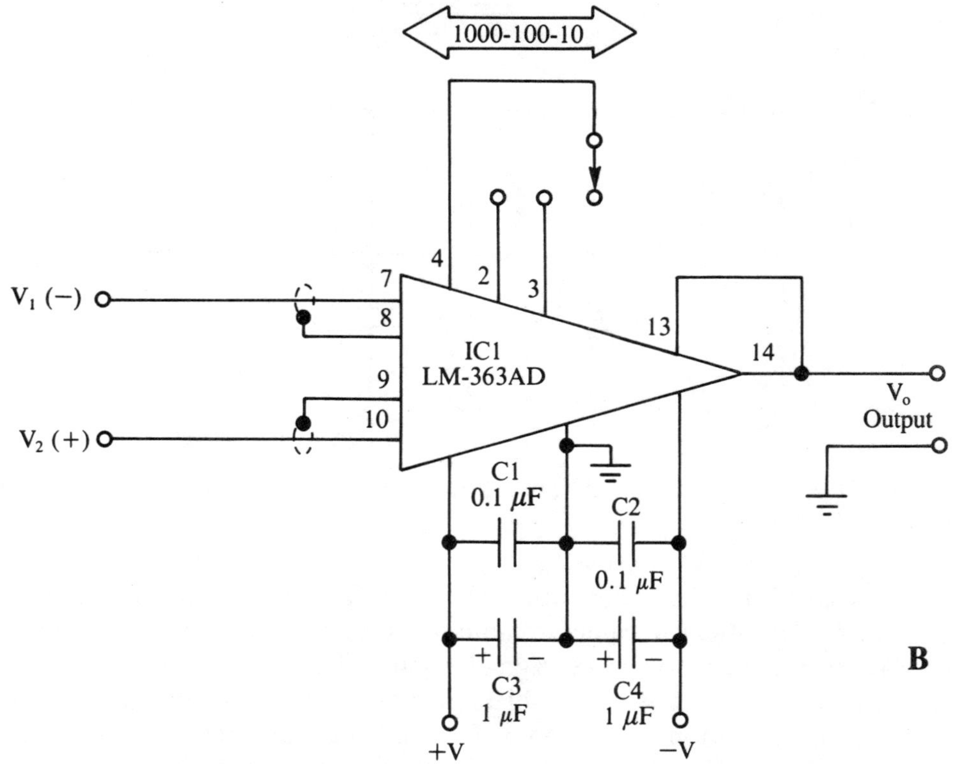

Figure 11-19. LM-363AD instrumentation amplifier: A) pinouts, B) circuit.

Unfortunately, the cancellation of interfering signals is not total. For example, imbalances might exist in the circuit that deteriorate the CMRR of the amplifier. These imbalances might be either internal or external to the amplifier circuit. Figure 11-20A shows a common scenario. In this figure, a differential amplifier is connected to shielded leads from the signal source, V_{in}.

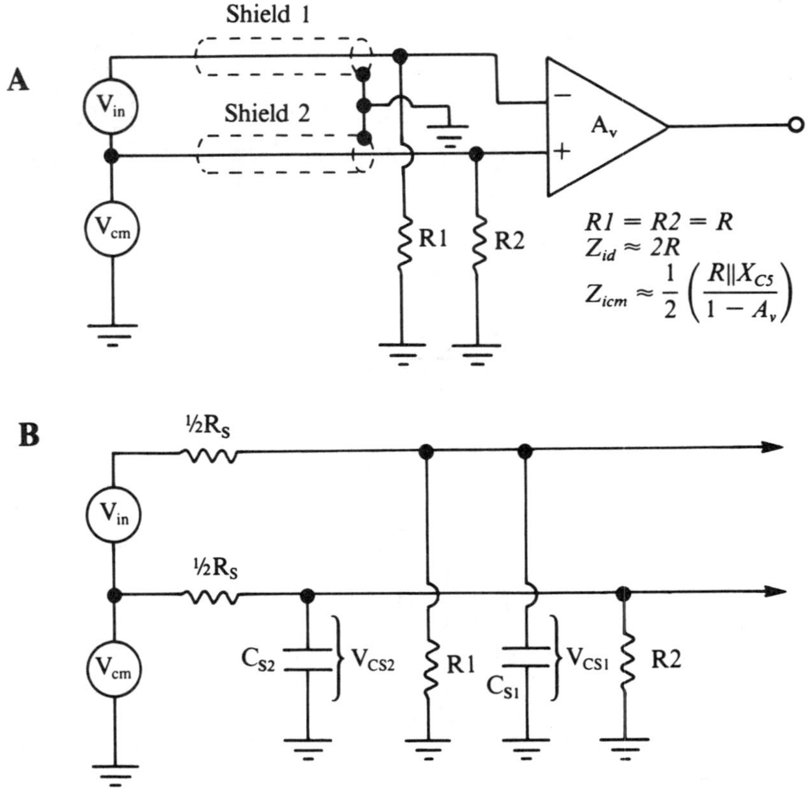

Figure 11-20. A) Usual shield connection of the instrumentation amplifier, B) equivalent circuit.

Shielded lead wires offer some protection from local fields, however, it is possible for shielded cables to manufacture a valid differential signal voltage from a common-mode signal! Figure 11-20B shows an equivalent circuit that demonstrates how a shielded cable pair can create a differential signal from a common-mode signal. The cable has capacitance between the center conductor and the shield conductor that surrounds it. In addition, the input connectors and the amplifier-equipment internal wiring also exhibit capacitance. These capacitances are lumped together in the model of Fig. 11-20B as C_{S1} and C_{S2}. As long as the source resistances and shunt resistances are equal and the two capacitances are equal, the circuit balance is not a problem. However, inequalities in any of these factors (which are commonplace) create an unbalanced circuit, in which common-mode signal V_{cm} can charge one capacitance more than the other. As a result, the difference between the capacitance voltages, V_{CS1} and V_{CS2}, is seen as a valid differential signal.

A low-cost solution to the problem of shield-induced artifact signals is shown in Fig. 11-21A. In this circuit, a sample of the two input signals is fed back to the shield, which in this situation is not grounded. In a *guard shield*, the amplifier output signal drives the shield. Either double shields (one on each input line) as shown or a common shield for the two inputs can be used.

An improved guard-shield example for the instrumentation amplifier is shown in Fig. 11-21B. In this case, a single shield covers both input lines, but it is possible to use separate shields. In this circuit, a sample of the two input signals is taken from the junction of resistors R8 and R9, and it is fed to the input of a unity-gain buffer/driver guard amplifier (A4). The output of A4 drives the guard shield.

Perhaps the most common approach to guard shielding is the arrangement shown in Fig. 11-21C. In this case, two shields are used; the input cabling is double-shielded insulated wire. The guard amplifier drives the inner shield, which serves as the guard shield for the system. The outer shield is grounded at the input end in the normal manner, and it serves as an EM1-suppression shield.

Summary

The instrumentation amplifier is a differential amplifier whose inputs are first amplified by noninverting followers. This increases the input impedance, increases the gain, and provides a means by which the gain can be easily adjusted. This circuit is often used to amplify small differential voltages, such as those which come from an unbalanced bridge circuit. Additional circuitry may be used to increase common-mode rejection and to adjust frequency response for ac amplifiers.

Questions

1. What is an instrumentation amplifier?
2. How is common-mode rejection adjusted in an instrumentation amplifier?

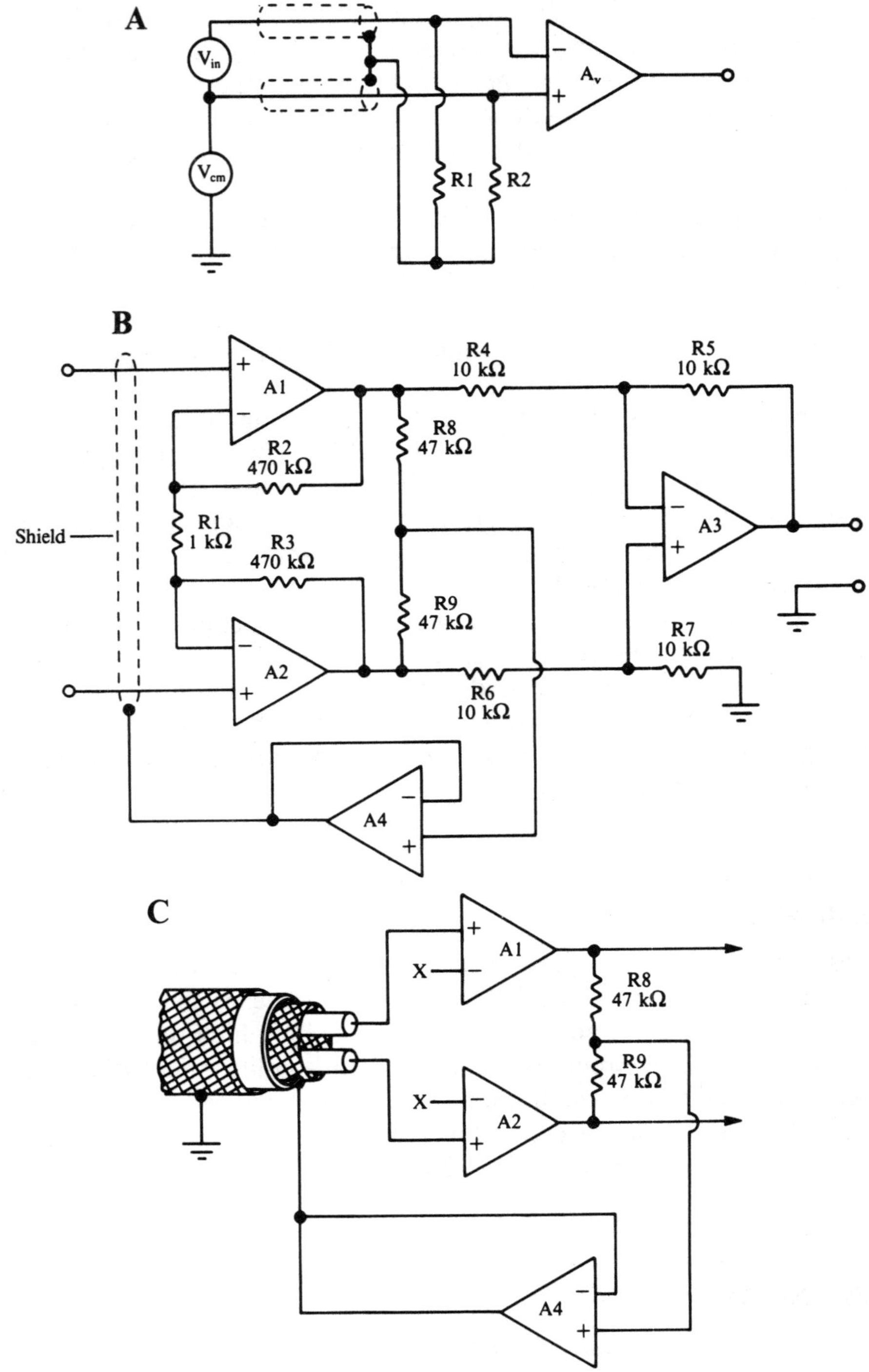

Figure 11-21. A) Guard-shield connection, B) using driver amplifier (A4), C) double-shielded method.

3. What advantages does the instrumentation amplifier have when used to detect out-of-null voltages in a Wheatstone bridge?
4. How can an instrumentation amplifier be modified to serve as an ac amplifier?
5. How should the ac-coupled instrumentation amplifier shown in Fig. 11-12A be modified for use as a low-pass filter?
6. Name two commercial instrumentation amplifier ICs.
7. Using the characteristics of the op amp, derive the formula in Eq. (11-2).

Problems

1. For the circuit in Fig. 11-2, R_1 can vary from 100 to 1000 ohms. If $R_2 = R_3 = 1000$ ohms, $R_4 = R_5 = 10$ kohms, and $R_6 = R_7 = 100$ kohms, calculate the range for the gain of the circuit.
2. Repeat Problem 1 for R_1 varying from 1 kohm to 5 kohms.
3. For the circuit in Fig. 11-8B, $R_1 = R_3 = 1$ kohm, $R_2 = 10$ kohms, $R_4 = 10.5$ kohms. If V+ = 12 volts, what is V_o?
4. For the circuit in Fig. 11-8B, $R_1 = R_3 = 5$ kohms, $R_2 = 20$ kohms, $R_4 = 19.5$ kohms. If V+ = 12 volts, what is V_o?
5. For the circuit in Fig. 11-12A, the desired breakpoint frequency is 200 Hz. Calculate the necessary value for C_1 and C_2.
6. For the circuit in Fig. 11-12A, the desired breakpoint frequency is 1 kHz. Calculate the necessary value for C_1 and C_2.

12 NONLINEAR OP-AMP CIRCUITS

AFTER COMPLETING THIS chapter, you will be able to:

- ☐ Explain ideal versus real diodes and the circuits involving them
- ☐ Discuss zener diode circuits
- ☐ Describe rectifier circuits

The op amp is also very well suited for numerous nonlinear applications. In this chapter, various nonlinear op-amp circuits are explored. Of particular interest are circuits in which pn-junction diodes are used: precise rectifiers, bounded value circuits, clippers/clampers, etc. Some of the following experiments will require a low-cost oscilloscope.

Review of the pn-junction diode

The pn diode is the oldest type of solid-state electronic component available. Indeed, naturally occurring diodes of *galena crystals* (lead sulphide, PbS) were used prior to World War I as the demodulator (a.k.a. detector) in crystal radio receivers. During World War II, radar research led to the development of the 1N34, 1N60, and 1N63 germanium video-detector diodes and the 1N21 and 1N23 microwave diodes.

Two different types of solid-state pn diodes exist. Although they are considerably different from one another, it has become a common (erroneous) practice to consider them the same. Figures 12-1A and 12-1B show the differences. Both consist of p-type and n-type semiconductor materials, which are intimately close.

In the *point-contact pn diode* (Fig. 12-1A), the main body of the semiconductor material consists of one type material (shown here as p-type), with the alternate type (here n-type) diffused into the bulk material. A large metallic electrode connects the p-type end to an external electrode. The

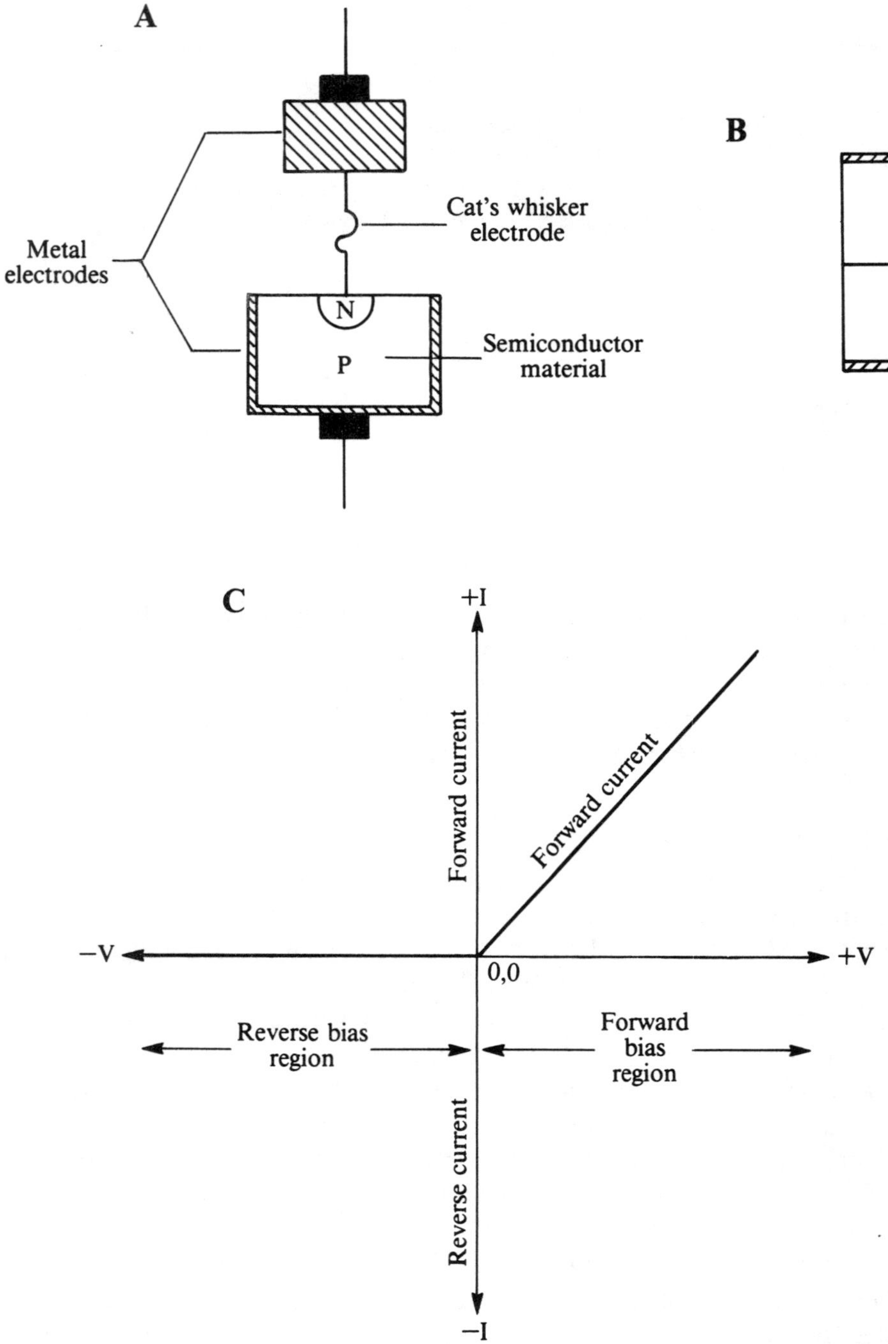

Figure 12-1. A) Point-contact diode structure, B) pn-junction diode, C) transfer function.

n-type material is connected to its electrode through a "cat's-whisker" contact, which is reminiscent of the crystal sets from early radio.

The *pn-junction diode* (Fig. 12-1B), on the other hand, consists of a bulk semiconductor with n-type impurities at one end and p-type impurities at the other end. Metallic-end electrodes connect the semiconductor material to the outside world. The modern diode is of the pn-junction type, although a few point-contact types are still used. Germanium diodes, such as 1N34 and 1N60, tend to be point-contact types, as do the older microwave diodes (e.g., 1N21, 1N23). Modern silicon signal and switching diodes (e.g., 1N914, 1N4148) are pn-junction types. When this book mentions pn-junction diodes, you can assume that both types are meant. I am bowing to the modern usage—even if it is a little sloppy.

The standard "ideal" pn-junction diode has an *I-vs-V transfer characteristic*, such as Fig. 12-1C. Real diodes don't meet this ideal, but it at least makes the operation of the device clear. When the anode is positive, with respect to the cathode, the diode is forward-biased and it conducts current. Alternatively, when the anode is negative, with respect to the cathode, the diode is reverse-biased and no current flows.

Real diodes fail to meet the ideal in several important respects. Figure 12-2 shows a transfer characteristic for a practical nonideal diode. For the ideal diode, the reverse current is always zero. In real diodes, a minute leakage current (I_L) flows backward across the junction. A manifestation of this current can be seen by measuring the forward and reverse resistances of a pn-junction diode. The forward resistance is very low and the reverse resistance is very high, but not "infinite" as one might expect from an open circuit.

Another departure from the ideal in the reverse-bias region is the avalanche point (V_z), at which the reverse-current flow increases sharply. At this point, the reverse-bias voltage is too great and it causes a breakthrough. When carefully regulated, the breakdown potential is both sharply defined and is reasonably stable, except for a slight temperature dependence. In such cases, the device is a *zener diode*, and it is used as a voltage regulator.

In the forward-biased region, another anomaly departs from the ideal. In the ideal case, an ohmic relationship exists between current flow and applied forward voltage. Similarly, a linear relationship exists between applied forward voltage (V_f) and output voltage V_o. In real diodes, however, the ideal transfer characteristic is departed from. Between zero volts and a critical junction potential V_g), the characteristic curves are quite nonlinear. The actual value of this potential is a function of both the type of semiconductor material and the junction temperature. In general, V_g will be 0.2 to 0.3 volts for germanium (Ge) diodes (1N34, 1N60, etc.) and 0.6 to 0.7 volts

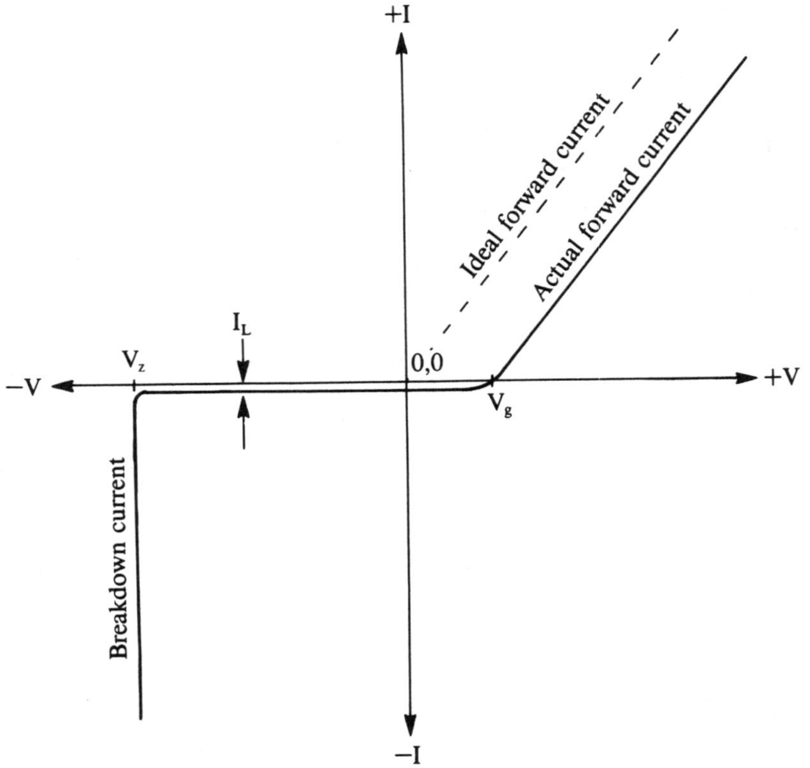

Figure 12-2. Zener-diode transfer function.

for silicon (Si) diodes (1N400x, 1N914, 1N418, etc.). In the 0-to-V_g region, the diode forward resistance is a variable function of V_f and T and the I-vs-V_f characteristic is logarithmic. Above V_g, the characteristic becomes nearly linear.

Experiment 12-1

This experiment examines the behavior of pn-junction diodes. The diodes selected are power-supply rectifiers, rather than signal diodes, because they are much more robust than small signal diodes. Select any of the 1N4000x series of diodes (i.e., 1N4001 through 1N4007). These diodes can tolerate forward currents up to 1 ampere. Two versions of the experiment can be attempted. One requires a dc milliammeter (or the mA scale of a digital multimeter, VOM, etc.), and the second requires an oscilloscope.

Procedure (meter version)

Connect the circuit of Experiment 12-1. This circuit consists of a diode (D1) in series with a 0-to-50 or 0-to-100 mA milliammeter (or the equivalent

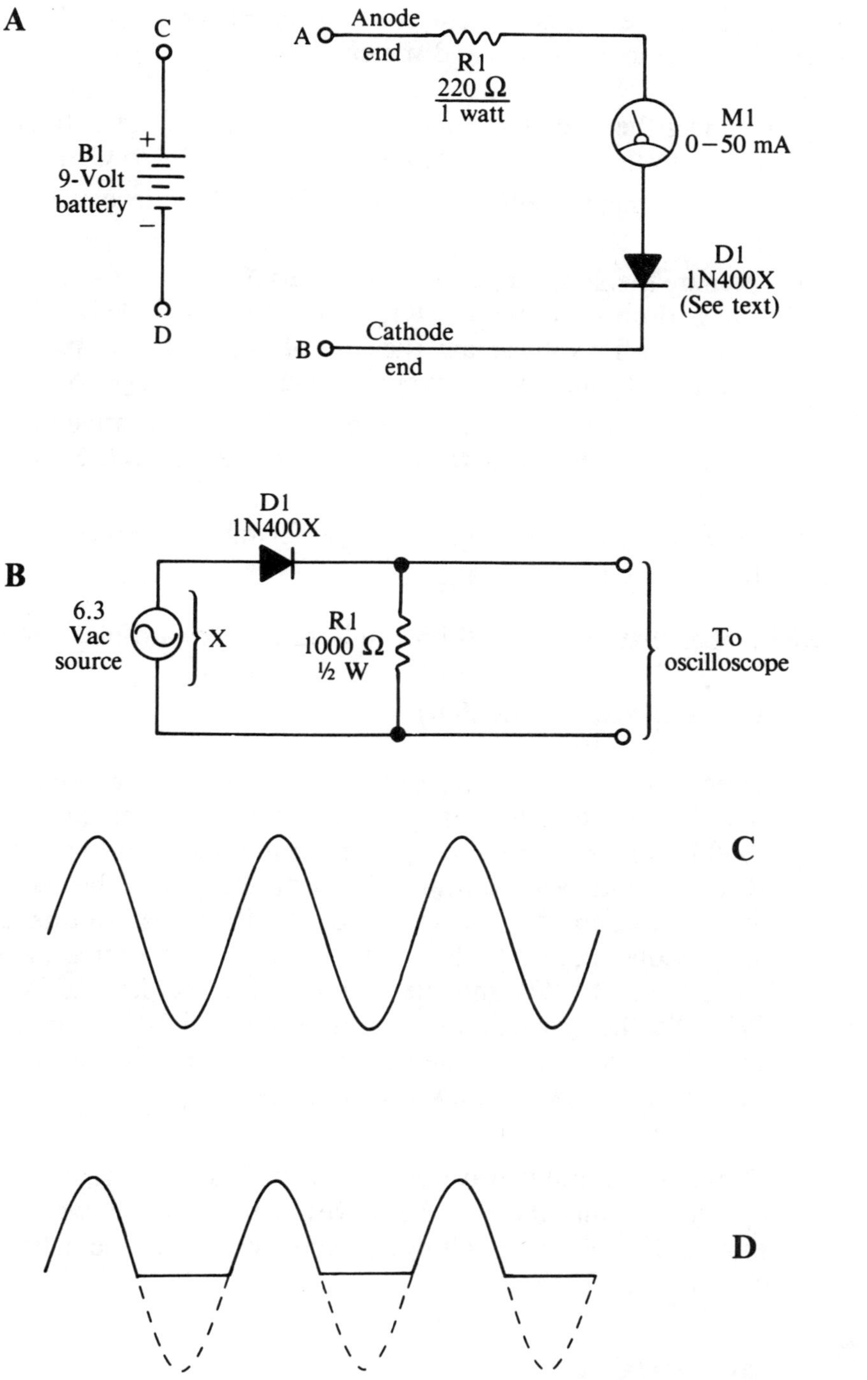

Experiment 12-1A. A) Circuit for Experiment 12-1, B) equivalent circuit, C) sine wave input, D) half wave rectified version.

scale on a multimeter), and a current-limiting resistor (R1). The resistor prevents damage to the diode and meter.

1. Connect the circuit so that the positive end of the battery is connected to the anode end of the diode circuit (A to C and B to D).
2. Take the meter reading. Is this the forward- or reverse-bias condition?
3. Measure the dc voltage across terminals A and B. Now, measure the voltage drop across resistor R1, meter M1, and diode D1. According to Kirchhoff's voltage law, the sum of the voltage drops across R1, M1, and D1 should be equal to the applied voltage (A to B).
4. Now, reverse the battery terminals so that the negative terminal of the battery is applied to the anode end of the circuit (i.e., A to D and B to C).
5. Take the current reading. Is this the forward- or reverse-bias condition?

Remember that current will flow only in the forward-bias condition.

Procedure (oscilloscope version)

1. Connect the circuit of Experiment 12-2. The 6.3-Vac source should be a filament transformer (any current rating). You can also use a 1000 Hz or less sine-wave signal generator and adjust the output for 10 to 20 volts peak-to-peak. The vertical input of the oscilloscope should be connected across resistor R1. If you have a dual-channel oscilloscope, connect CH-1 across the 6.3-Vac source (across the point marked "X") and the output of the rectifier (across R1) to CH-2. Set the 'scope position controls so that 6.3-Vac is in the upper half of the screen and the output waveform is in the lower half. Adjust the 'scope to show two or three complete cycles of the ac waveform.
2. Compare the input waveform with the half-wave rectified output waveform. Only the positive halves of the input ac waveform are shown. The negative half cycles were clipped by the action of the diode.

Precise diodes

A *precise-diode* (also called *ideal-rectifier*) *circuit* combines an active device, such as an operational amplifier, with a pair of solid-state pn-junction diodes to essentially "servo-out" the errors of the nonideal diode with feedback. Two advantages are obtained from this arrangement. First, the circuit will rectify very small ac signals between zero volts and V_g (i.e., $0 < V < V_g$, about 0.65 volts). Second, the rectification will be more linear than is the case with just the diode alone, even in the diode's ohmic range.

Figure 12-3 shows the basic circuit for the inverting halfwave precise rectifier. A basic assumption in this circuit is that load impedance R_L is purely resistive, and therefore contains no energy production or storage elements. The circuit is essentially an inverting-follower amplifier with two pn-junction diodes (D1 and D2) added. Halfwave rectification occurs because the circuits offer two different gains that are dependent on the polarity of the input signal. For positive values of V_{in}, the gain (V_o/V_{in}) is zero; for negative values of V_{in}, the voltage gain is R_f/R_{in}.

Consider operation of the circuit for positive values of V_{in}. The noninverting input (+In) is grounded, so it is held at zero volts. By the properties of the ideal op amp, consider the inverting input (−In) as if it is also grounded ($V_a = 0$). Remember that this concept is a virtual ground. Thus, differential voltage V_d is zero.

When $V_{in} > 0$ (i.e., when it is positive) current $I_1 = +V_{in}/R_{in}$. In order to maintain the equality $I_1 + I_3 = 0$ because of Kirchhoff's Current Law (KCL), op-amp output voltage V_b swings negative, but is limited by the D1 junction potential to V_g (about 0.6 to 0.7 volts). With $V_b < 0$, even by only

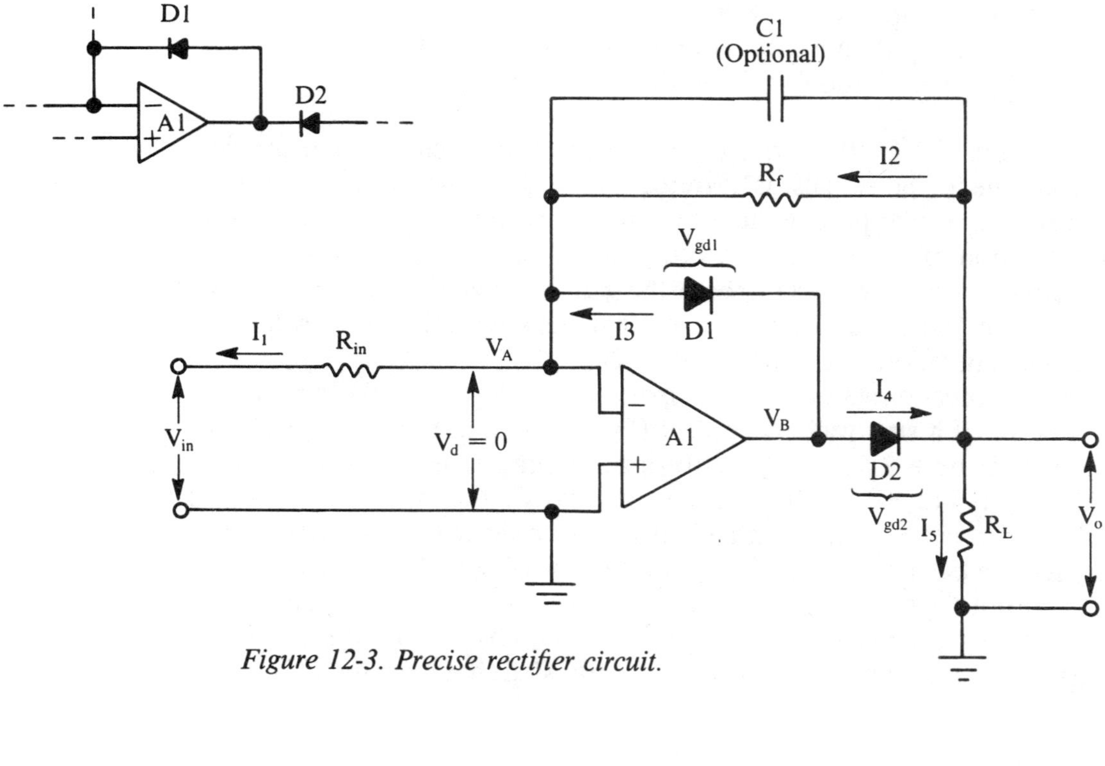

Figure 12-3. Precise rectifier circuit.

0.6 to 0.7 volts, diode D2 is reverse-biased and therefore cannot conduct. Currents I_2, I_4, and I_5 are zero. Thus, for positive values of V_{in}, output voltage V_o is zero.

Now, consider operation for $V_{in} < 0$. Under this input condition, op amp output voltage V_b swings positive, forcing diode D1 to become reverse-biased, and D2 to conduct. In order to conserve KCL (as before), for this case ($I_1 + I_2 = 0$), current I_2 will have the same magnitude, but opposite direction relative to I_1. Because $V_{in}/R_{in} = -V_o/R_f$, voltage gain $A_v = V_o/V_{in}$ reduces to $-R_f/R_{in}$, as is appropriate for an inverting amplifier. Thus, the gain for negative input voltages ($V_{in} < 0$) is $-R_f/R_{in}$, and for the positive-input voltage ($V_{in} > 0$), it is zero. That difference provides halfwave rectification.

The voltage drop across diode D2 is about +0.6 to +0.7 volts. It is "servoed-out" by D2 being in the negative-feedback loop around A1. Voltage V_b is correspondingly higher than V_o, in order to null the effects of V_{gd2}.

The precise rectifier is capable of halfwave rectifying very low level input signals. The minimum signal allowed is given by:

$$V_{in} > \frac{V_g}{A_{vol}} \tag{12-1}$$

Where:

V_{in} is the input signal voltage
V_g is the diode junction potential (0.6 to 0.7 volts)
A_{vol} is the open-loop gain of the amplifier

In Eq. (12-1), the term A_{vol} is the open-loop gain, which for dc and low-frequency ac signals, is extremely high. However, at some of the frequencies at which precise diodes operate, the input frequency is a substantial fraction of the gain-bandwidth product, so A_{vol} will be less than might otherwise be true. For example, if the gain-bandwidth product is 1.2 MHz, the gain at 100 Hz is 12,000. But at 1000 Hz, a typical frequency for precise rectifier operation, the gain is only 1,200.

The precise-rectifier circuit operation is shown by the waveforms in Fig. 12-4. If a sine wave is applied (Fig. 12-4A), output voltage V_o will be zero from time T_1 to T_2 (positive-input voltage), and V_b will rest at $-V_g$ (about -0.6 to -0.7 volts). Between T_2 and T_3, the input is negative, so V_o will be a positive voltage with a halfwave sine shape (Fig. 12-4B). However, notice the behavior of V_b, the op-amp output (Fig. 12-4C). From T_1 to T_2, the output rests at $-V_g$, but at T_2 it snaps to a value of $2V_g$. The halfwave sine shape rests on top of the $+V_g$ offset caused by V_{gd2}. Figure 12-4D shows this same situation in the form of the transfer characteristic (V_o-vs.-V_{in}).

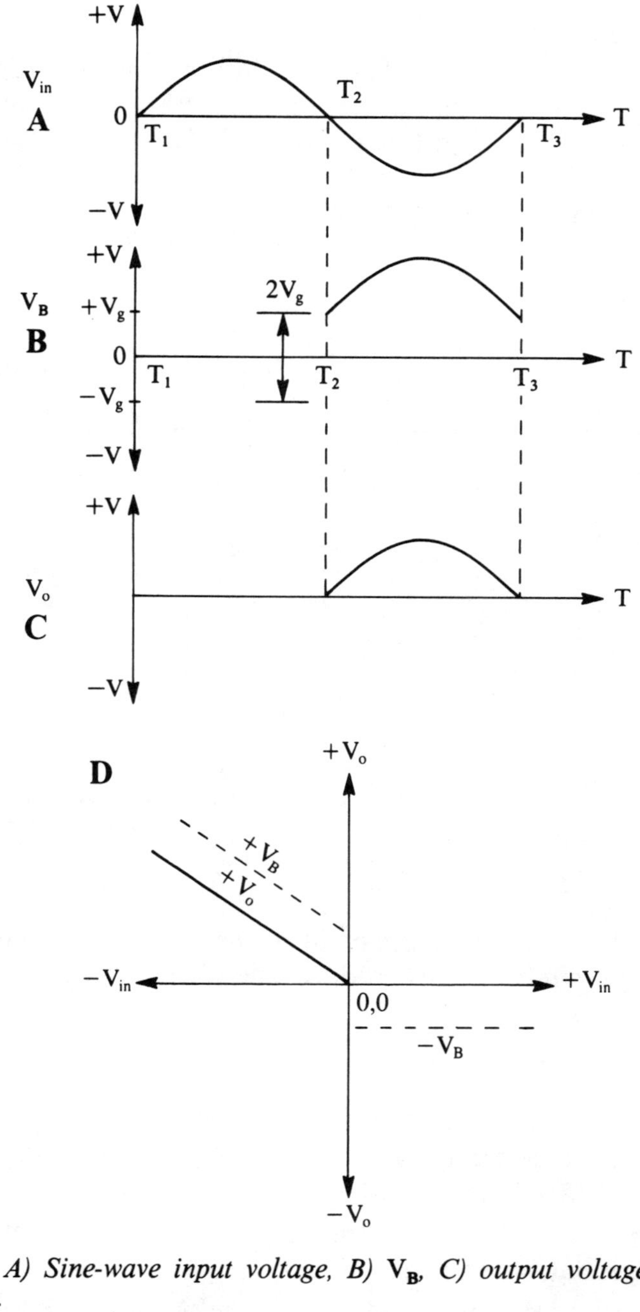

Figure 12-4. A) Sine-wave input voltage, B) V_B, *C) output voltage, D) transfer characteristic.*

The circuit of Fig. 12-3 will rectify and invert negative peaks of the input signal. In order to accommodate the positive peaks, only reverse the polarity of diodes D1 and D2.

Experiment 12-2

Select either a 741 or CA-3140 op amp in the 8-pin miniDIP package. Connect the circuit of Experiment 12-2. All resistors are ¼-watt carbon composition or metallic-film types. The ±12-Vdc power supplies are bypassed with 0.1-μF capacitors. Place these capacitors as close as possible to the body of the operational amplifier.

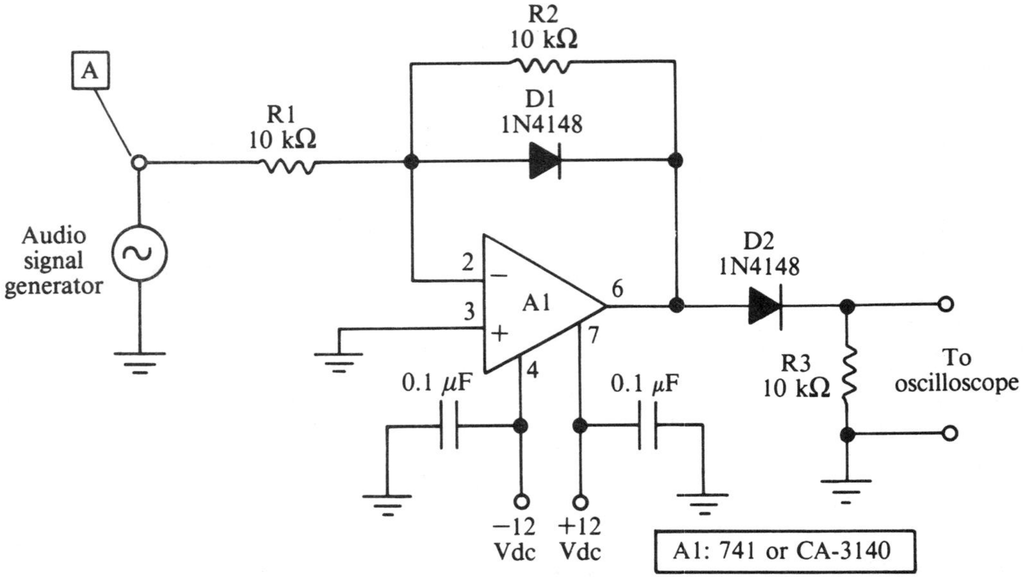

Experiment 12-2. Circuit for Experiment 12-2.

The signal source is an audio oscillator or audio signal generator (same as a *function generator*), set to produce sine-wave output signals.

Procedure

1. Adjust the output of the signal source for a 2-volt peak-to-peak signal at some frequency between 100 and 1000 Hz. The oscilloscope will facilitate this adjustment.
2. Adjust the oscilloscope to trigger on and lock to two or three cycles of the input waveform.
3. Turn on the dc power and connect the signal source to the input of the circuit (point A) and ground.
4. Using the oscilloscope's input probe, examine the output of the circuit. You should see the same kind of halfwave rectified waveform as was seen in the oscilloscope version of Experiment 12-1.

5. If you have a dual-trace oscilloscope, place the input signal on one channel and the output signal on the other so that they can be compared on a common time base.
6. Disconnect the signal source from point A and turn off the ±12-Vdc power supply.
7. Reverse the polarity of the diodes D1 and D2 (i.e., turn them around backward in the circuit).
8. Turn on the dc power and apply the signal source to point A.

The output of a precise diode (ideal rectifier) is a pulsating dc waveform, as seen in Experiment 12-2. If you need to create a dc level that is proportional to the average level (rms value) of the output waveform, time-average the pulsating dc (Fig. 12-5). In this circuit, the precise diode is followed by either a low-pass filter or an electronic integrator (basically the same thing!) that has a low enough time constant that it will produce a time-averaged output. Basically, the experimental value of the time constant should start at five times the period of the applied waveform; test from that point.

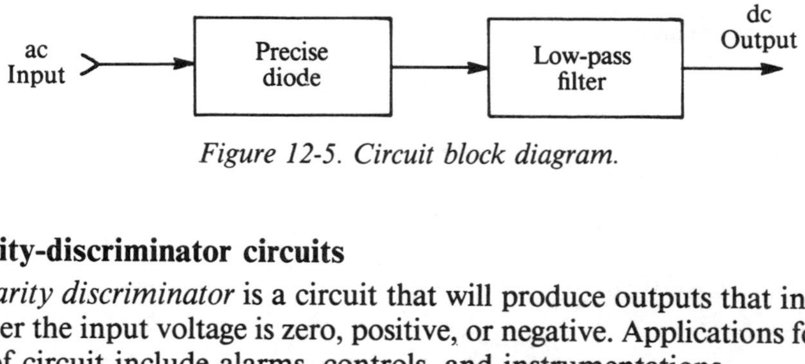

Figure 12-5. Circuit block diagram.

Polarity-discriminator circuits

A *polarity discriminator* is a circuit that will produce outputs that indicate whether the input voltage is zero, positive, or negative. Applications for this type of circuit include alarms, controls, and instrumentations.

Figure 12-6A shows a typical polarity-discriminator circuit. The basic configuration is the inverting-follower op-amp circuit, but with two negative feedback circuits. Each feedback path contains a diode, but the diodes are connected in the opposite polarity. The polarity of the output potential will determine which diode conducts and which is reverse-biased.

Consider first the case where input signal V_{in} is positive. In this case, current I_{in} flows away from the summing junction, toward the source with a magnitude of $+V_{in}/R_{in}$. The output terminal of the op amp will swing negative and cause diode D1 to be reverse-biased and D2 to be forward-biased. Current I_1 is zero and I_2 is equal to V_{o2}/R_f. Output voltage V_{o2} is negative and has a value of $V_{o2} = V_o - 0.6$ volts; output voltage V_{o1} is zero.

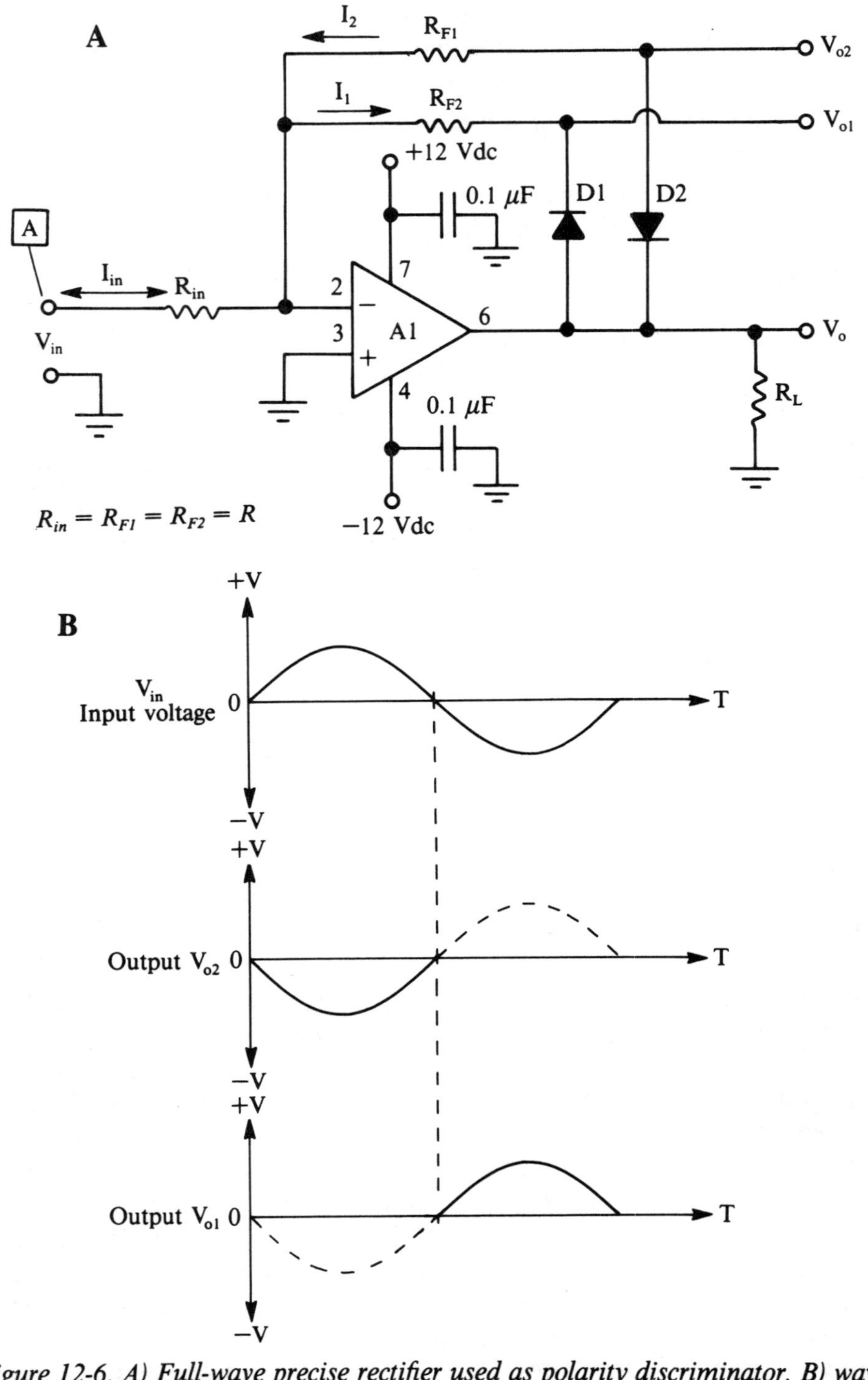

Figure 12-6. A) Full-wave precise rectifier used as polarity discriminator, B) waveforms, C) transfer function.

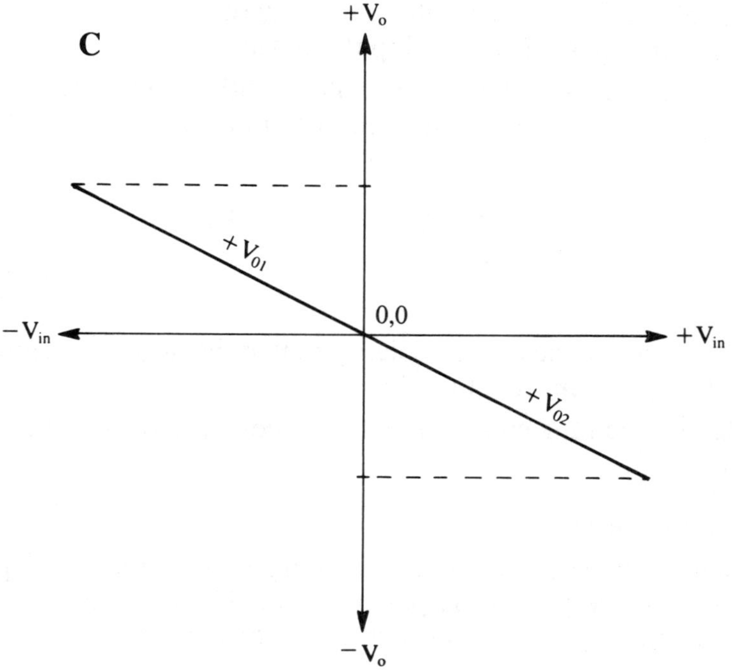

Now consider the opposite case where V_{in} is negative. The current flows away from the source, toward the summing junction. The output terminal of the op amp swings positive and causes diode D1 to become forward-biased, while D2 is reverse-biased. Current I_2 is zero and I_1 is V_{o1}/R_f. In this case, V_{o1} is positive and V_o is zero.

The operation of this circuit can be seen in the waveforms (shown in Fig. 12-6B) and in the transfer characteristics (shown in Fig. 12-6C).

Experiment 12-3

This experiment examines the behavior of the polarity-discriminator circuit of Fig. 12-6A using 10 kohm resistors for R_{in}, R_{f1}, R_{f2}, and R_L. Select an operational amplifier, either 741 or CA-3140 in the miniDIP package (the pinouts in Fig. 12-6A are for the miniDIP configuration). The ±12-Vdc power-supply terminals of the op amp are bypasses with 0.1-μF capacitors; place the capacitors as close as possible to the body of the op amp.

Procedure (dc version)

1. Connect the circuit as required.
2. Turn on the dc power, then connect a 1.5-volt dry cell as an input

signal. At first, connect the positive terminal of the battery to the input (point A). Measure input voltage V_{in}.

3. **Trial 1** Measure the output voltage with a voltmeter at points V_o, V_{o1}, and V_{o2}. Compare these readings and write them on a table such as:

Trial	V_{in}	V_o	V_{o1}	V_{o2}
1				
2				

4. Now, reverse the battery polarity so that the negative terminal of the battery is connected to point A.
5. Repeat the four measurements and compare them with the results of step 3.

Procedure (ac version)

This version of the experiment uses an ac signal source for the input signal and an oscilloscope for the output indicator. If you have a two-channel oscilloscope, put V_{o1} and V_{o2} on each of the two channels.

1. Adjust the oscilloscope to lock on two or three cycles of the ac input signal, with a potential of 1 to 4 volts peak-to-peak.
2. Turn on dc power, and apply a 1-to-4-Vac 100-to-1000-Hz sine-wave signal.
3. Examine and compare the two outputs on the oscilloscope. If you don't have an oscilloscope camera, try to sketch the output wave-forms on the same time base.

Fullwave precise rectifier

The fullwave rectifier uses both halves of an input sine wave. Remember that the halfwave rectifier removes one polarity of the sine wave; the fullwave rectifier preserves it. Figure 12-7 shows the relationships present in a fullwave-rectifier circuit. Figure 12-7A shows the input sine-wave and the pulsating dc output of a fullwave rectifier. Notice that the negative halves of the sine waves are flipped over and appear in the positive-going direction. The characteristic function for the fullwave rectifier is shown in Fig. 12-7B. Because the output voltage is always positive, regardless of whether the input signal is positive or negative, the fullwave rectifier is an *absolute-value circuit*. The output voltage will be either:

$$V_o = k/V_{in}/ \qquad (12\text{-}2)$$

or,

$$V_o = -k/V_{in}/ \qquad (12\text{-}3)$$

depending on the direction of the diodes within the circuit.

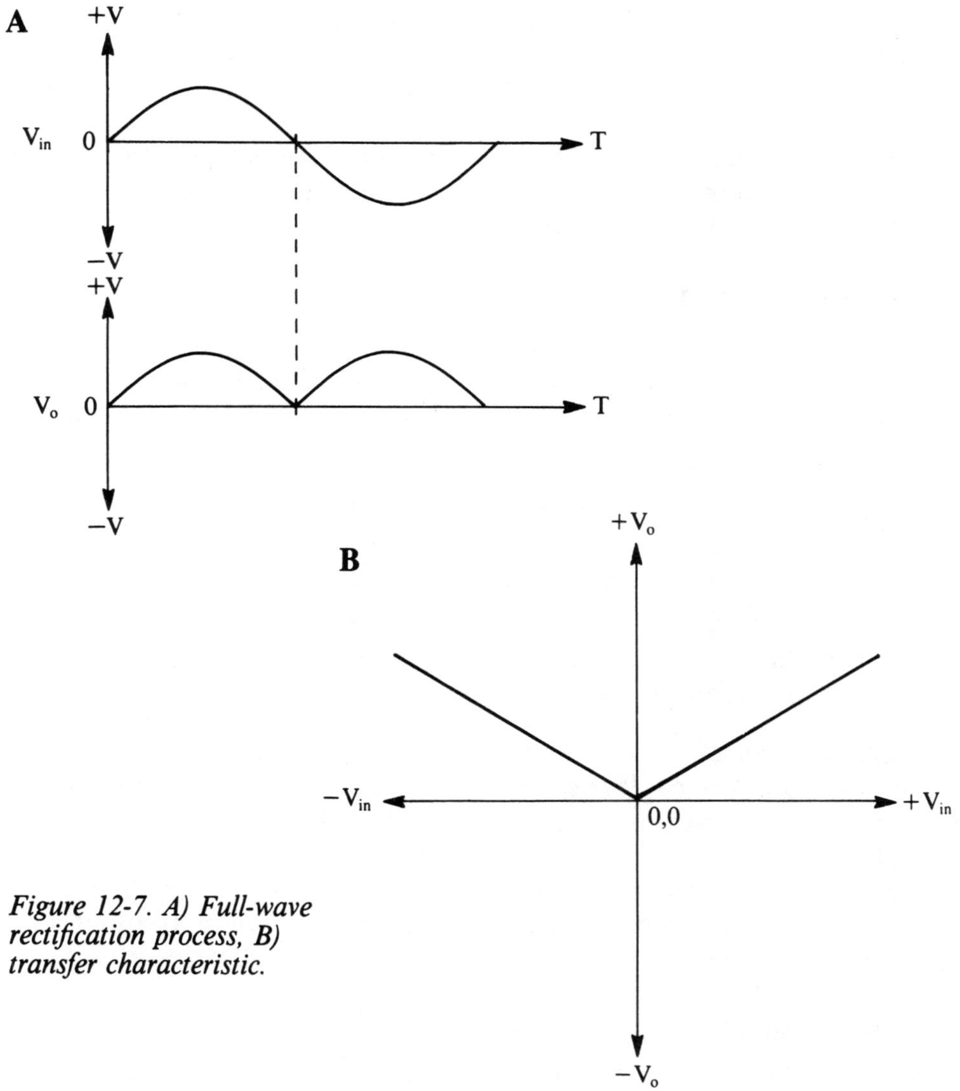

Figure 12-7. A) Full-wave rectification process, B) transfer characteristic.

Although the fullwave rectifier has major applications in dc power supplies, it is the absolute-value feature that makes the fullwave rectifier important for instrumentation and other related purposes.

Several methods exist to create a precise fullwave rectifier (a.k.a. absolute value amplifier) and some of these are shown in Fig. 12-8. The first circuit (Fig. 12-8A), is based on the polarity-discriminator circuit of Fig. 12-6. The two outputs, V_{o1} and V_{o2}, are applied to the inputs of a dc differential amplifier.

A second approach is shown in Fig. 12-7B. In this circuit, a pair of oppositely connected diodes are applied to the inputs of a simple dc differential amplifier. This approach is not as highly regarded because the diodes in the input stage are not in the feedback loop. Thus, the voltage drops (V_g) are not "servoed-out."

The third alternative is a pair of operational amplifiers. Still another possibility is to connect two precise-diode circuits, each with opposite polarity (the two versions discussed earlier), in parallel.

Experiment 12-4

Connect the circuit of Fig. 12-8C with two 742 operational amplifiers. Connect 0.1-μF capacitors to each of the dc power-supply terminals (not shown in Fig. 12-8C); $-$V is pin 4, and $+$V is pin 7 on each. All resistors are 10 kohms.

1. Adjust the output of an audio signal generator to produce 4 volts peak-to-peak, from 100 to 1000 Hz; use a sine-wave output.
2. Adjust the oscilloscope vertical input, trigger, and time-base controls to display two or three cycles of the applied sine-wave ac signal.
3. Compare the output waveform V_o with the input waveform V_{in}. The output waveform should be a fullwave rectified output.

This experiment can be performed with a dc voltmeter to measure input- and output-signal levels, and a 1.5-volt dry cell as the signal source. Connect a dc generator, such as Experiment 12-4. Measure the input and output voltages with a dc voltmeter and plot the V_o-to-V_{in} transfer function of the circuit for both settings of the DPDT switch. Use at least 10 approximately evenly spaced settings of the potentiometer to select 10 input voltages for best accuracy.

Zero-bound and dead-band circuits

A zero-bound circuit is one in which the output voltage is limited so that it will be nonzero for certain input-voltage values and zero for all other input voltages. The term does not mean that the values of V_{in} are in any way constrained, but rather that allowable output voltages are constrained. The output of a zero-bound circuit indicates when the input signal exceeds a certain threshold and by how much.

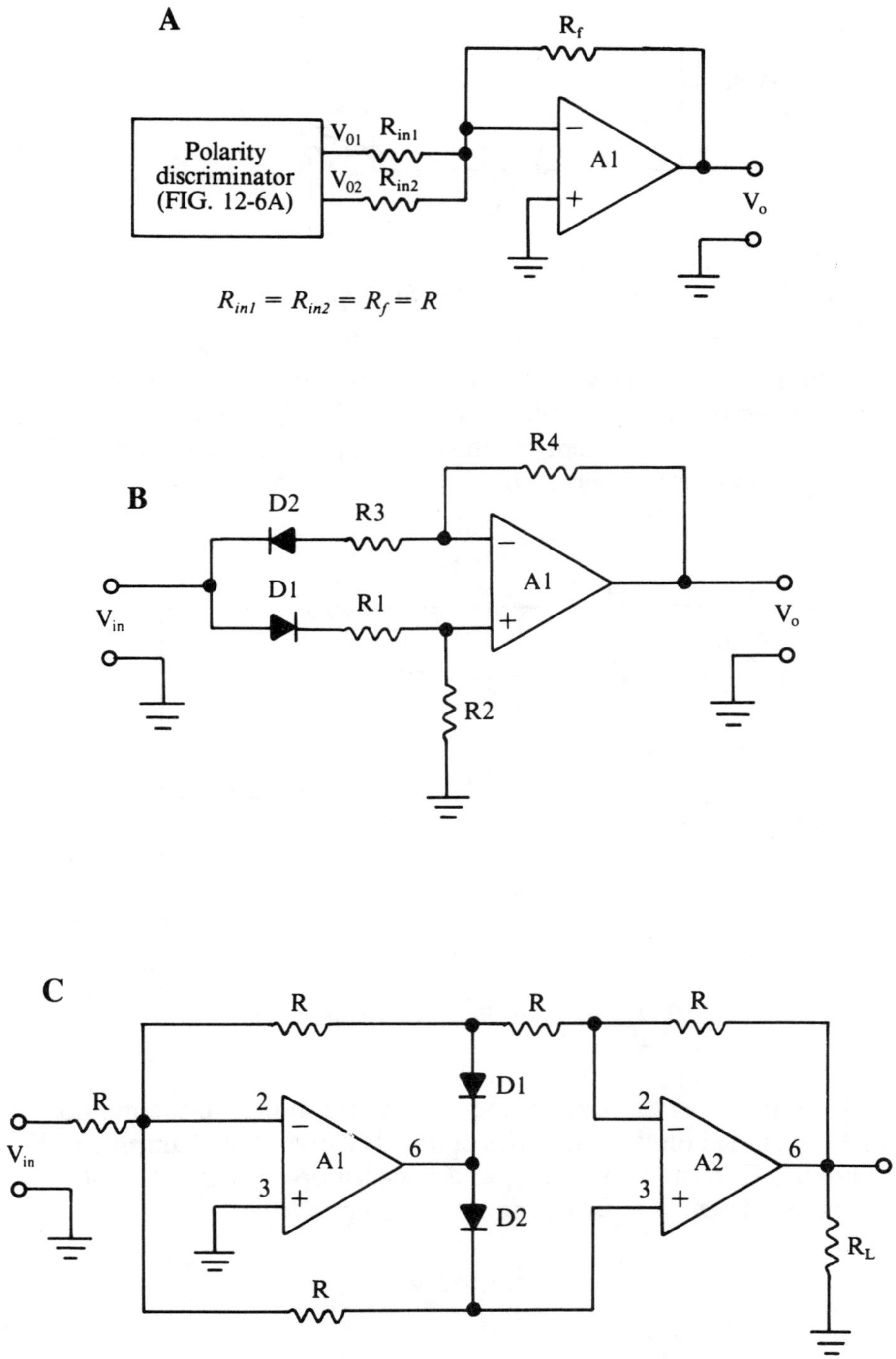

Figure 12-8. A) Polarity-discriminator precise-rectifier circuit, B) Diode version, C) active op-amp version.

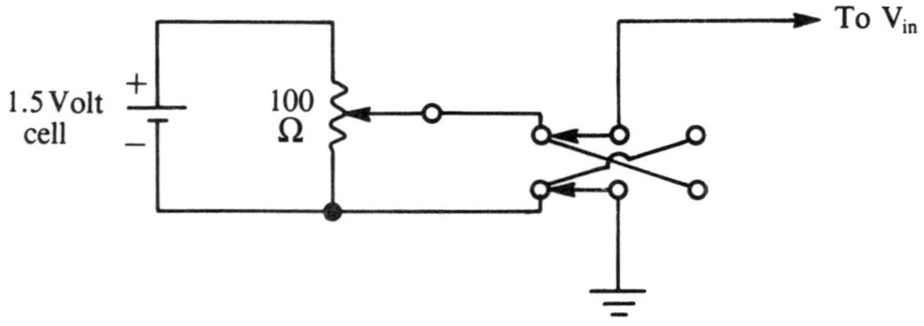

Experiment 12-4. Circuit for Experiment 12-4.

Figure 12-9 shows a zero-bound amplifier circuit. This circuit is based on the halfwave precise-rectifier circuit of Fig. 12-3 and it functions in exactly the same way, except for the extra input reference current, I_{ref}. I_{ref} offsets the trip point, at which the input voltage takes effect.

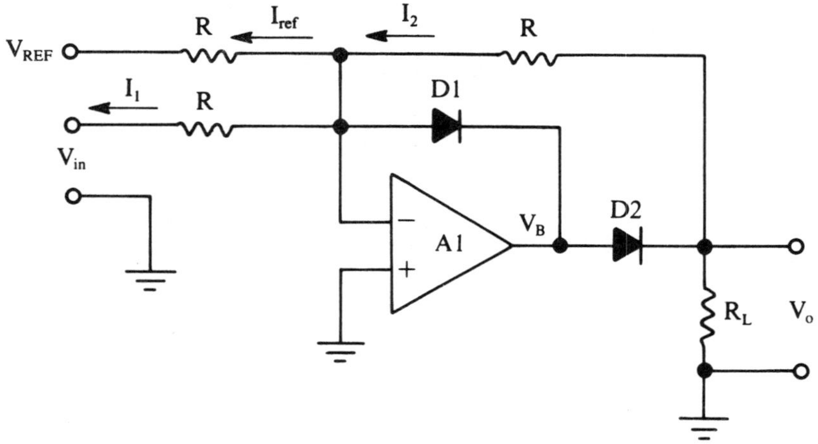

Figure 12-9. Bounded value circuit.

To understand this circuit, use an analysis similar to the method used before (i.e., based on the properties of the ideal operational amplifier). From Kirchhoff's Current Law (KCL) and because op-amp inputs neither sink nor source current, the following relationship is true:

$$I_1 + I_{ref} = 0 \tag{12-4}$$

or,

$$I_1 + I_{ref} = -I_2 \tag{12-5}$$

We also know that:

$$I_1 = \frac{V_{in}}{R} \tag{12-6}$$

$$I_{ref} = \frac{V_{ref}}{R} \tag{12-7}$$

$$I_2 = \frac{V_o}{R} \tag{12-8}$$

Thus,

$$\frac{V_{in}}{R} + \frac{V_{ref}}{R} = -\frac{V_o}{R} \tag{12-9}$$

and after multiplying both sides by R:

$$V_{in} + V_{ref} = V_o \tag{12-10}$$

Thus, the output voltage is still proportional to the input voltage, but it is offset by V_{ref}. The transfer characteristics for this circuit are shown in Fig. 12-10. In Fig. 12-10A, V_{ref} is negative, but in Fig. 12-10B, V_{ref} is positive. In both cases, the transfer curve is offset by the reference-signal potential.

Consider the operation of the circuit in Fig. 12-9 under two conditions: $V_{in} > 0$ and $V_{in} < 0$. First assume that $V_{ref} = 0$. For the positive input ($V_{in} > 0$), the output of the operational amplifier (A1) swings negative (the circuit is an inverter) and causes diode D2 to be reverse-biased and D1 to be forward-biased. Output voltage V_o is zero in this case. The output voltage is bound to zero for all values of $V_{in} > 0$.

Experiment 12-5

Perform Experiment 12-4 again with the circuit of Fig. 12-9 so that V_{ref} is 1-Vdc. Perform this experiment twice: one time with each polarity of V_{ref}. Also, perform this experiment with a 1-to-4-volts peak-to-peak 100-to-1000-Hz sine wave as the signal source. A two-channel oscilloscope is needed to perform this experiment at its best.

Dead-band circuits

In a dead-band circuit, two zero-bound circuits work together to produce a summed output. Figure 12-11A shows the transfer characteristic of such a

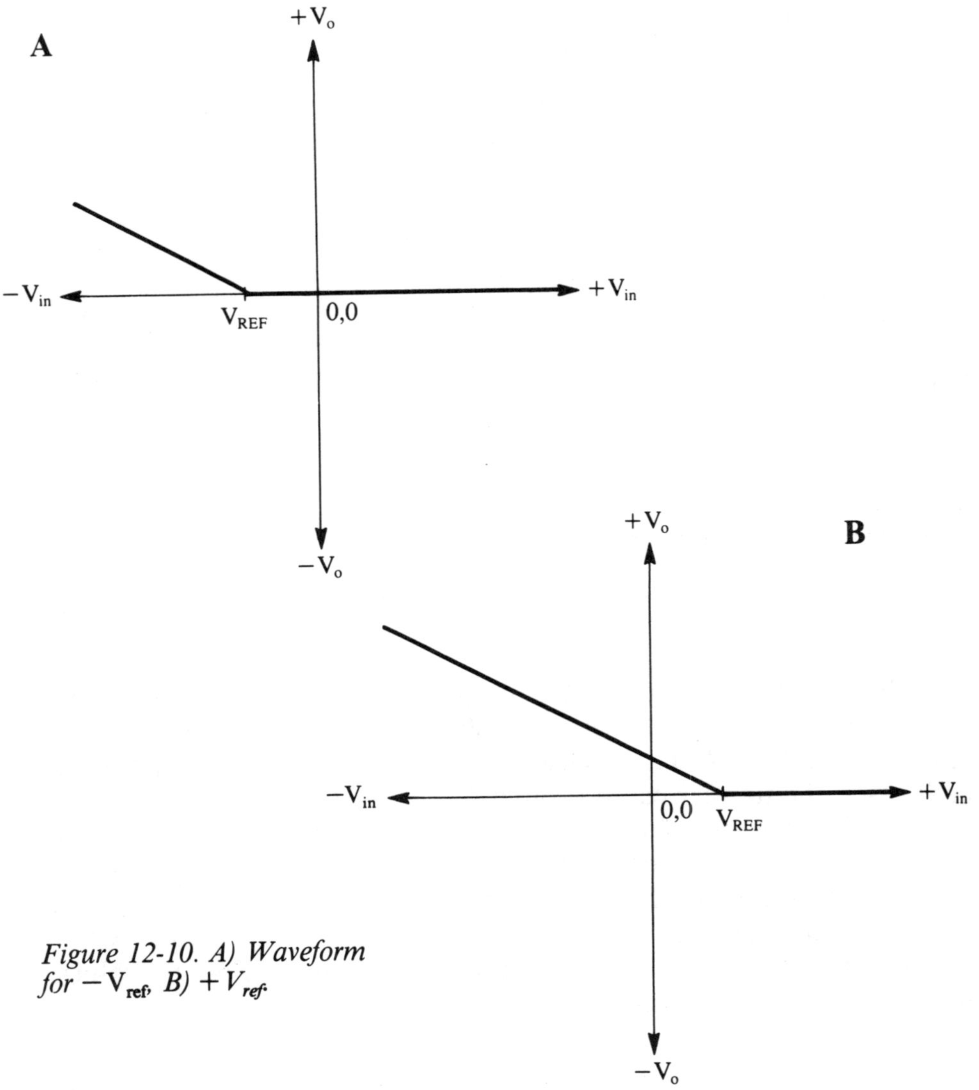

Figure 12-10. A) Waveform for $-V_{ref}$*, B)* $+V_{ref}$*.*

circuit. Two different threshold values are shown in this curve. The circuit will output signals only when the input signal is less than the lower threshold ($V_{in} < -V_{th}$) or greater than the upper threshold ($V_{in} > +V_{th}$). This behavior is relative to the sine-wave input signal in Fig. 12-11B. The output will be zero for all values of input within the shaded zone.

Remember that the output voltage will not suddenly snap to a high value above the threshold potential. Rather, it will be equal to the difference between the peak voltage and the threshold voltage. Assuming unity gain for both reference voltage and input-signal voltage, the output peaks will be $[(+V_p - (+V_{th})]$ and $[(-V_p - (-V_{th})]$.

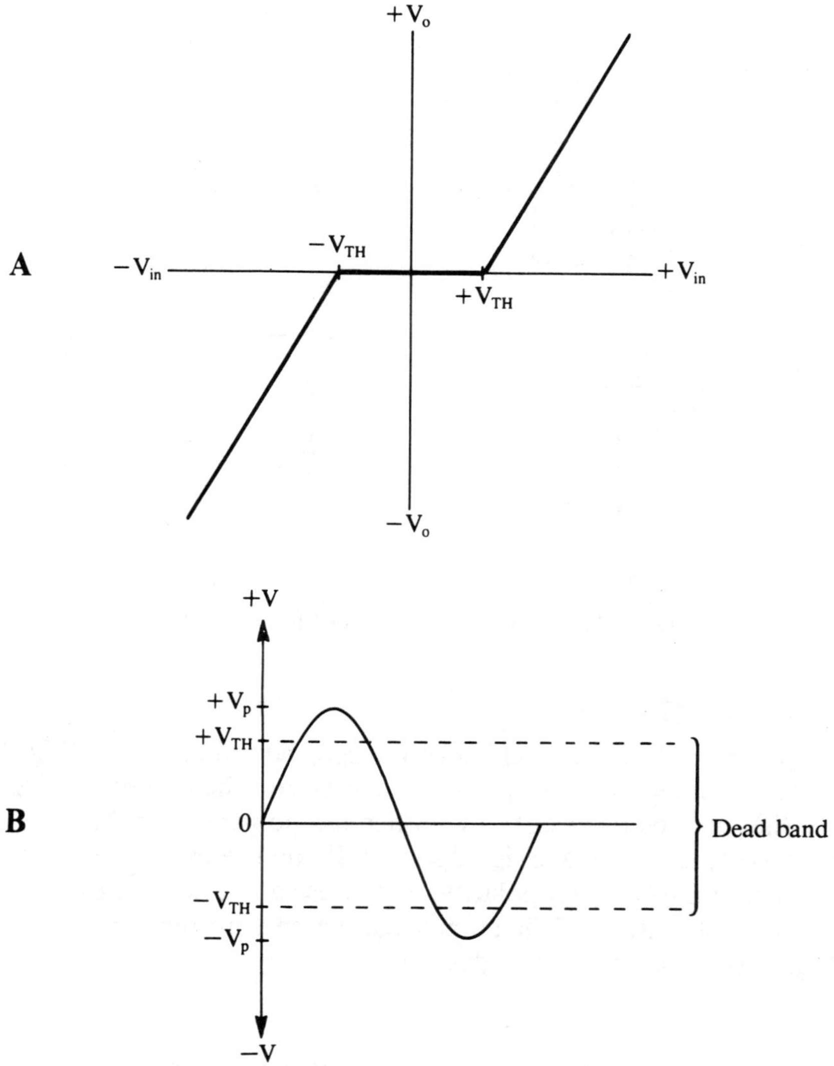

Figure 12-11. Zero-bounded operation: A) transfer characteristic, B) action on sine wave.

The dead-band amplifier circuit consists of a pair of zero-bound circuits that are summed together (Fig. 12-12). Both zero-bound circuits are similar to Fig. 12-9. The first circuit uses diodes of the same polarity that were used in Fig. 12-9; and the second circuit uses reverse-polarity diodes. In the case of the first circuit, the V + dc power supply is used as V_{ref}; in the second, the V − dc supply is used as V_{ref}. In both cases, the magnitude is the same, but the polarities are reversed.

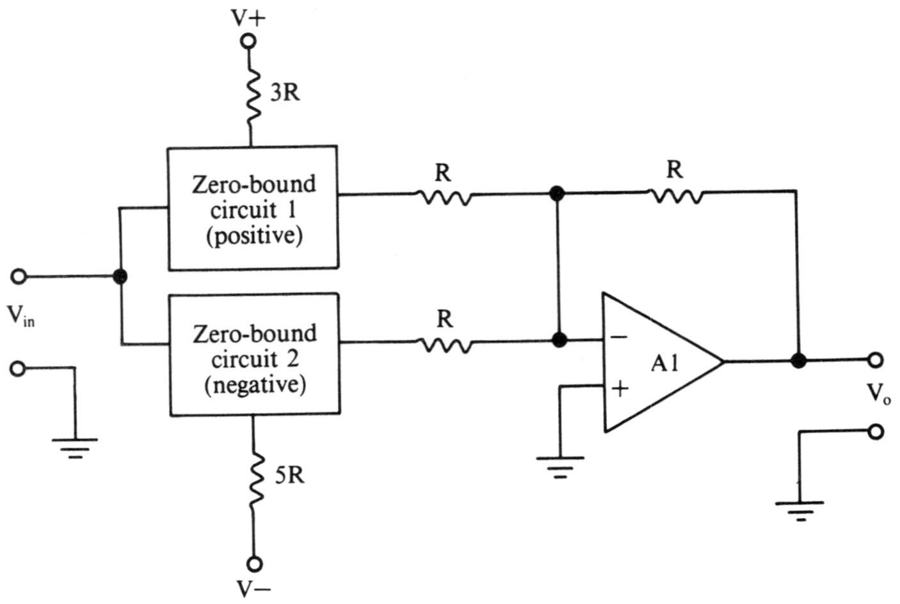

Figure 12-12. Dual-polarity zero-bound circuit.

Experiment 12-6

Connect the circuit of Fig. 12-12 with the zero-bound circuits of Fig. 12-9. Make one of the circuits with positive polarity and the other with negative polarity (i.e., diodes reversed). Use either the dc generator that was used earlier or use an ac sine-wave signal source. If you use the dc source, take at least 10 data points at each polarity. If you use an ac source, examine the waveform at the output of the circuit, compared with the input sine-wave signal, at various signal amplitudes.

Summary

An ideal diode develops a very large current with a forward-bias voltage, and zero current with a reverse-bias voltage. Real diodes, however, require a minimum junction potential for forward bias and exhibit a small reverse current when reverse-biased. The zener diode exhibits a large reverse current once a breakdown voltage, or zener voltage, is reached. These diodes can be combined with op amps to produce amplifying rectifiers and polarity-discriminator circuits. Another diode-amplifier circuit, known as a *zero-bound circuit*, allows a nonzero output only when the input is outside a certain range. This nonzero output will have the same shape as the input but is reduced by some dc value. A dead-band circuit combines two zero-bound circuits to provide an output clipped for two different values.

Questions

1. What is the difference between an ideal diode and a real diode?
2. Sketch the transfer function for **a.** an ideal diode **b.** a real diode
3. What is a polarity-discriminator circuit? What applications does it have?
4. What is a zero-bound circuit?
5. What is a dead-band circuit? How can zero-bound circuits be combined to form a dead-band circuit?

Problems

1. For the circuit in Fig. 12-9, V_{in} is a 10-volt peak-to-peak sine wave, and all the resistors are equal. If V_{ref} is +3 volts dc, sketch the output voltage.
2. Repeat Problem 1 for a reference voltage of −2 volts dc.
3. For the circuit in Fig. 12-12, the input voltage is a 12-volt peak-to-peak sine wave, the positive reference voltage is +4 volts, and the negative reference voltage is −4 volts. Sketch the output voltage.
4. Repeat Problem 3 for a positive reference voltage of +3 volts and a negative reference voltage of −2 volts.

13 OPERATIONAL-TRANSCONDUCTANCE AMPLIFIERS

AFTER COMPLETING THIS chapter, you will be able to:

- □ Understand amplifier transfer functions
- □ Discuss operational transconductance amplifiers and circuits

Commonly people erroneously assume that op amps are the only flexible linear IC amplifiers on the market. Although some concede that other linear ICs exist on the market, they are often believed to be special-purpose devices or limited-use devices. However, several other types of flexible ICs are available. This chapter looks at the *operational-transconductance amplifier* (*OTA*) and some of its applications.

Before getting into the specifics of transconductance amplifiers, however, you need to look at amplifier transfer functions. This book has covered at least one type of transfer function all along, so don't be too concerned about the fancy names. The concepts are quite simple.

Amplifier transfer functions

The *transfer function* of any electronic network is the ratio of its output characteristic to its input characteristic. Perhaps the most familiar transfer function is the one that this book has already dealt with: the *voltage amplifier*. In the voltage amplifier, the transfer function is the ratio of the output signal voltage (V_o) to the input signal voltage (V_{in}). The transfer function for a voltage amplifier (V_o/V_{in}) is also its *voltage gain* (A_v), so the transfer function of the voltage amplifier can be written in the forms:

$$A_v = \frac{V_o}{V_{in}} \tag{13-1}$$

or,

$$A_v = \frac{\Delta V_o}{\Delta V_{in}} \tag{13-2}$$

Similarly, the transfer function for a current amplifier is:

$$A_I = \frac{I_o}{I_{in}} \tag{13-3}$$

or,

$$A_i = \frac{\Delta I_o}{\Delta I_{in}} \tag{13-4}$$

These two amplifier classes (voltage and current) are familiar to most readers, or at least the concepts are easily visualized. Perhaps less familiar, however, are the *transresistance amplifiers* and the *transconductance amplifiers*. Resistance (*R*) is the quotient of voltage (*V*) and current (*I*), so a transresistance amplifier is one that relates an output voltage to an input current:

$$R_m = \frac{\Delta V_o}{\Delta I_{in}} \tag{13-5}$$

Conductance (current/voltage) is the inverse of resistance, so the gain of the OTA is expressed in units of conductance. In other words, the OTA gain relates an output current to an input voltage:

$$G_m = \frac{\Delta I_o}{\Delta V_{in}} \tag{13-6}$$

Where:

G_m is the transconductance (in *Siemens*, formerly *mhos*)
I_O is the output current change

V_{in} is the input voltage change
Δ indicates a small change in the related quantity

OTA devices

The OTA-equivalent circuit is shown in Fig. 13-1. The input circuit is similar to the input circuit of the operational amplifier because they have differential voltage inputs (−In and +In). The input voltages are: differential signal V_d (i.e., $V_2 - V_1$), and common-mode signal V_{cm}. The output side of the amplifier, however, is a current source that produces an output current, I_o, which is proportional to the gain and the input voltage. The current gain (A_{gm}) of this circuit is a function of the transconductance (I_o/V_{in}) and the load resistance (R):

$$A_{gm} = G_m R \tag{13-7}$$

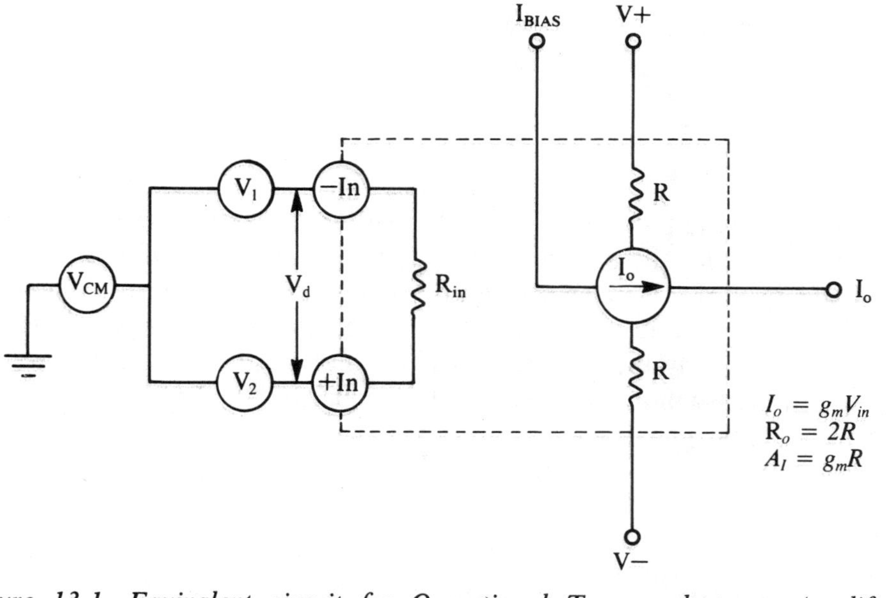

Figure 13-1. Equivalent circuit for Operational Transconductance Amplifiers (OTA).

Where:

A_{gm} is the gain
G_m is the transconductance (I_o/V_{in})
R is the load resistance (one-half of the output resistance R_o)

Perhaps the most common commercial versions of the OTA are the RCA CA-3080, CA-3080A, and CA-3060 devices. The CA-3080 devices are available in the eight-pin metal IC package with the pinouts that are shown in Fig. 13-2. The CA-3080 devices will operate over dc power-supply voltages, from ±2 to ±15 volts, with adjustable power consumption of 10 to 30 milliwatts. The gain range is zero to the product G_mR. The input voltage spread is ±5 volts. The bias current can be set to as high as 2 mA.

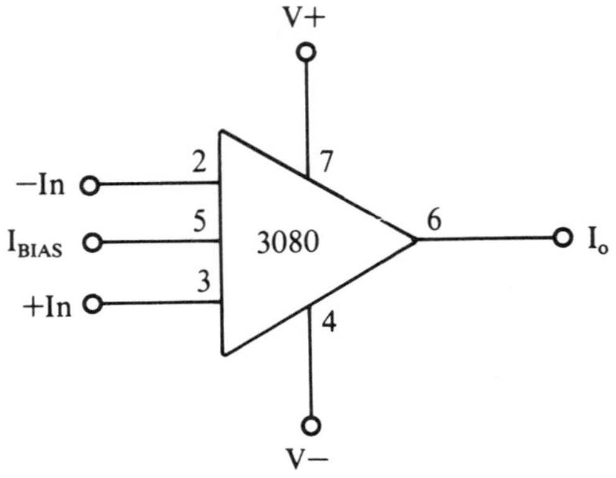

Figure 13-2. CA-3080 OTA device.

Note that the pinouts for the CA-3080 are industry-standard op-amp pinouts, except for the bias current (I_{bias}) that is applied to pin 5: V− is on pin 4; V+ is on pin 7; inverting input (−In) is on pin 2; noninverting input (+In) is on pin 3; and the output is on pin 6. The operating parameters of the OTA are set by the bias current (I_{bias}). For example, on the CA-3080 device, the transconductance is 19.2 times higher than the bias current:

$$G_m = 19.2 I_{bias} \qquad (13\text{-}8)$$

Where:

I_{bias} is the bias current in milliamperes (mA)

Figure 13-3A shows the relationship between transconductance and the bias current, I_{bias}, in graphical form. Figure 13-3B shows the relationship between bias current and both the input resistance, R_{in}, and the output resistance, R_o.

In many actual design cases, you will know the required value of G_m from knowledge of I_o/V_{in}, so the G_m required can be set by adjusting the bias current. In those cases, the I_{bias} is found by rewriting the previous expression:

$$I_{bias} = \frac{G_m}{19.2} \tag{13-9}$$

The CA-3080 output resistance of the device is also a function of the bias current:

$$R_o = \frac{7.5}{I_{bias}} \tag{13-10}$$

Where:

R_o is the output resistance in megohms
I_{bias} is the bias current in milliamperes (mA)

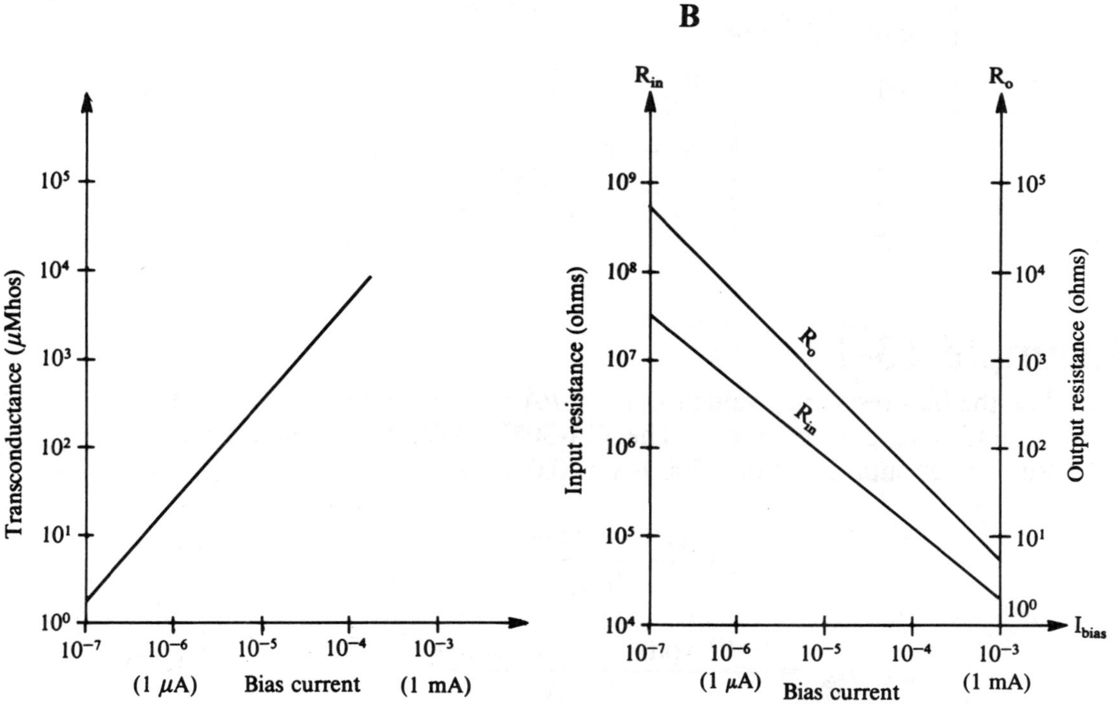

Figure 13-3. A) Transfer function for bias current, B) relationship of input resistance and output resistance to bias current.

The OTA bias current is set in most circuits by a resistance between the bias input, pin 5, and a voltage source (Fig. 13-4). Unless a higher degree of either stability or isolation is required, the bias resistor can be connected to the V+ dc power-supply line. The value of the bias resistor is found from:

$$R_{bias} = \frac{(V+) - (V-)}{I_{bias}} \tag{13-11}$$

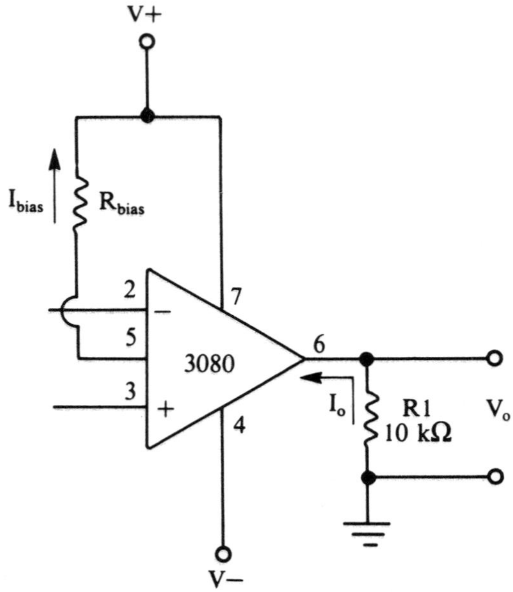

Figure 13-4. Bias current and output resistors used on the CA-3080.

Example 13-1

Calculate the bias resistance value for a 500-μA bias current that is derived from a +12-Vdc power supply. The CA-3080 OTA is powered from ±12-Vdc power supplies. Note: 500 μA = 0.0005 A.

$$I_{bias} = \frac{(V+) - (V-)}{I_{bias}} \tag{13-12}$$

$$I_{bias} = \frac{(+12 \text{ volts}) - (-12 \text{ volts})}{0.0005 \text{ A}} \tag{13-13}$$

$$R_{bias} = \frac{24 \text{ volts}}{0.0005 \text{ A}} = 48{,}000 \text{ ohms} \tag{13-14}$$

Voltage amplification from the IC OTA

The OTA is a current-output device, but it can be used as a voltage amplifier when one of the circuit strategies of Fig. 13-5 are used. The simplest method is the resistor load shown in Fig. 13-5A. Because the output of the OTA is a current (I_o), you can pass this current through a resistor (R1) to create a voltage drop. The value of the voltage drop (and the output voltage, V_o) is found from Ohm's law:

$$V_o = I_o R_1 \tag{13-15}$$

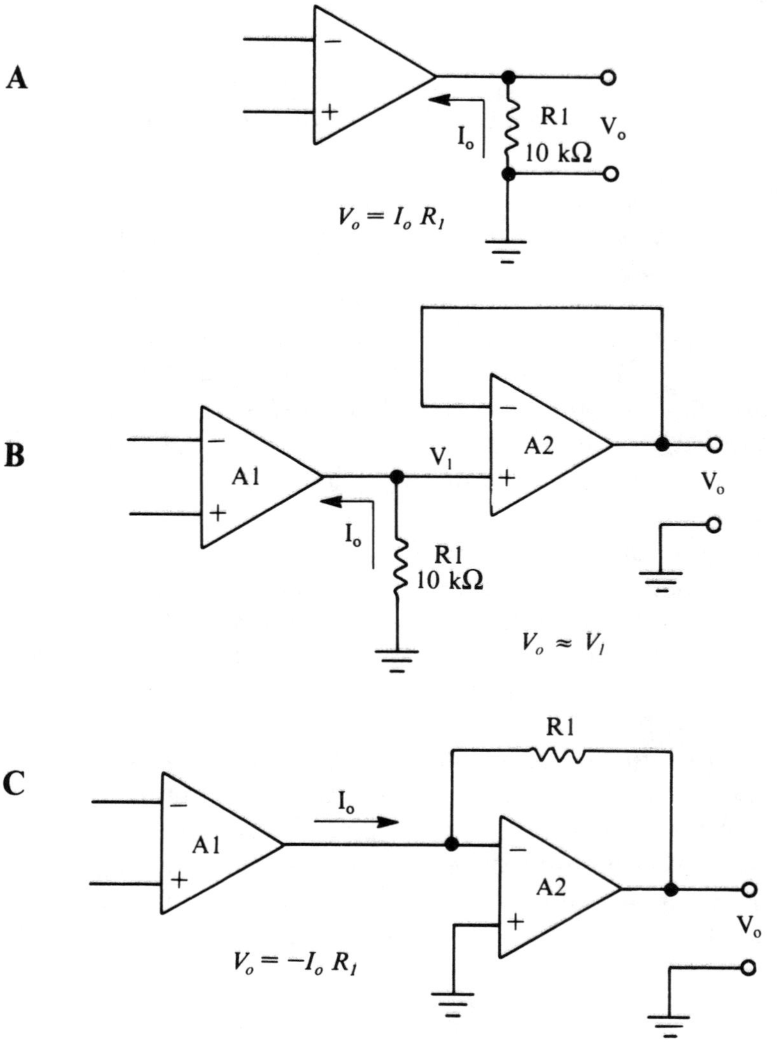

Figure 13-5. A) Voltage output from OTA, B) buffered version, C) gain version.

A problem with this circuit is that the source impedance is very high—equal to resistance R_1. In the example shown in Fig. 13-5A, the output impedance is 10 kohms. This problem can be overcome by adding a unity-gain noninverting operational amplifier, such as A2, in Fig. 13-5B. The output voltage in this case is the same as for the nonamplified version: $I_o R_1$, but the output impedance is very low.

Another form of low-impedance output circuit is shown in Fig. 13-5C. This form uses the inverting-follower configuration of the op amp (A2). The output voltage is the product of the OTA output current (I_o) and the op-amp feedback resistance (R_1):

$$V_o = I_o R_1 \qquad (13\text{-}16)$$

Figures 13-5B and 13-5C both have an output impedance that is equal to the op-amp output impedance, which is typically less than 100 ohms.

OTA compensation

The *bandwidth* of an amplifier refers to the width of the frequency-response spectrum between the upper and lower −6-dB points (−3-dB points in power amplifiers). Typically, the frequency response is specified in terms of the lower- and upper-frequency limits (F_1 and F_2), or the difference between them (BW $= F_2 - F_1$). For example, an amplifier might be said to operate over the range 30 to 3,000 Hz. So, the −6-dB points are at 30 and 3,000 Hz, and that the bandwidth is $F_2 - F_2 = 3{,}000 - 30 = 2{,}970$ Hz. In some amplifiers, the lower-end limit is dc, so such an amplifier is said to have a frequency response of dc to F_{upper}.

The *slew rate* of the amplifier is the time required to transit from a maximum-negative output to a maximum-positive output (i.e., its response to a high-amplitude input square wave). The units used for slew rate are volts per microsecond (V/μS). The slew rate is related to the bandwidth because the steepness of the output waveform (i.e., the fastness of the rise time) is related to the frequency response.

Both bandwidth and slew rate are set by the bias current and the compensation network connected across the output terminal (pin 6 on the CA-3080). Two networks are shown in Fig. 13-6. The simplest consists of a single capacitor that is shunted from the output terminal to ground. The other is a series RC network. If no network is used, the natural slew rate is on the order of 50 V/μS. When a single capacitor (C_L) is used, then the slew rate is:

$$S_r = \frac{I_{bias}}{C_L} \frac{V}{\mu S} \qquad (13\text{-}17)$$

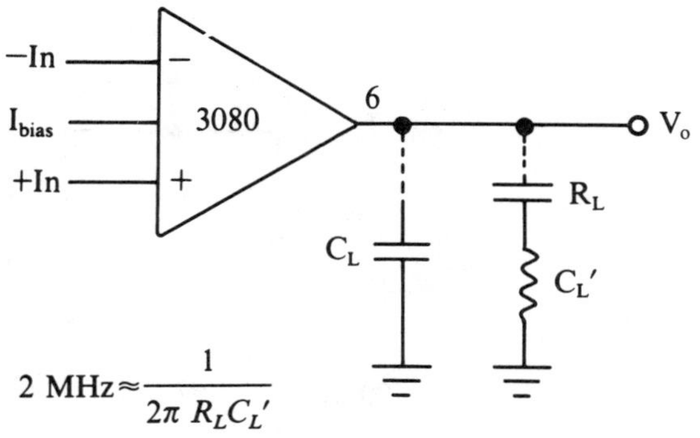

Figure 13-6. Frequency compensation for CA-3080.

Where the capacitance is in picofarads (pF) and the bias current is in microamperes (μA).

In the case of the RC network, set the value of the networks so that:

$$\frac{1}{2\pi R_L C_L'} = 2 \text{ MHz} \tag{13-18}$$

Experiment 13-1

This experiment is designed to show you how an operational transconductance amplifier works. It is an ac amplifier that requires an audio sine-wave signal source with a frequency range of 100 to 1,000 Hz and a variable-output voltage level that is controllable from 0 to 1 volt peak-to-peak. An oscilloscope or ac voltmeter can be used to measure the output voltage, V_o.

1. Measure the exact value of the 20-kohm resistor used for R_{B2}. This resistor forms part of the bias circuit (the other part being potentiometer R_{B1}. Write this value down, and keep it for later.
2. Set potentiometer R_{B2} to maximum value (100 kohms).
3. Connect the circuit of Experiment 13-1.
4. Turn on the dc power ($\pm$12 volts dc), and apply a 100 millivolt ac signal (100 to 1,000 Hz) to the input, V_{in}.
5. Measure the dc voltage drop across resistor R_{B1}.
6. Make the following calculation to find I_{bias}:

$$I_{bias} = \frac{V_{RB2}}{R_{B2}} \tag{13-19}$$

7. Measure the ac output voltage, V_o.

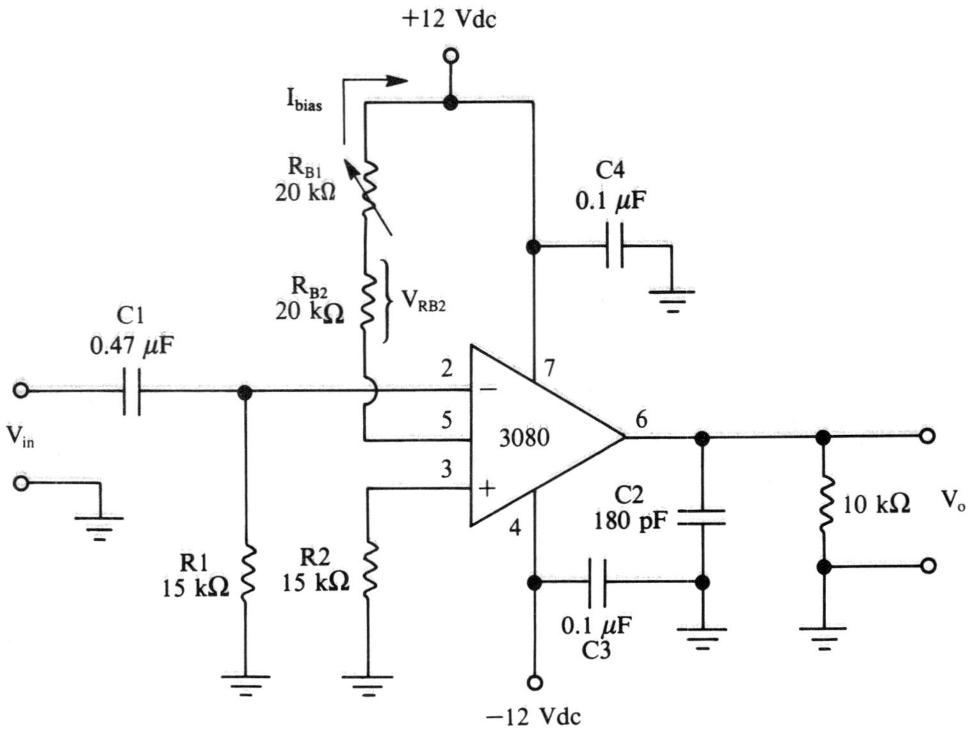

Experiment 13-1. Circuit for Experiment 13-1.

8. Measure the input ac voltage, V_{in}.
9. Calculate the voltage gain V_o/V_{in}.
10. Calculate the transconductance, g_m.
11. How do these results compare with the values calculated from the equations in this chapter?

Project 13-1
20-dB-gain voltage amplifier

Figure 13-7 shows the circuit for a gain-of-10 voltage amplifier that is based on the CA-3080 OTA. The gain is set by the feedback network (R_4/R_1) and the bias current is created by connecting resistor R3 from the bias terminal to the +12-Vdc power supply. The basic configuration of this amplifier is the inverting follower, which uses an ac-coupled input circuit.

The output current is converted to a voltage by the action of optional load resistance R_L and feedback resistance R_4. The feedback resistor is seen effectively as a load resistor because the inverting input is a virtual ground (the same concept as is in operational amplifiers). The maximum permitted

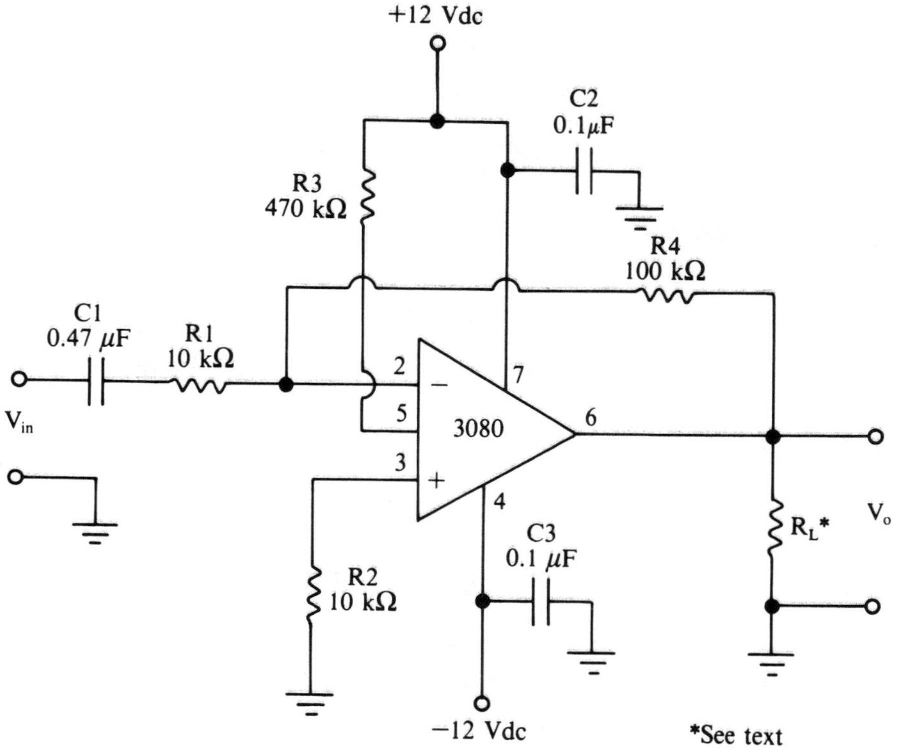

Figure 13-7. 20-dB amplifier.

output voltage is determined by the value of R_L. When R_L is infinite (i.e., when no resistor is present), the maximum output voltage is set by R_4 and is on the order of 10 volts. The maximum output voltage and the gain are reduced by shunting R_4 with a load resistor from pin 6 and ground. A minimum value of 10 kohms reduces the output voltage to 0.5 volts.

Project 13-2
40-dB-gain voltage amplifier

A 40-dB-gain amplifier has a voltage gain of 100. Unlike the previous project, this amplifier's gain is set entirely by the bias resistor and the output-load resistor (see Fig. 13-8). This mode of operation is normal to OTAs. Like the previous amplifier, however, this circuit is also an inverting follower that has an ac-coupled input circuit.

The bias current is set by 48-kohm resistor R4:

$$I_{bias} = \frac{(V+) - (V-)}{R_4}$$

$$I_{bias} = \frac{(+12 \text{ volts}) - (-12 \text{ volts})}{48{,}000 \text{ Ohms}}$$

$$I_{bias} = \frac{24}{48{,}000} = 5 \times 10^{-4} \text{ amperes}$$

The transconductance is found from:

$$G_m = 19.2\ I_{bias}$$

$$G_m = (19.2)\ (5 \times 10^{-4} \text{ amperes}) = 9.6 \times 10^{-3} \text{ mhos}$$

and the gain A_{gm} is:

$$A_{gm} = G_m R_3$$

$$A_{gm} = (9.6 \times 10^{-3})\ (10{,}000 \text{ ohms}) = 96 \approx 100$$

If you want to make the gain exactly 100, make R4 adjustable and set it to produce a gain of 100. This technique is used in the following adjustable-gain amplifier.

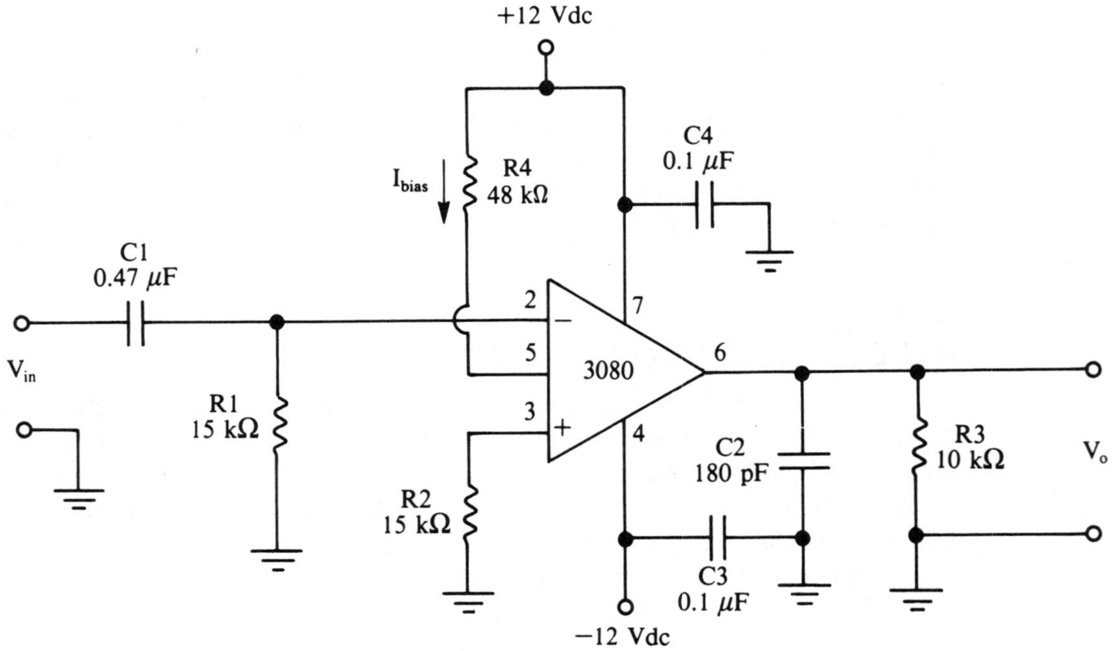

Figure 13-8. 40-dB amplifier.

The output terminal of this project has a capacitor to ground for slew-rate/frequency compensation. This 180-pF capacitor will set the slew rate to:

$$S_r = \frac{I_{bias}}{C_L}$$

$$S_r = \frac{500\ \mu\text{A}}{180\ \text{pF}}$$

$$S_r = 2.78\ \text{V}/\mu\text{S}$$

The bias current can be varied to produce other gains as is required.

Project 13-3 Variable-gain voltage amplifier

Figure 13-9 shows the circuit for a variable-gain voltage amplifier. This circuit offers gain from 14 to 40 dB, depending on the setting of 500-kohm potentiometer, R5. Also, a means for nulling the dc-output offset voltage is provided. The circuit is essentially the same as the previous circuits, except that the bias current is adjustable. Therefore, I am not listing the calculations once again.

Adjust the dc-output offset voltage by setting R5 to its minimum value (i.e., zero ohms), then adjust potentiometer R4 (the offset-null control) for zero-volts dc output when the input ac signal is zero. In order to make this adjustment, either short the input to ground, or connect it to an audio signal generator that is turned off.

Project 13-4 Voltage-controlled gain amplifier (VCGA)

An amplitude-modulator circuit, such as Fig. 13-10, can be used to make CW signal into an AM signal, as in a signal generator or a radio transmitter. Or, it can be used to make a voltage-controlled variable-gain amplifier.

Because the gain of an OTA is controlled by the bias current, this current can also be used to amplitude modulate the ac-carrier signal (V_c) that is applied to the inverting input. You can vary the bias current by varying resistor R1 (as was done in the variable gain amplifier above). Alternatively, R1 can remain fixed as long as you vary the modulating voltage V_m.

The dc-output offset control is adjusted as in the previous case, but modulating voltage V_m is set to $+12$ Vdc and V_c is set to zero.

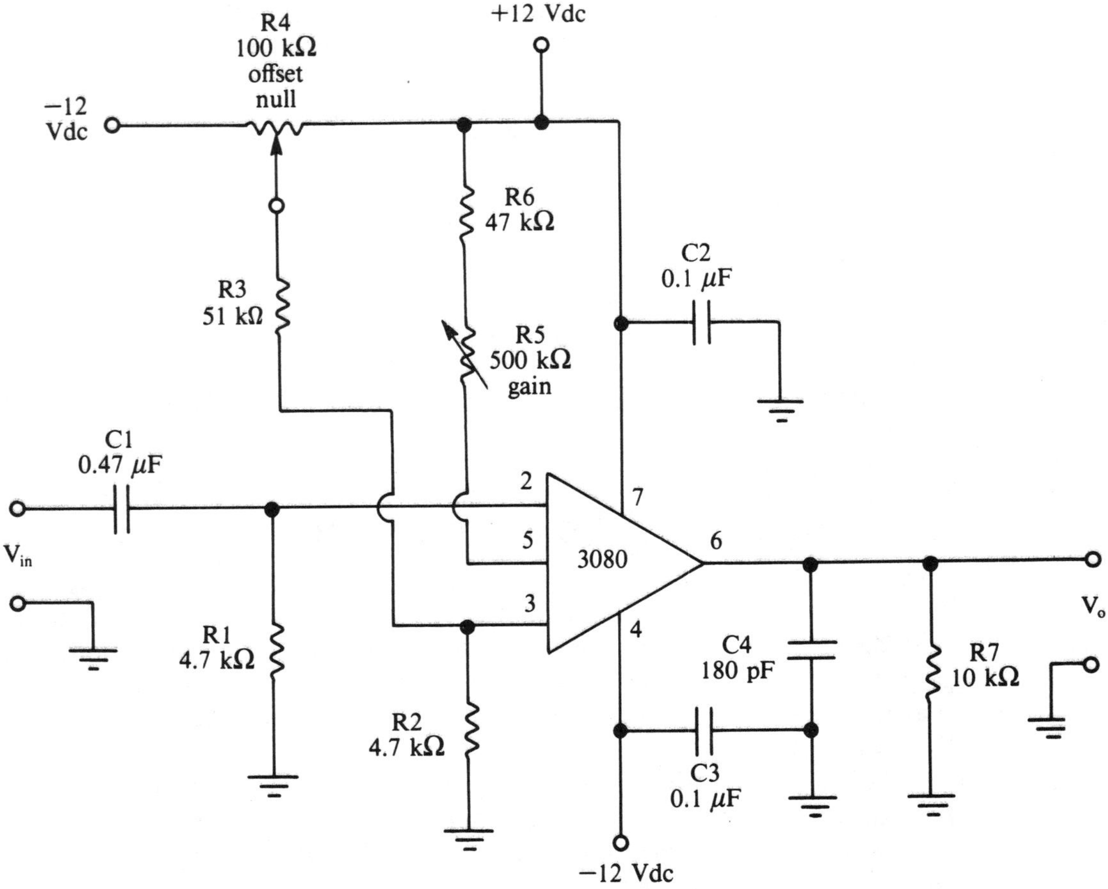

Figure 13-9. Variable-gain amplifier.

Project 13-5
40-dB-gain dc differential amplifier

The differential amplifier (Fig. 13-11) has two inputs, each of which has identical gain, but opposite polarity. The inverting input (−In) produces an output signal that is 180 degrees out of phase with the input signal, but the noninverting input produces an output signal that is in-phase (0 degrees) with the input signal.

Differential amplifiers are used for two main reasons. First, the input-signal source has a natural-output configuration that is either differential or push-pull. Second, when a high probability of receiving a noise level from external electrical or magnetic fields exists. For example, when an input line is subjected to the ubiquitous 60-Hz ac power-line fields.

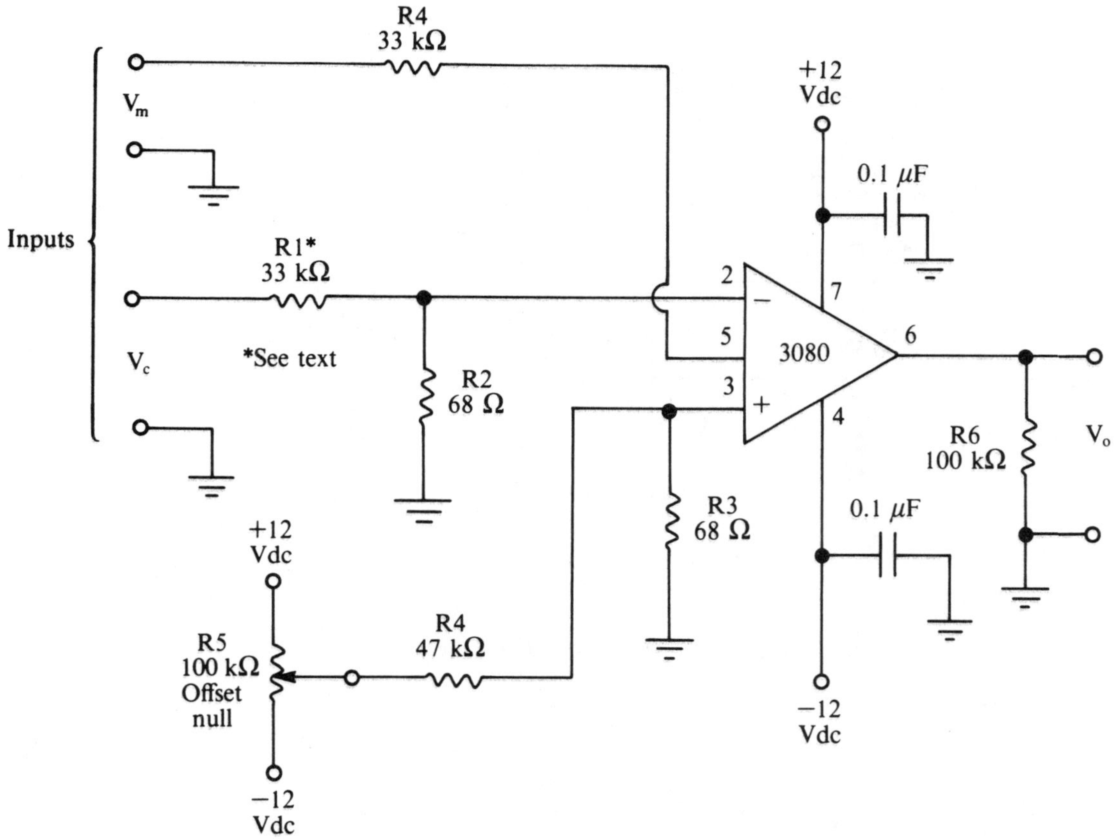

Figure 13-10. OTA amplitude modulator.

The voltage gain of this circuit is set by the bias resistor, (R3). Therefore, it is similar to the previous 40-dB single-ended amplifier.

Project 13-6 Wideband ac amplifier

A wideband amplifier is used either for high-fidelity reproduction of complex waveforms or to pass video signals. Figure 13-12 shows the circuit for a single-ended OTA amplifier that is based on the CA-3080, which will serve as a wideband amplifier.

The input impedance of this circuit is set to approximately 50 ohms by resistor R1. This impedance is used because it is the standard-system impedance for RF systems (75 ohms in television-antenna systems). In wideband and RF circuits, the system impedances should be matched in order to reduce the standing-wave-ratio (SWR) errors that can occur.

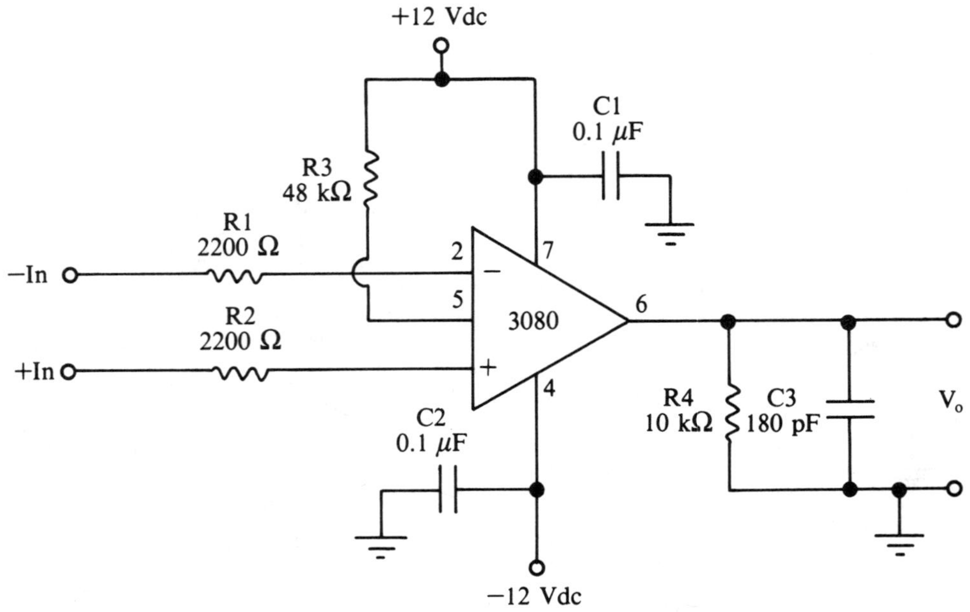

Figure 13-11. 40-dB differential OTA amplifier.

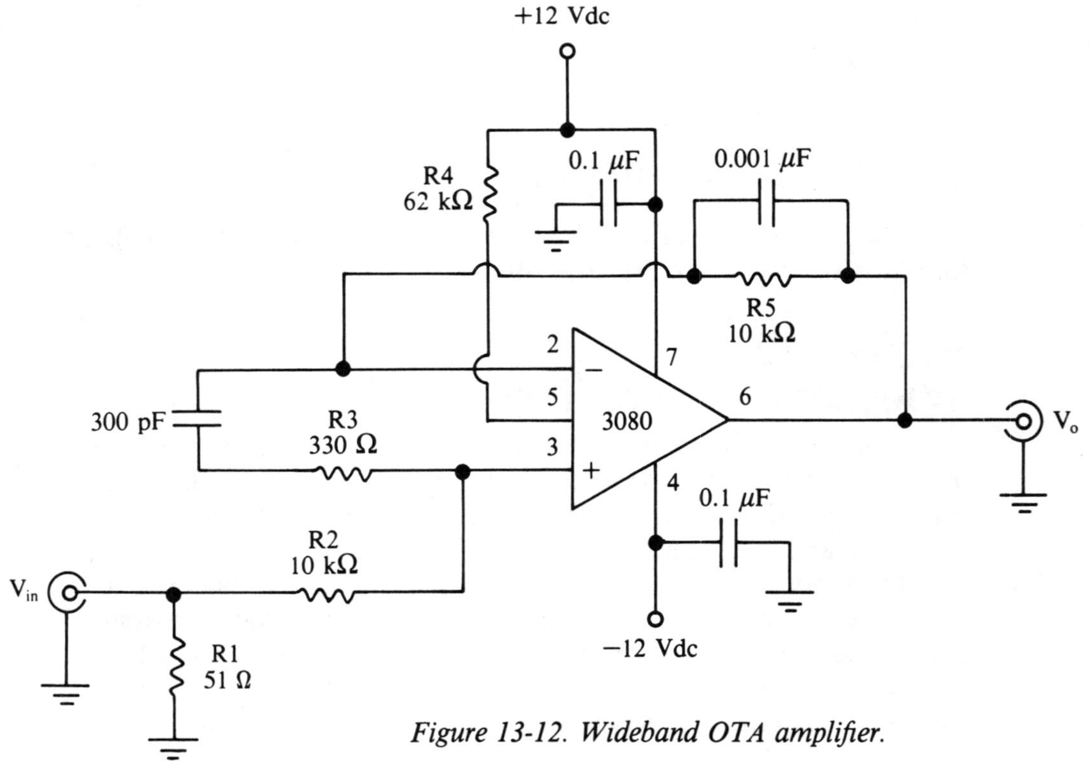

Figure 13-12. Wideband OTA amplifier.

Summary

The operational transconductance amplifier is an op amp type of device that takes a differential input voltage and converts it to an output current. The transfer function, or gain, is the change in output current divided by the change in input voltage. On some variable devices, an input bias current can be used to adjust the gain. Placing a resistor in the output circuit, in parallel with the load, will create voltage amplification. For circuits in which the input changes rapidly, a capacitor placed in the output circuit improves frequency response.

Questions

1. What is a transfer function?
2. What is transresistance? transconductance?
3. How can the transconductance amplifier be converted to a voltage amplifier? What precautions need to be taken?
4. What is slew rate?

Problems

1. A change in input voltage of 0.2 volt causes a change in output current of 50 mA. What is the transconductance?
2. A change in input voltage of 0.05 volt causes a change in output current of 20 mA. Calculate the transconductance.
3. For the CA-3080 OTA, the bias current is 1.5 mA. What is the output current for an input voltage of 0.15 volt?
4. For the CA-3080 OTA, the bias current is 1.2 mA. What is the output current for an input voltage of 0.08 volt?

14 CURRENT-DIFFERENCE AMPLIFIERS

AFTER COMPLETING THIS chapter, you will be able to:

- ☐ Explain the function of the current-difference amplifier
- ☐ Describe how the CDA operates as an inverting-follower
- ☐ Describe how the CDA operates as a noninverting-follower

Still another form of linear IC amplifier is the *current-difference amplifier* (*CDA*), also called the *Norton Amplifier*. It is, like the OTA (chapter 13) another nonoperational linear-IC amplifier that performs much like the op amp. The CDA has features that make it uniquely useful for certain applications. The CDA is more useful than the op amp in circuits that process ac signals, but are limited to a single-polarity dc power supply. For example, in automotive electronics equipment that are limited to a single +12-to-+14.4-Vdc battery power supply and use the car chassis for negative common return.

In other cases, the linear IC amplifier is just a minor feature of the circuit, which usually operates from a single dc power supply. It might be wasteful in such circuits to provide a power supply to use the operational amplifier in the usual bipolar manner. You would either need to bias the operational amplifier with an external-resistor network or provide a second dc power supply.

The normal CDA circuit symbol is shown in Fig. 14-1. This symbol looks much like the regular op-amp symbol, except that a current source is placed along the side that is opposite the apex. This symbol is typically used for several products, such as the LM-3900 device, which is a quad Norton amplifier. You can sometimes find schematics where the op-amp symbol is used for the CDA, but that is technically incorrect.

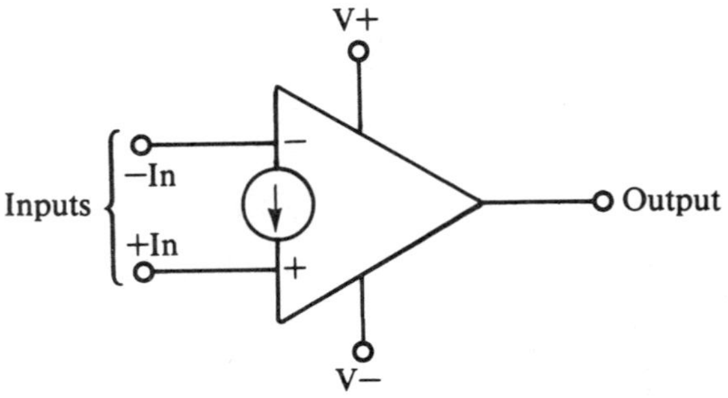

Figure 14-1. Current-difference amplifier (CDA) symbol.

CDA circuit configuration

The input circuit of the CDA differs radically from the operational amplifier. Remember that the op amp used a differential-input common-emitter amplifier that is driven from a constant-current source to supply the collector-emitter current. The CDA is quite different, however, as can be seen in Fig. 14-2. The overall circuit of a typical CDA is shown in Fig. 14-2A and an alternate form of the input circuit is shown in Fig. 14-2B. Transistor Q7 in Fig. 14-2A forms the output transistor and Q5 is the driver. Both the npn output transistor and the pnp driver transistor operate in the emitter-follower configuration. Transistors Q4, Q5, Q6, and Q8 are connected as current sources. The input transistor is Q3 and it operates in the common-emitter configuration. The base of Q3 forms the inverting (−In) input for the CDA.

The noninverting input of the CDA is formed with a "current-mirror" transistor, Q1 (Q1 in Fig. 14-2A is "diode connected" and it serves exactly the same function as diode D1 in Fig. 14-2B). The dynamic resistance offered by the current-mirror transistor (Q2) is given by:

$$r = \frac{26}{I_b} \qquad (14\text{-}1)$$

Where:

r is the dynamic resistance of Q2 (in *ohms*)
I_b is the base bias current of Q3 (in *milliamperes*)

Equation (14-1) is used only at or near normal room temperature because I_b will vary with wide temperature excursions. For most common applications, however, the room-temperature version of the equation will suffice. Data sheets for specific current-difference amplifiers give additional details for amplifiers that must operate outside of the relatively narrow temperature range (25 to 30°C) that is specified for the simplified equation.

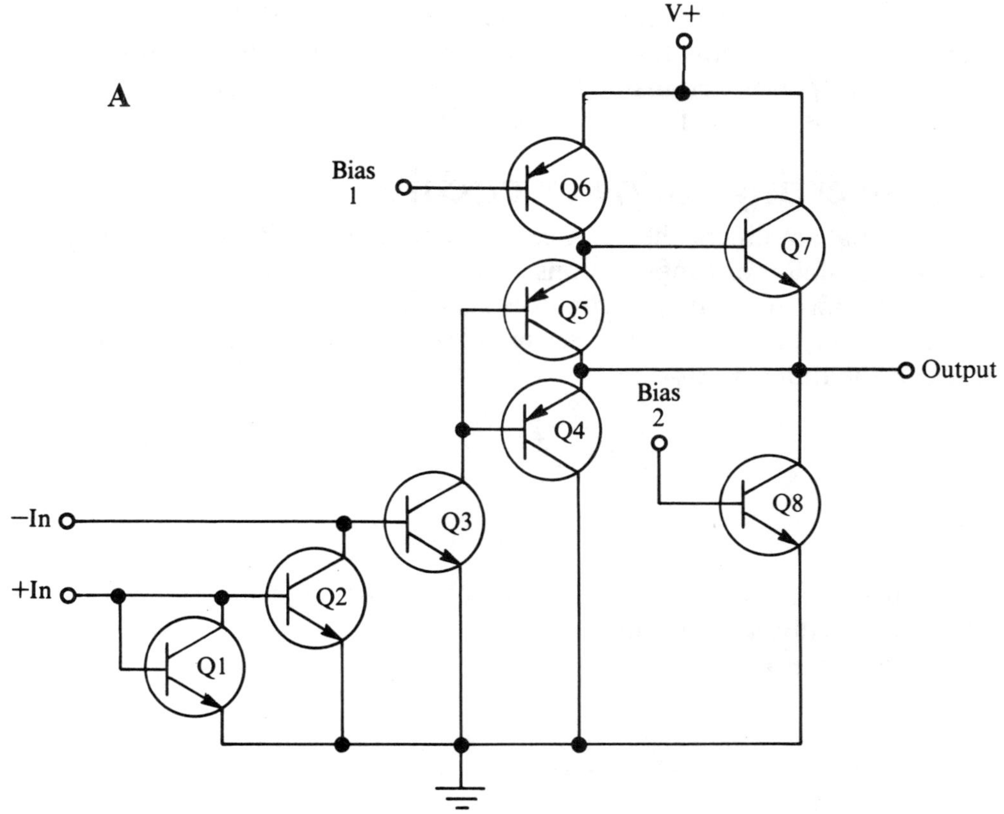

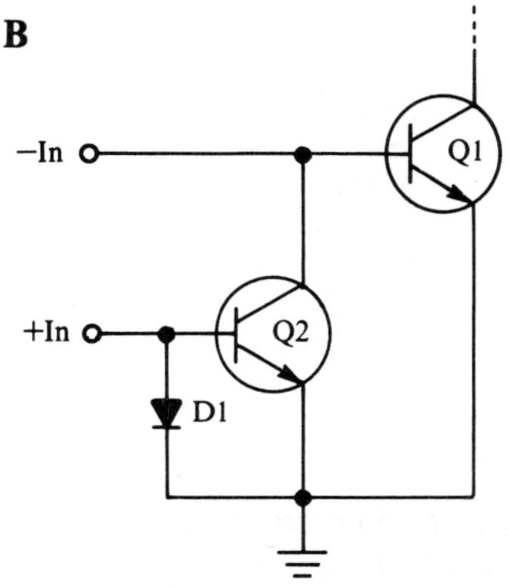

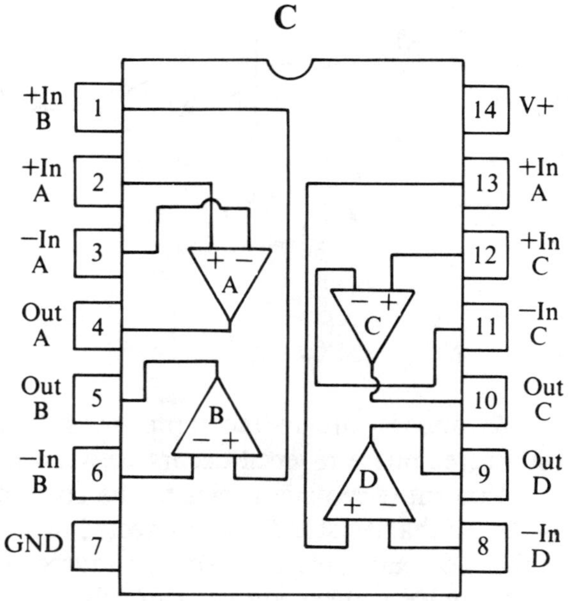

Figure 14-2. A) Internal circuit for CDA, B) current mirror for differential operation, C) quad CDA (LM-3900).

Figure 14-2C shows a common CDA device, the LM-3900. This chip is a 14-pin DIP that contains four CDA devices that operate from a single ground and a dc power terminal. Other than the power lines, however, the four devices (A, B, C, and D) are completely independent.

CDA inverting-follower circuits

Like operational amplifiers, the CDA can be configured in either inverting- or noninverting-follower configurations. The inverting follower is shown in Fig. 14-3. In many respects, this circuit is very similar to op amp circuits. The voltage gain of the circuit is set approximately by the ratio of the feedback to the input resistor:

$$A_v = -\frac{R_2}{R_1} \tag{14-2}$$

Where:

A_v is the voltage gain
R_2 is the feedback resistance
R_1 is the input resistance

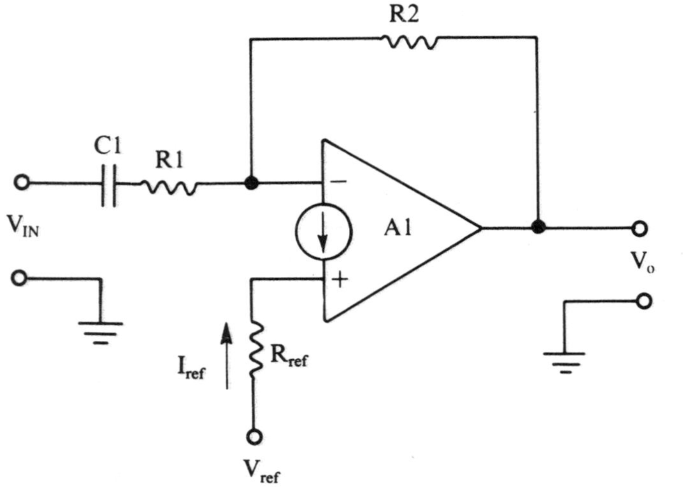

Figure 14-3. Inverting CDA circuit.

R_1 and R_2 are in the same units. The minus sign indicates that a 180-degree phase reversal occurs between input and output signals.

You must provide a bias to the current-mirror transistor (Q2 in Fig. 14-2A) so that resistor R_{ref} is connected in series with the noninverting input of the CDA and reference-voltage source, V_{ref}. In many practical circuits, the reference-voltage source is merely the V+ supply that is used for the

CDA. In other cases, however, some other potential might be required. Alternatively, the required reference current might be regulated more tightly (or with less noise) than the supply voltage. Ordinarily, the reference current is set to some convenient value between 5 and 100 μA. For V+ power-supply values of +12 Vdc, for example, it is common to find a 1-megohm resistor for R_{ref}. In that case, the input reference current is:

$$I_{ref} = \frac{12}{1{,}000{,}000} = 12\ \mu\text{A}$$

A constraint placed on CDAs is that the input resistance (R_1) used to set gain must be high compared with the value of the current-mirror dynamic resistance. The CDA becomes nonlinear (i.e., distorts the input signal) if the input resistor value approaches the current-mirror resistance, r. In that case, the voltage gain is not $-R_2/R_1$, but rather:

$$A_v = -\frac{R_2}{R_1 + r} \tag{14-3}$$

Where:

A_v is the voltage gain
R_2 is the feedback resistance
R_1 is the input resistance
r is the current-mirror resistance

Equation (14-3) essentially reduces to Eq. (14-2) when you force R_1 to be much larger than r. This goal is easily achieved in most circuits because r is typically tiny.

The output voltage of the CDA will exhibit an offset potential, even when the ac input signal is zero. This potential is given by:

$$V_o = -\left(\frac{V_{ref}R_2}{R_{ref}} + 1\right) - \left(\frac{R_2}{R_{ref}}\right)\Phi \tag{14-4}$$

Where:

V_o is the output potential (in *volts*)
V_{ref} is the reference potential (in *volts*—usually V+)
R_2 is the feedback resistance (in *ohms*)
R_{ref} is the current-mirror bias resistance (in *ohms*)
Φ is a temperature-dependent factor (0.70 volts for room temperature)

The capacitor in series with the input circuitry limits the low-end frequency response. The -3-dB cutoff frequency is a function of the value of this capacitor and the input resistance, R_I. This frequency, F, is given by:

$$F = \frac{1{,}000{,}000}{2\pi R_I C_I} \tag{14-5}$$

Where:

F is the lower end -3-dB frequency (in *hertz*)
R_I is the resistance (in *ohms*)
C_I is the capacitance (in *microfarads*)

In some CDA circuits, a capacitor is also in series with the output terminal. The purpose of that capacitor is to prevent the dc offset that is inherent in this type of circuit from affecting following circuits. The output capacitor will also limit the low-end frequency response. The same form of equation, Eq. (14-5), is used to determine this frequency, but the input resistance of the load is R.

As is often the case with equations that are presented in electronics books, Eq. (14-5) is not necessarily in the most useful form. In most cases, you will know the input resistance (R_I) from the application. It is typically not less than 10 times the source impedance and it forms part of the gain equation. The driving source impedance and voltage gain tend to determine the value of R_I. The required low-end frequency response is usually determined from the application. You generally know (or can find) the frequency spectrum of input signals. From the lower limit of the frequency spectrum, you can determine the value of F. Thus, you will determine F and R_I from considerations other than the circuit. You, therefore, need a version of Eq. (14-5) that calculates capacitance C_I from knowledge of R_I and F. Using the elementary algebra:

$$C_I = \frac{1{,}000{,}000}{2\pi R_I F} \tag{14-6}$$

all terms as defined for Eq. (14-5).

Noninverting amplifier circuits

The noninverting-amplifier CDA configuration is shown in Fig. 14-4. This circuit retains the reference-current bias that was applied to the noninverting input, but it rearranges some of the other components. As in the case of

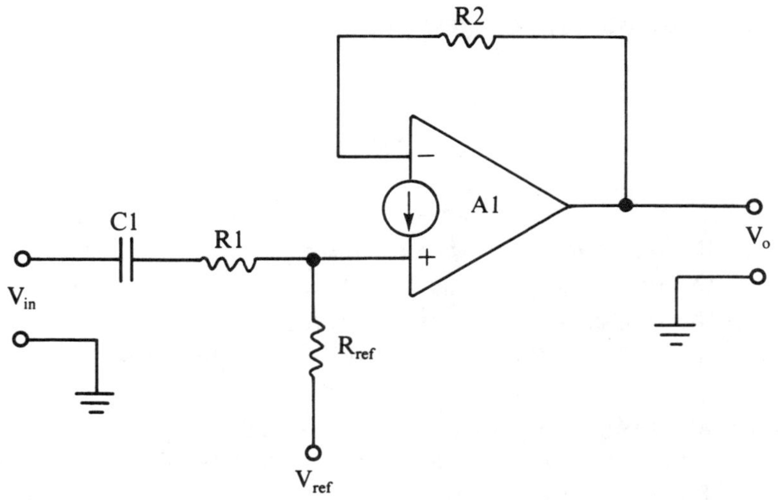

Figure 14-4. Noninverting CDA circuit.

the inverting amplifier configuration, the noninverting amplifier uses R_2 to provide negative feedback between the output terminal and the inverting input. Unlike the inverting CDA circuit, however, input resistor R1 is connected in series with the noninverting input. The gain of the noninverting CDA amplifier is given by:

$$A_v = \frac{R_2}{\left(\dfrac{26\ R_1}{I_{ref(mA)}}\right)} \tag{14-7}$$

Where:

A_v is the voltage gain
R_1 and R_2 are the resistances (in *ohms*)
I_{ref} is the bias current (in *milliamperes*)

The reference current, I_{ref}, is set to a value between 5 and 100 μA. Unlike the situation in the inverting amplifier, the value of this current is partially responsible for setting the gain of the circuit. Some clever designers have even used this current as a limited-gain control for some CDA stages. The value of the resistor that provides the reference current (R_{ref}) is set by Ohm's Law, considering the required value of reference current and reference voltage, V_{ref}. In most common applications, the reference voltage is merely one of the supply voltages. The value of R_{ref} is determined from:

$$R_{ref} = \frac{V_{ref}}{I_{ref}} \tag{14-8}$$

Where:

R_{ref} is the reference resistance (in *ohms*)
V_{ref} is the reference potential (in *volts*)
I_{ref} is the reference current (in *amperes*)
(note: 1 μA = 0.000001 ampere)

The value of input impedance is approximately equal to R_1, provided that R_1 is much higher than the dynamic resistance of the current mirror inside the CDA (typically the case). As was true in the inverting-follower case, input capacitor C1 sets the low-end frequency response of the amplifier. The −3-dB frequency is given by exactly the same equation as for the inverting case—see Eqs. (14-5) and (14-6).

Figure 14-5 shows a modification of the noninverting-follower circuit that allows for a noisy reference source. This type of amplifier circuit might be used where the reference-voltage dc power supplies are electrically noisy. Such noise could come from other stages in the circuit or from outside sources.

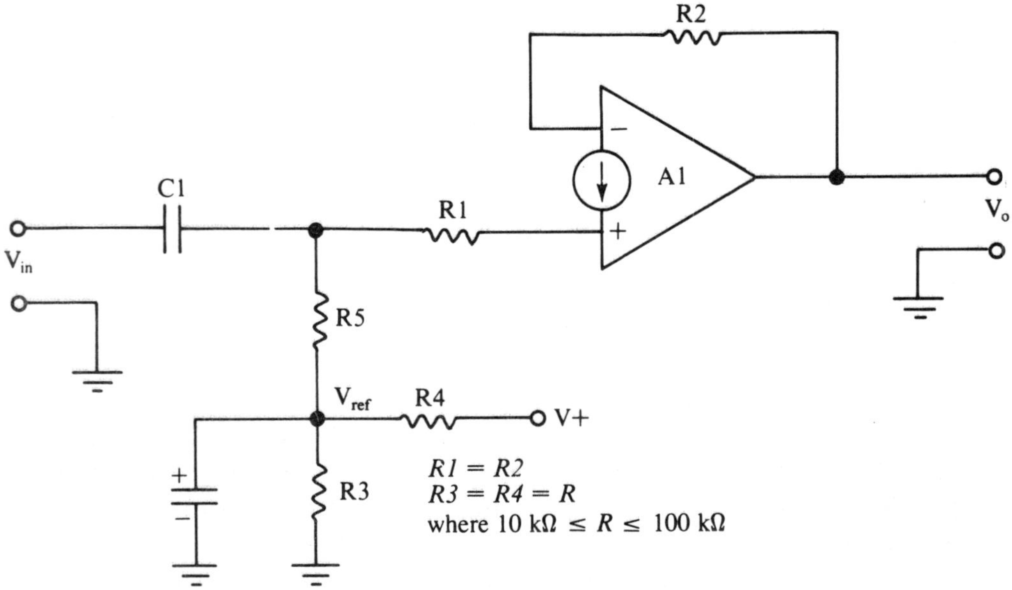

Figure 14-5. Single power-supply operation.

The purpose in Fig. 14-5 is to form a reference voltage from a resistor-voltage divider circuit that consists of R3 and R4. The value of V_{ref} will be:

$$V_{ref} = \frac{(V+)\,R_3}{R_3 + R_4} \tag{14-9}$$

Where:

V_{ref} is the reference potential (in *volts*)
$V+$ is the supply potential (in *volts*)
R_3 is the resistance (in *ohms*)

An inspection of Eq. (14-9) will reveal that $V_{ref} = (V+)/2$ when $R_3 = R_4$, which is usually the case in practical circuits. The reference current is:

$$I_{ref} = \frac{V_{ref}}{R_1} \tag{14-10}$$

Super-gain amplifier

The voltage gain has a practical limit when using standard resistor values and standard circuit configurations (a similar problem also exists for operational amplifiers). Figure 14-6 shows a means to overcome the limitations. This super-gain amplifier forms a noninverting follower that is similar to the earlier circuit, except that feedback resistor R2 is driven from an output-voltage divider network, rather than directly from the output terminal of the CDA. The voltage gain of the circuit of Fig. 14-6 is given by:

$$A_v = \left(\frac{R_2}{R_1}\right)\left(\frac{R_3 + R_4}{R_3}\right) \tag{14-11}$$

Capacitor C1 is set using the same Eq. (14-6) that was used previously and C2 is set to have a capacitive reactance of $R_4/10$ at the lowest frequency of operation (the low-end -3-dB point).

Using bipolar dc power supplies

The current-difference amplifier is designed primarily for single-polarity power-supply circuits. In most cases, the CDA will operate with a V+ dc power supply, has one side grounded. However, you can operate the CDA in a circuit (such as Fig. 14-7) with a bipolar dc power supply. The reference resistor, R_{ref}, is connected from the noninverting input to ground. The V− and V+ power supplies are each ground referenced at equal potential. Thus, the 5-to-100-μA bias current is found from $(V+)/R_{ref}$.

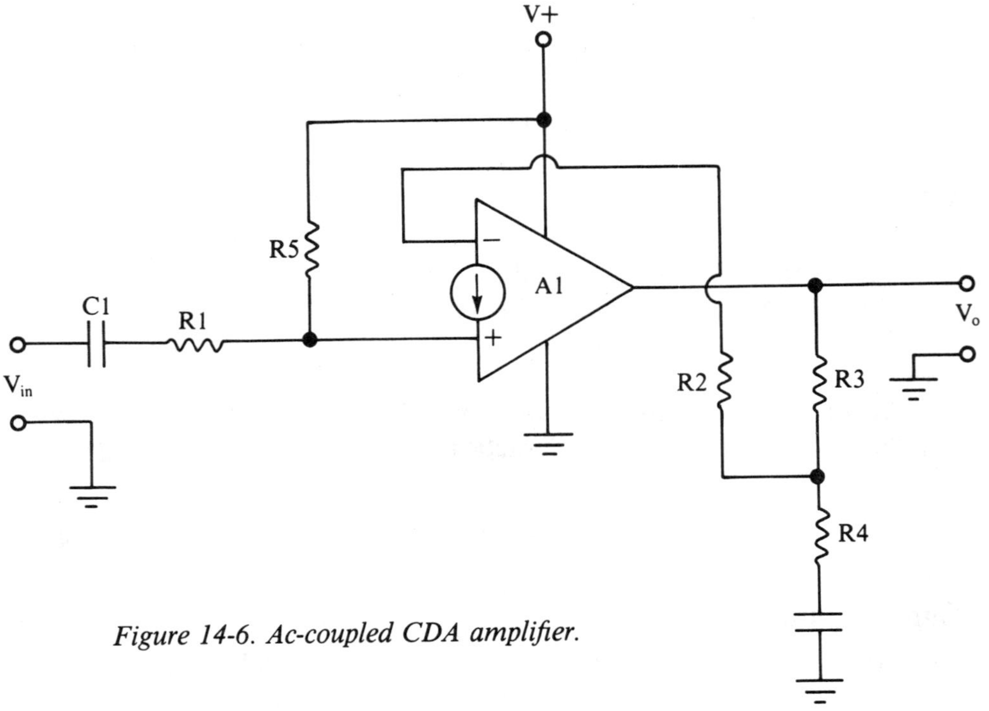

Figure 14-6. Ac-coupled CDA amplifier.

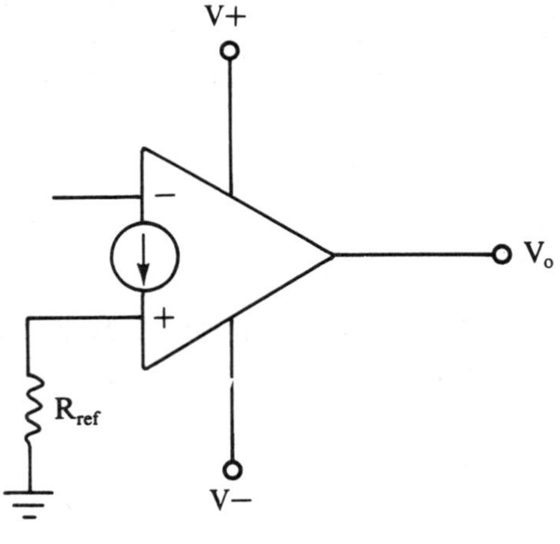

Figure 14-7. Dual supply operation.

Applications projects

The CDA device is very useful and has almost as many applications as the operational amplifier. Unlike the op amp and the OTA, however, the CDA can operate from a single dc power supply. Thus, it is popular in automotive and portable-equipment designs. This section looks at some representative applications in the form of small projects.

Project 14-1
20-dB-gain ac amplifier

An inverting amplifier is needed that has an ac gain of 100 (i.e., 20 dB) and an input impedance of at least 10 kohms. The LM-3900 CDA was selected for the application. The circuit for this project is shown in Fig. 14-8. The pinouts can be selected from the four internal devices (Fig. 14-2C).

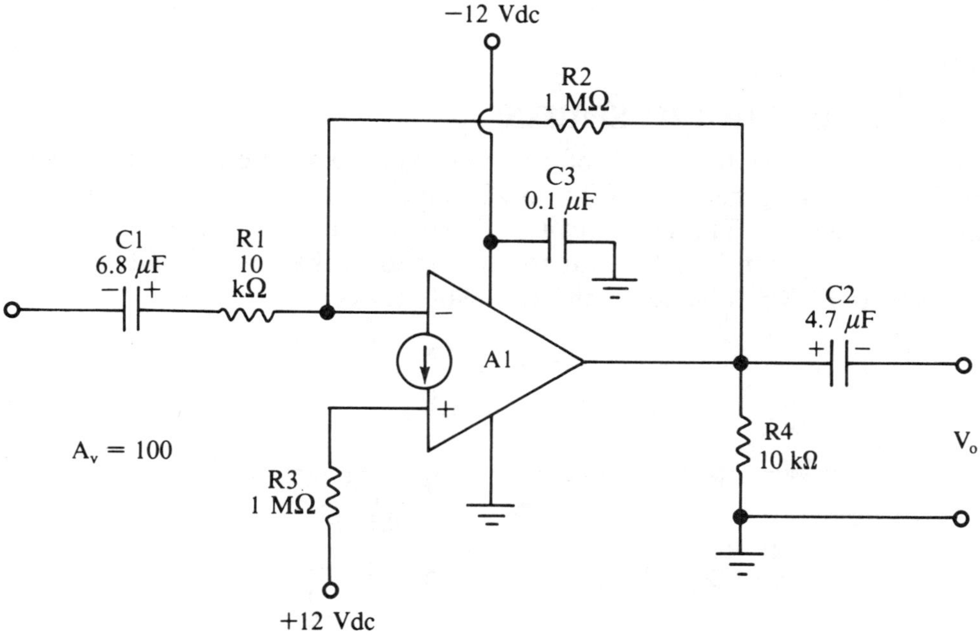

Figure 14-8. Inverting gain-of-100 CDA circuit.

The reference resistor is set to 1 megohm. The gain (A_v) is set by the ratio of the feedback and input resistors: $-R_2/R_1$. Because the input impedance needs to be at least 10 kohms, R_1 = 10 kohms. The gain is 100, so the feedback resistor must be:

$$R_2 = A_v R_1 \tag{14-12}$$

$$R_2 = (100)\ (10{,}000 \text{ ohms}) = 1{,}000{,}000 \text{ ohms} \tag{14-13}$$

The dc power supplies are set to ±12 volts, a common value.

The input and output terminals are capacitor coupled. These capacitors and their associated resistors, form high-pass filters that set the lower-end −6-dB points in the frequency-response characteristic. In each case, the frequency is:

$$F = \frac{1}{2\pi RC} \tag{14-14}$$

The last capacitor in the circuit (C3) is used to decouple the dc power-supply line. This capacitor is a 0.1-μF unit, it must be mounted as close as possible to the body of the LM-3900.

Project 14-2 Variable-voltage source

Figure 14-9 shows the circuit of a variable-voltage source (VVS) that will produce output potentials of 0.6 to 20 volts, depending on the setting of potentiometer R1. This circuit can also be used as a fixed voltage source by replacing the potentiometer with a fixed resistor. Like other circuits in this chapter, the VVS is based on the LM-3900 device.

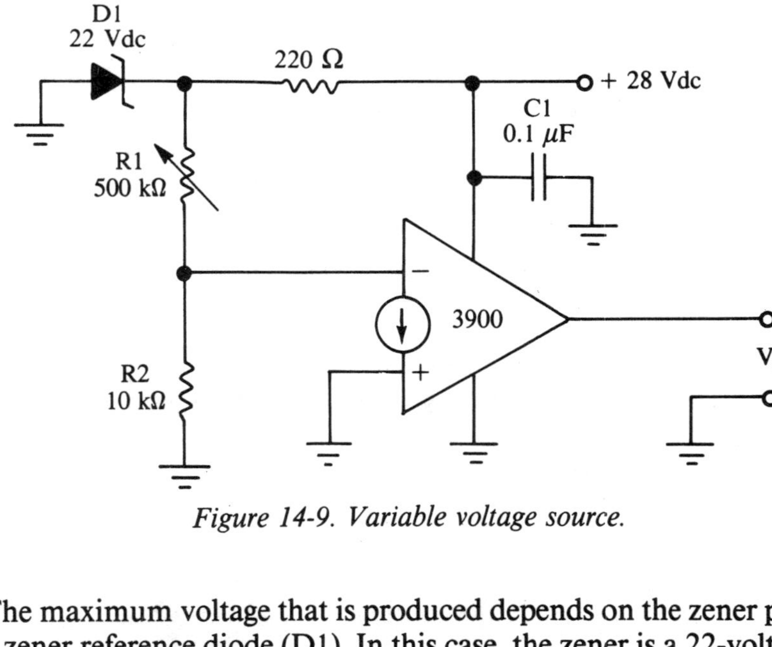

Figure 14-9. Variable voltage source.

The maximum voltage that is produced depends on the zener potential of the zener reference diode (D1). In this case, the zener is a 22-volt unit, so the maximum output voltage will be between 20 and 22 volts. Lower values of maximum output voltage can be accommodated with a diode of lower zener potential for D1. The advantage of this change is that the resolution is improved (i.e., the voltage change per turn of potentiometer R1).

Project 14-3
Over/under temperature alarms

The CDA can be used as a voltage comparator, so it is also useful for certain alarm circuits that use a voltage comparison at the heart. Figures 14-10A and 14-10B show two such circuits: Fig. 14-10A is an overtemperature alarm and Fig. 14-10B is an undertemperature alarm.

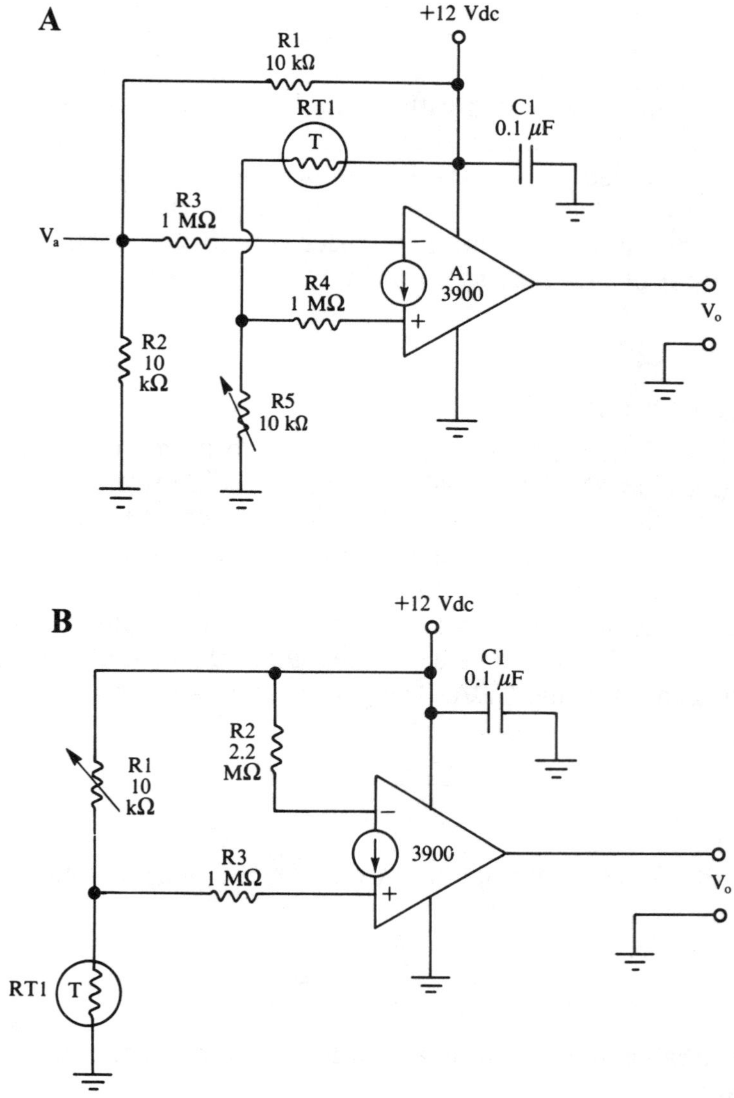

Figure 14-10. A) Over-temperature alarm, B) undertemperature alarm.

In Fig. 14-10A, two reference circuits are provided. First is a 2:1 voltage divider that consists of R1 and R2, which produces voltage V_a of ½(V+), about 6 volts. This potential produces an input current through resistor R3, which is set to 1,000,000 ohms (partly to avoid loading the voltage divider).

The other reference circuit is a voltage divider that consists of a thermistor, RT1, and a potentiometer, R5. The potentiometer sets the trip point, at which the output changes to indicate the alarm condition. The thermistor should be a negative-temperature-coefficient type, in which the resistance will be approximately 10 kohms in the temperature range of interest.

The circuit can be calibrated with a liquid (such as water), into which both the thermistor and a mercury thermometer are immersed. The water is either heated or chilled to the desired temperature (which takes a little time), or is left to equillabrate at room temperature, if that is the goal. Let both the thermistor and the thermometer reach equilibrium, then adjust R5 so that the output of the CDA snaps to high at that point.

Project 14-4
CDA audio mixer

An audio mixer is a circuit that will combine two or more audio signal sources into one channel. Figure 14-11 shows an audio mixer that is based on the current-difference amplifier. The LM-3900 can be used for this project and you can select the pinouts from the diagram in Fig. 14-2C.

The crux of this circuit is the three input networks, principally R3, R4, and R5. These resistors are connected to different input sources (labelled *Input 1*, *Input 2*, and *Input 3*). These three resistors are joined into one point at the inverting input of the CDA. The gain is approximately:

$$A_v = \frac{-R_2}{R_x} \tag{14-15}$$

Where R_x is the value of any one input resistor, the output voltage is:

$$V_o = -R_2\left(\frac{V_{in1}}{R_3} + \frac{R_{in2}}{R_4} + \frac{V_{in3}}{R_5}\right) \tag{14-16}$$

If R2 is variable, the potentiometer will serve as a master gain control for the audio mixer.

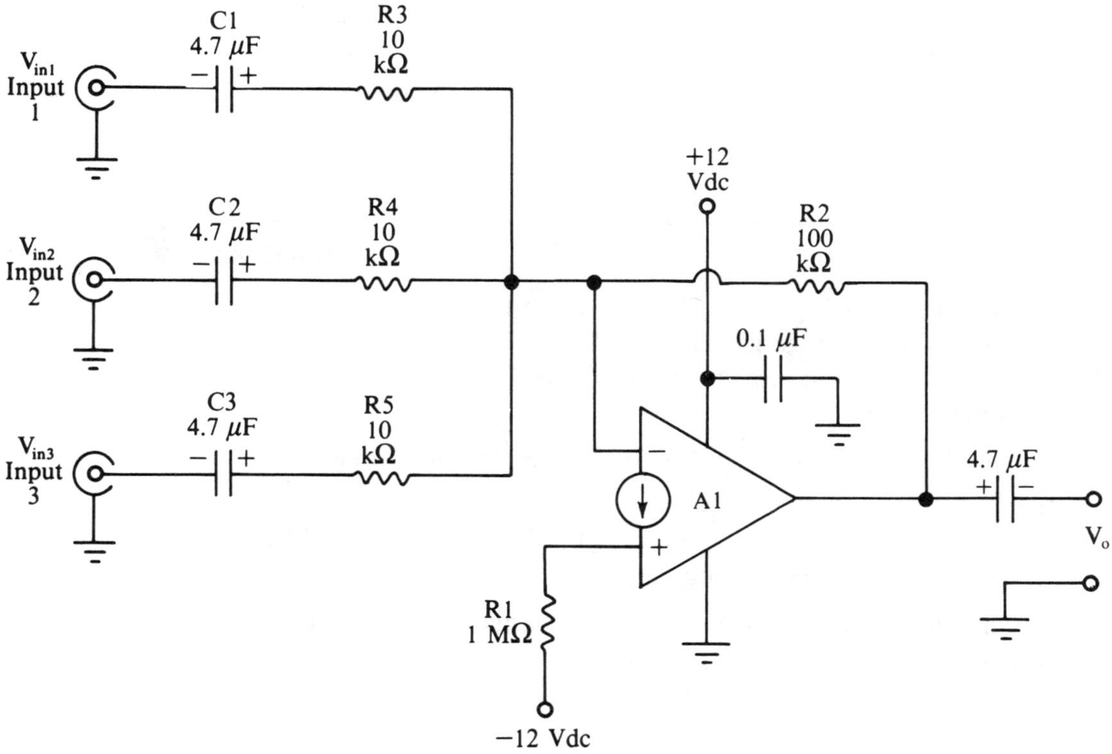

Figure 14-11. Three-channel audio CDA mixer.

Summary

Just as the standard op amp converts a voltage difference to an output voltage, the CDA takes the difference between input currents and converts it to an output voltage. One of the two inputs is usually a reference current. The CDA can be used as an inverting-follower or as a noninverting-follower, both with and without frequency compensation. One of its main advantages is that it can easily be used with a single-polarity power supply.

Questions

1. What is a current-difference amplifier? What is another name for this type of amplifier?
2. How can the current-difference amplifier be used as an inverting-follower amplifier?

3. Why must the input resistors be much larger than the dynamic gain of the device?
4. How does the reference current in a noninverting amplifier affect the gain?

Problems

1. For the circuit in Fig. 14-3, R_I = 100 kohms and R_2 = 1 Mohm. **a.** What is the voltage gain for the circuit? **b.** If the supply voltage is +12 volts and the reference resistor is 500 kohms, what is the reference current?
2. Repeat Problem 1 for R_I = 500 kohms, R_2 = 750 kohms, and R_{ref} = 1 Mohm.
3. For the circuit in Fig. 14-3, R_I = 100 kohms and C_I = 0.22 μF. Calculate the breakpoint frequency.
4. For the circuit in Fig. 14-3, R_I = 500 kohms and C_I = 0.033 μF. Calculate the breakpoint frequency.

15 LINEAR IC AMPLIFIER APPLICATIONS

AFTER COMPLETING THIS chapter, you will be able to:

- □ Discuss IC reference-voltage sources
- □ Describe IC audio amplifiers
- □ Explain light-sensitive IC amplifier circuits

Most of this book is about electronic theory, especially in regard to linear ICs. Experiments and some projects are provided to help you learn the electronic theory that underlies the devices being studied. No experiments are in this chapter, because it is a change of pace that is designed to give you some useful circuits. Some of these circuits can be used by themselves as mini-projects and others are intended as part of a larger circuit. In all of these, however, the emphasis is on useful circuits that are based on linear IC devices.

Reference voltage source

Figure 15-1 shows the circuit for a reference-voltage source that is based on a zener diode and an operational amplifier. The output voltage (V_{adj}) will be a stable potential with a value set by the potentiometer, R3.

The zener diode is a pn-junction diode that is heavily reverse-biased. When the reverse-bias potential hits a critical point (the *zener voltage*), the diode breaks down and the voltage across the diode does not increase further. This voltage is relatively stable, so it can be used as a reference potential. In Fig. 15-1, the zener diode (D1) is a 6.8-volt device (6.8 volts is a common value), biased by the +12-Vdc power supply. Resistor R5 is a 4.7-kohms ¼-watt metallic-film or carbon-composition type.

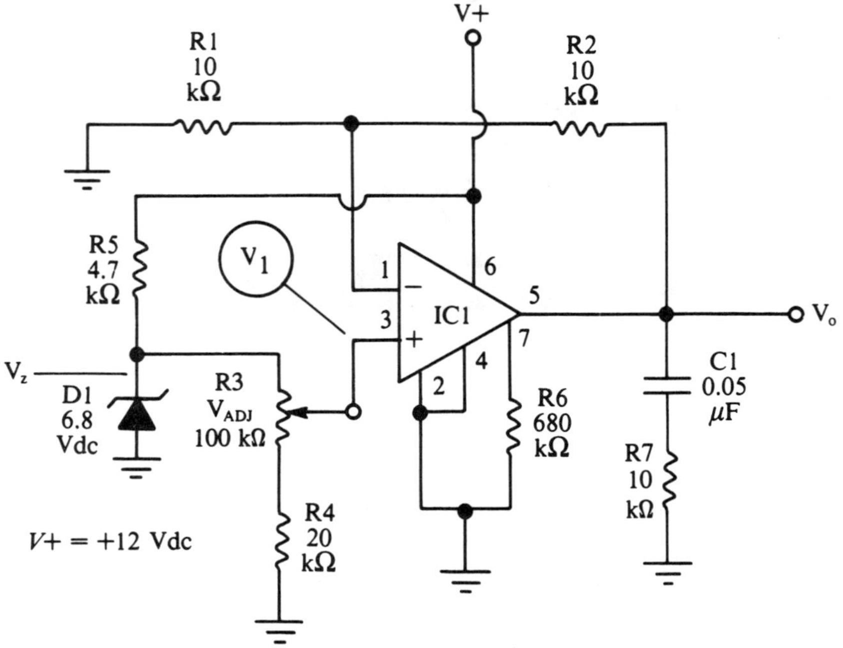

Figure 15-1. Precision dc reference source.

The zener voltage (V_z) is applied to a resistor voltage divider that consists of R3 (a potentiometer) and R4 (a fixed resistor). The input voltage to the operational amplifier (V_1) is selected by the wiper of the potentiometer. Voltage V_1 is adjustable from about +1.13 to +6.8 volts.

The op amp (IC1) is connected in the noninverting-follower-with-gain configuration. The gain is given by:

$$A_v = \left(\frac{R_2}{R_1} + 1\right) \qquad (15\text{-}1)$$

Because $R_1 = R_2 = 10$ kohms, the gain of the operational amplifier in this circuit is two. So, the voltage at the output of op amp V_o is two times V_1.

IC1 is an LM-13080 high-frequency op amp that is operated from a single-polarity dc power supply. This device is a high-gain-bandwidth product unit, so a snubbing or compensation circuit is needed to prevent high-frequency oscillations. This is the function of C1 and R7 from the output terminal to ground.

600-ohm audio amplifiers

Commercial audio and broadcasting equipment use a balanced-audio circuit that has a standardized impedance of 600 ohms. By *balanced*, it is meant that it has two audio (hot) lines and a ground. Transformers for these circuits are center-tapped and the center tap is grounded.

The *volume unit* (*VU*) is a power-decibel system of measurement for audio levels. A volume unit is defined in terms of a power level (1 milliwatts) dissipated in a 600-ohm load. In other words, the 0-VU reference level is 1 mW in 600 ohms, which equates to a voltage of 0.775 volts rms. To find higher signal levels:

$$VU_{pwr} = 10 \text{ LOG}\left(\frac{P_{mW}}{1}\right) \tag{15-2}$$

or, if millivolts are intended, replace the *1* in the denominator with *0.775*, and the *10* with *20*.

A 600-ohm line-receiver amplifier is shown in Fig. 15-2A. This circuit is based on the operational-amplifier noninverting-follower configuration. The input circuit consists of a special audio transformer, T1, that has a 600-ohms center-tapped primary winding and either 600 ohms (in which case it is a 1:1 transformer) or higher secondary impedance.

The voltage gain of the amplifier is found from:

$$A_v = \left(\frac{R_2}{R_1} + 1\right)\left(\frac{N_s}{N_p}\right) \tag{15-3}$$

Where:

N_p is the number of turns in the primary winding of T1
N_s is the number of turns in the secondary winding of T1

Note: The ratio N_p/N_s is often specified in audio-transformer specification sheets. Turn this ratio upside-down to find the ratio needed in this equation.

With the values shown, the gain of the noninverting op amp is on the order of 6.75, but this value can be increased or decreased by adjusting the ratio of R_2/R_1.

The power supply is a ±12-Vdc regulated supply. These terminals are each bypassed by two capacitors, 0.1 μF and 4.7 μF. The higher-value capacitor decouples low audio frequencies and the 0.1-μF units prevent high-frequency oscillations. The 0.1-μF capacitor must be mounted as close as possible to the body of the LM-301 operational amplifier.

Frequency compensation for the LM-301 operational amplifier is provided by a lead capacitor (C5), a 30-pF ceramic-disk type. This capacitor alters the gain of the op amp at higher frequencies, and thus tailors the frequency response of the amplifier.

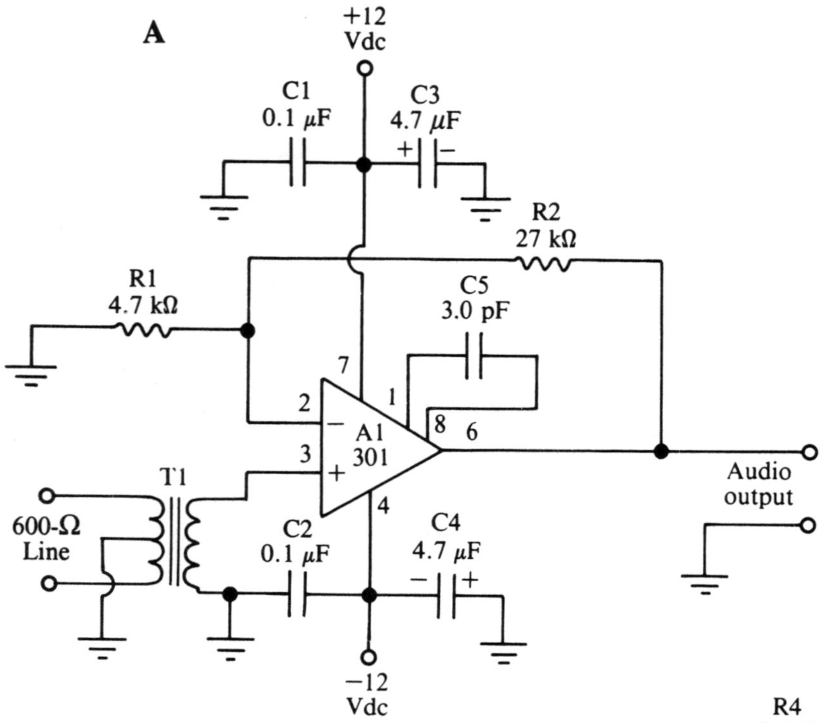

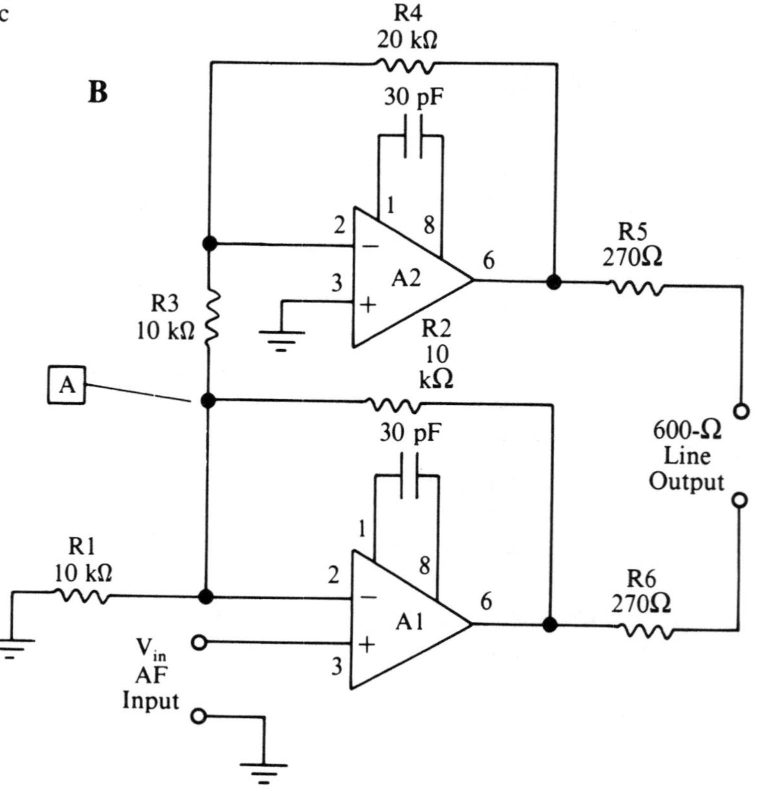

Figure 15-2. A) 600-ohm line-receiver amplifier, B) 600-ohm line-driver amplifier.

A 600-ohm line-driver amplifier is shown in Fig. 15-2B. This circuit is used at the other end of the 600-ohm balanced line as the previous circuit. The input is a standard single-ended audio signal and the output is the push-pull signal that is required for the balanced audio line. The 270-ohm resistors keep the impedance at the 600-ohm level when combined with the internal output impedances of the operational amplifiers.

Audio amplifier stage

The basis for this audio stage is the easily obtained LM-386 IC audio-subsystem chip. This chip is easily obtained in replacement semiconductor lines from Radio Shack and from a number of mailorder sources. Although the low-power level leaves something to be desired, the LM-386 has a better behavior and is easier to obtain than certain higher-powered audio-subsystem chips.

Figure 15-3 shows the basic circuit for the LM-386 chip when it is used as an audio stage for a radio receiver, a small hand-held PA amplifier, or a headphones amplifier. Notice that the circuit is extremely simple—it basically just has input, output, ground, and V+ connections. The circuit can provide two levels of gain. If capacitor C4 is used, the gain is ×200, and if the capacitor is deleted the gain is ×20. Normally, the gain will be set to ×20 in receiver projects, unless the design uses little gain ahead of the detector (which might be the case in the simplest superhets or in direct-conversion designs).

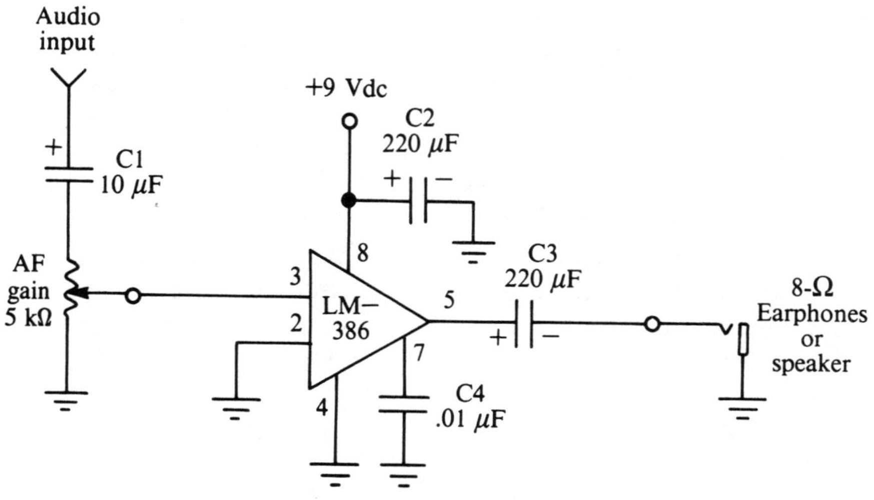

Figure 15-3. LM-386 audio amplifier stage.

The LM-386 is relatively well-behaved in projects, so an ordinary layout will produce good results. However, because of the high gain, this IC might oscillate mercilessly if proper layout is not followed. Be sure to keep the inputs and outputs physically separated and don't wind the ground connections all over the wiring board. Also, make sure that the V+ bypass capacitor is used and has sufficient value (as shown).

The circuit of Fig. 15-3 is designed to produce 250 milliwatts into an 8-ohm load, such as a pair of earphones or a small loudspeaker. Although the power level is relatively low, the circuit will produce a startlingly high volume for small radio projects.

Photosensors

Perhaps the most modern electronic light sensor is a pn junction with special photodiodes and phototransistors. Certain types of diodes, when the pn junction is illuminated, will increase the reverse leakage-current level. The diode is normally reverse-biased with a current-limiting resistance in series. The same principle applies to a class of npn or pnp phototransistors. In these devices, collector-to-emitter current flows when the base region is illuminated. These devices are the heart of optoisolator and optocoupler ICs and they are used in various instrumentation sensor applications.

Figures 15-4A and 15-4B show one use for photodiodes and phototransistors. The current flow through the device under light conditions is the *transducible event*, so you must connect them in a circuit that makes use of that property. The inverting-follower op-amp circuit will do that trick neatly. In Fig. 15-4A, the output voltage (V_o) is equal to the product of the input current and the feedback resistance ($I_L \times R_1$). In the case of Fig. 15-4A, a zero control is added, which can also be added to the other circuit as well. However, the zero control is not necessarily needed in either circuit.

Figure 15-4B shows a circuit that can be used when the light signal is either varying or is modulated. An example of the former is used in medical electronics in the form of pulse meters. These devices, called *photoplethysmographs*, use a red LED on one side of a thumb, finger, and ear lobe, and use a phototransistor on the other. When blood pulses through the tissue, it slightly changes the optical density to red light. This change can be used to take the pulse waveform. Modulated light sources are used in some alarms and other applications to help the circuit distinguish between light from a pulsed emitter and ambient light. It also helps in burglar-alarm circuits, where the trespasser tries to defeat the alarm by shining a light into the sensor. If the source emitter is pulsed, shining a constant light or a light that is pulsed at the wrong frequency, will not help him at all.

The circuit of Fig. 15-4B consists of a phototransistor (Q1) that drives an ac-coupled noninverting-follower op-amp circuit. The sensitivity of Q1

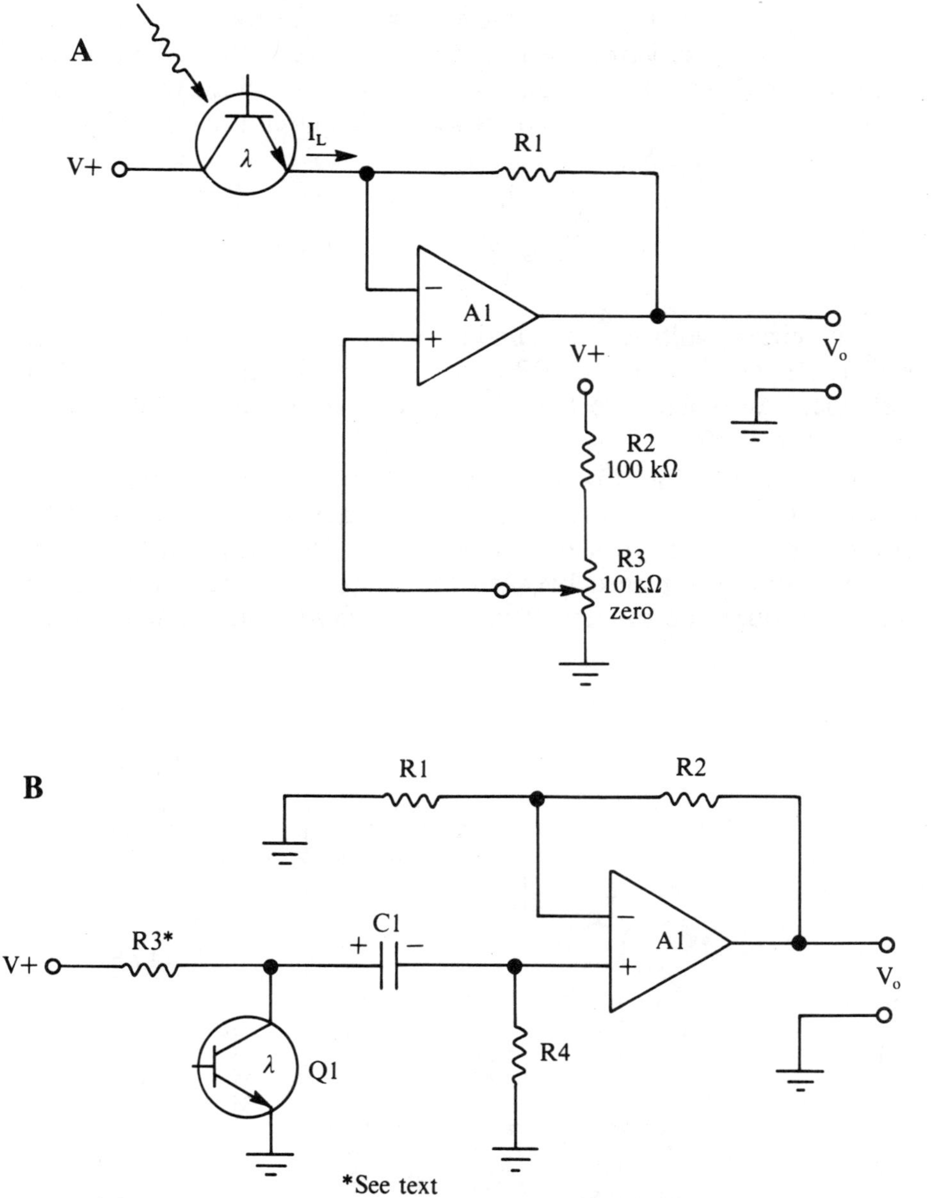

Figure 15-4. A) Inverting amplifier for a phototransistor, B) noninverting amplifier for a phototransistor.

is determined in some measure by the resistor that is in series with the collector and V+ supply (i.e., R3). A high value, 100 kohms to 1 megohms, will increase sensitivity, but a low value, 10 to 100 kohms, will increase operating speed.

The collector of Q1 has a large dc offset, with variation due to stimulus light levels being superimposed on the dc component. The capacitor (C1) strips off the dc component, allowing only the variations to pass on to the amplifier. The low frequency response of this circuit is dependent on the capacitor value and R_4:

$$F_{-3dB} = \frac{1}{2\pi R_4 C_1} \tag{15-4}$$

The circuit voltage is a function of the two resistors in the negative-feedback network, R1 and R2, and is found from Eq. (15-4). For the particular resistances selected, $R_1 = 1$ kohms and $R_2 = 100$ kohms, the voltage gain is 101.

The circuit of Fig. 15-5 is a practical photoplethysmograph that I built. A red LED is positioned so that it shines on a photodarlington transistor (several appropriate types are available from Digi-Key and Radio Shack). When a thumb is placed between the source and the sensor, the blood pulsing through the tissue will produce a varying output signal from the

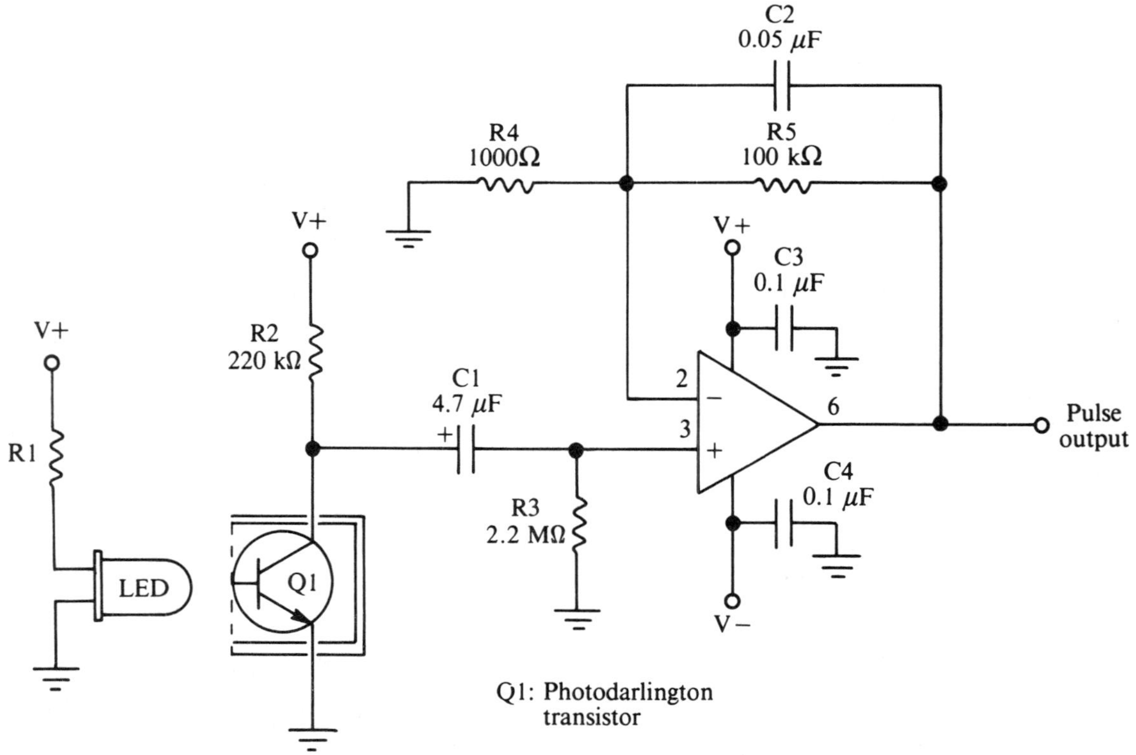

Figure 15-5. Photoplethysmograph circuit.

photodarlington transistor Q1. This signal is amplified by A1 (which can be any op amp, even a 741) and it becomes an analog output signal that has the shape of the heart's waveform.

The LED source and the sensor must be mounted inside a relatively light, tight housing. I used the large alligator-clip rubber-boot insulator because the opening is about the right size to insert over a finger or thumb. I cut small holes into the side of the boot for the sensor and the LED, and I used black electrical tape to prevent the wires and parts from coming loose.

Summary

The linear IC is the basis for many larger circuits, because of its high input impedance and its ability to use external components to control gain. The fixed reference voltage uses an op amp together with a zener diode to provide a stable voltage. The balanced audio amplifier uses a transformer for isolation and to provide possible additional gain. Photosensitive devices, which often have small changes in resistance when exposed to light, can have these changes amplified by op amps as well.

Questions

1. How can a zener diode provide a constant reference voltage for an op amp input?
2. For the circuit in Fig. 15-1, what effect will moving the potentiometer needle up have on the output?
3. What is meant by a balanced-audio circuit?
4. What is meant by VU?
5. For the circuit in Fig. 15-2A, what effect does increasing the turns ratio (N_p/N_s) have on the gain?
6. What is a photoplethysmograph?

Problems

1. For the circuit in Fig. 15-6, R_1 = 5.1 kohms, R_2 = 250 kohms, and R_3 = 100 kohms. If the zener voltage is 4.7 volts, what is the range of voltages for V_x?

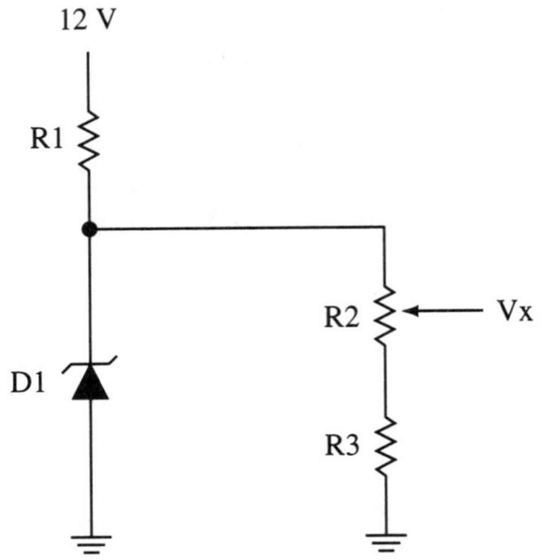

Figure 15-6.

2. Repeat Problem 1 for R_1 = 10 kohms, R_2 = 500 kohms, R_3 = 51 kohms, and a zener voltage of 8.2 volts.
3. The output power for a 1-milliwatt input is 50 mW. What is that in VU?
4. One milliwatt of input power provides an output of 75 mW. What is the gain in VU?
5. For the circuit in Fig. 15-3, what is the breakpoint frequency? Assume the maximum resistance for the AF gain potentiometer.
6. Calculate the breakpoint frequency for the circuit in Fig. 15-3 if the potentiometer is increased to 10 kohms.
7. For the circuit in Fig. 15-4A, V+ = 12 volts, R_1 = 10 kohms, and the potentiometer needle is at the top of R_3. **a.** If the dark resistance of the phototransistor is 100 kohms, what is the output voltage? **b.** What is the output voltage if the resistance is 80 kohms?
8. Repeat Problem 7 if the potentiometer needle is in the middle of R_3.

16 RC NETWORKS

AFTER COMPLETING THIS chapter, you will be able to:

- □ Understand charging and discharging in RC circuits
- □ Discuss trip-point circuits

Most of the waveform generator and oscillator circuits in this book depend on the characteristics of the simple resistor-capacitor (RC) network for proper operation. Understanding these networks is key to understanding the circuits that use them. As a result, this brief chapter is provided as a review of RC-network dc-circuit theory.

Consider Fig. 16-1A. Assuming that the initial condition is as shown, switch S1 is in position A, and is thus open circuited. There is initially no charge stored in capacitor C (i.e., $V_c = 0$). If switch S1 is moved to position B, however, voltage V is applied to the RC network. The capacitor begins to charge with current from the battery and V_c begins to rise towards V (see curve V_{cb} in Fig. 16-1B). The instantaneous capacitor voltage is found from:

$$V_c = V(1 - e^{-T/RC}) \qquad (16\text{-}1)$$

Where:

V_c is the capacitor voltage
V is the applied voltage from the source
T is the elapsed time (in seconds) after charging begins
R is the resistance (in *ohms*)
C is the capacitance (in *farads*)

The product RC is called the *time constant* (τ) of the network. If R is rated in *ohms*, and C is in *farads*, then the product RC is in *seconds*. The capacitor voltage rises to approximately 63.2 percent of the final "fully charged" value after *1RC*, 86 percent after *2RC*, and >99 percent after

5RC. A capacitor in an RC network is considered "fully charged" after five time-constants (>99.97%).

If switch S1 in Fig. 16-1A is next set to position C, the capacitor will begin to discharge through the resistor. In the discharge condition:

$$V_c = Ve^{-T/RC} \tag{16-2}$$

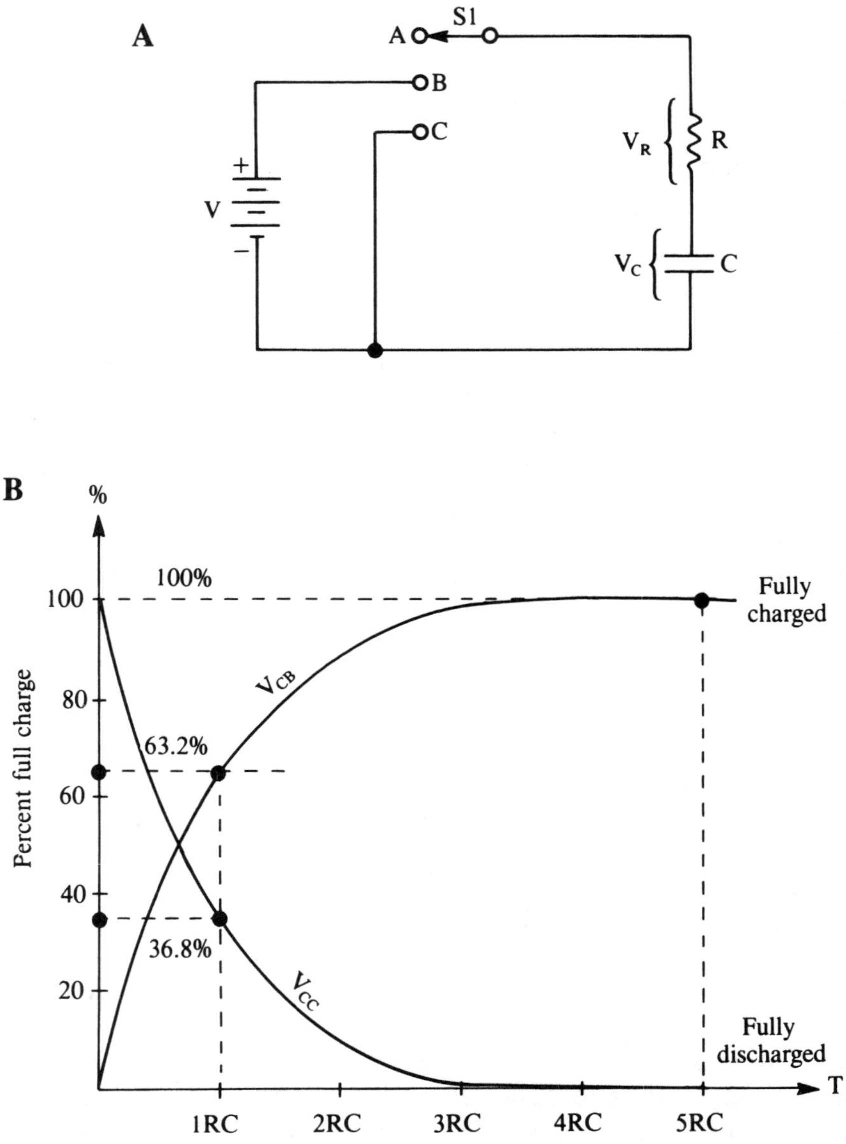

Figure 16-1. A) Resistor capacitor (RC) charge/discharge test circuit, B) charge/discharge curves, C) curve for charging from one potential to another.

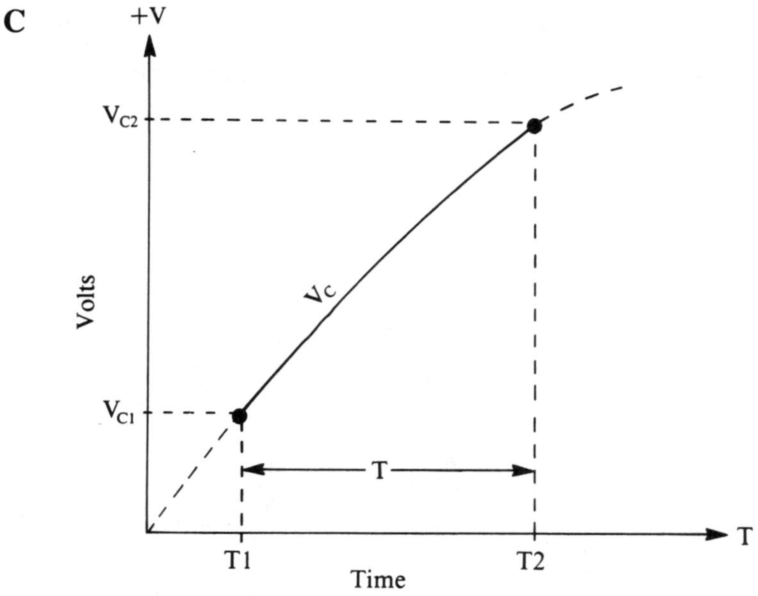

Voltage V_c drops to 36.8 percent of the full charge level after one time constant (*1RC*), and to very nearly asymptotic to zero after *5RC*. A capacitor in an RC network is considered fully discharged after five times constants (see the safety note at the end of the chapter).

Next, consider Fig. 16-1C. This graph represents a situation that is commonly encountered in waveform-generator circuits. In this graph, the capacitor is required to charge from the initial condition (V_{C1}), which might be 0 volts, to a final condition (V_{C2}), which might be the fully charged *5RC* point, in a specified time interval, *T*. The question asked by a circuit designer is: "What RC time constant will force V_{C1} to rise to V_{C2} in time *T*? Assume that $V_{C1} < V_{C2} < V$:

$$V - V_{C2} = (V - V_{C1})\, e^{-T/RC} \tag{16-3}$$

$$\frac{V - V_{C2}}{V - V_{C1}} = e^{-T/RC} \tag{16-4}$$

$$Ln\left(\frac{V - V_{C2}}{V - V_{C1}}\right) = \frac{-T}{RC} \tag{16-5}$$

or, rearranging terms:

$$RC = \frac{-T}{Ln\left(\dfrac{V - V_{c2}}{V - V_{C1}}\right)} \tag{16-6}$$

Experiment 16-1

An RC network is connected to a +12-Vdc source. What RC product will permit voltage V_c to rise from +1 to +4 Vdc in 200 mS?

Note: $V = +12$ Vdc, $V_{C2} = +4$ Vdc, and $V_{C1} = +1$ Vdc.

Solution:

$$RC = \frac{-T}{Ln\left(\frac{V - V_{c2}}{V - V_{c1}}\right)} \tag{16-7}$$

$$RC = \frac{-\left(200\text{ ms}\left(\frac{1\text{ s}}{1000\text{ ms}}\right)\right)}{Ln\left(\frac{12-4}{12-1}\right)}$$

$$RC = \frac{-0.200\text{ Sec}}{Ln\left(\frac{8}{11}\right)}$$

$$RC = \frac{-0.200\text{ Sec}}{Ln(0.727)}$$

$$RC = \frac{(-0.200\text{ Sec})}{(-0.319)} = 0.627\text{ Seconds}$$

Equation (16-3) can be used to derive the timing or frequency-setting equations of many different RC-based waveform-generator circuits. The key voltage levels will usually be device trip points or critical values that are set by the design of the circuit.

Experiment 16-2

This experiment is used to explore the capacitor-charging properties of RC networks. A low-voltage source (9-volt battery or equivalent dc power supply) is used to charge a 100-μF capacitor. Select a capacitor with a voltage rating of 15 wvdc or higher. Select three fixed resistors of 10 kohms, 100 kohms, and 1 megohms. A 0-to-10 Vdc voltmeter is used to measure the voltage across the capacitor (V_c).

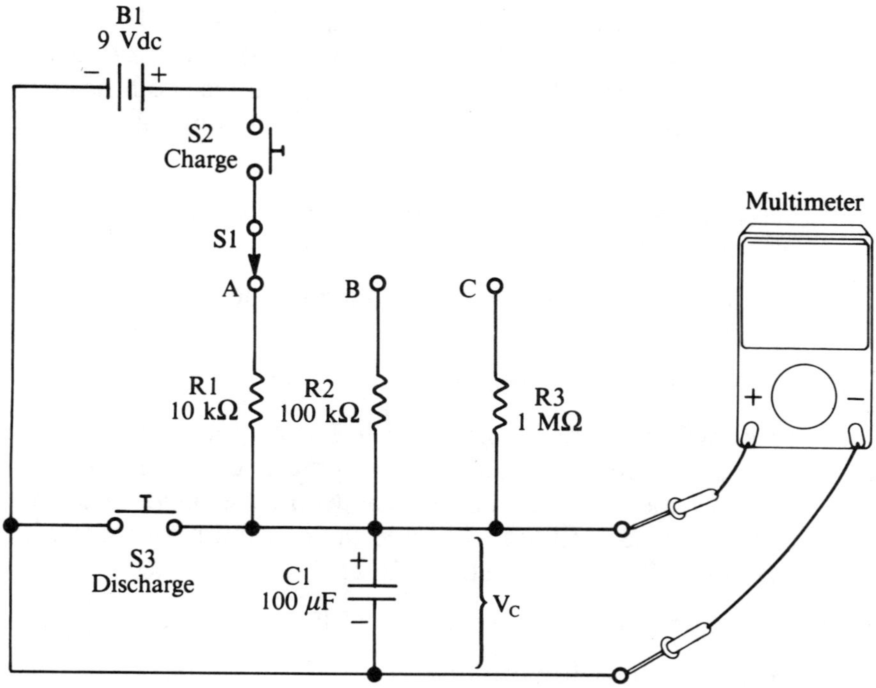

Experiment 16-1. Circuit for Experiment 16-1.

1. Measure the value of R1 and record the results.
2. Likewise, measure the values of R2 and R3 and record the results.
3. Calculate the following time constants:

$$RC_1 = R_1C_1 \text{ seconds}$$
$$RC_2 = R_2c_1 \text{ seconds}$$
$$RC_3 = R_3C_1 \text{ seconds}$$

4. Connect the components into the circuit of Experiment 16-1. Use pushbutton SPST switches for S2 and S3, and a single-pole, three-position (SP3P) slide or rotary switch for S1.
5. Press S3 several times to ensure that capacitor C1 is discharged.
6. Record V_{C1} (it should be zero).
7. Set switch S1 to position A.
8. While watching the dc voltmeter, press switch S2 and hold until V_{C1} no longer increases, then release S2. How fast was it? Nominally, it would be one second, which is a little fast to see. However, at longer time constants, you will be able to measure the time with the second hand on a watch.

9. Notice the meter reading. If the capacitor is ideal (it isn't), the voltage will remain at the final value for an indefinite period. In real circuits, stray "leakage" resistances will bleed off the charge in a relatively short time. How does the final value of V_{C1} compare with the battery voltage of B1?
10. Next, press S3 several times to ensure that the capacitor voltage returns to zero.
11. Use the second hand of a watch to record the charging times in the next two readings.
12. Set switch S1 to position B.
13. Press switch S2 and hold it until V_{C1} no longer increases, then release S2. How long did it take to reach the final value? Is this final value similar to the final voltage in the previous step? How does the time required to charge the capacitor compare with the calculated RC time constant, RC2?
14. Now, discharge C1 by pressing and holding S3 several times.
15. Move switch S1 to position C.
16. Press S2 and hold it down until the voltage across C1 no longer increases, then release S2. Notice the voltage across C1. How long did it take to reach this voltage? How does that time compare with RC3?
17. Your calculations and observations probably differed by a small amount. Part of this error is because the actual value of C1 is not 100 μF, but rather is close to that value $\pm$ a small tolerance. Also, the accuracy of your time measurements is not perfect.

This experiment can suggest a method for measuring the capacitance of a test capacitor. By measuring the resistance and the battery voltage very accurately and noting the time that is required to reach a fully charge value (or some other point), the capacitance can be calculated.

Trip-point circuits

A *trip-point circuit* remains in one state until a certain input voltage level is reached, then it abruptly switches to an alternate state. Trip-point circuits are at the heart of certain waveform-generator circuits that are studied in subsequent chapters.

The op amp *voltage-comparator circuit* is an example of a trip-point circuit. A voltage comparator compares two input voltages (V_1 and V_2) and

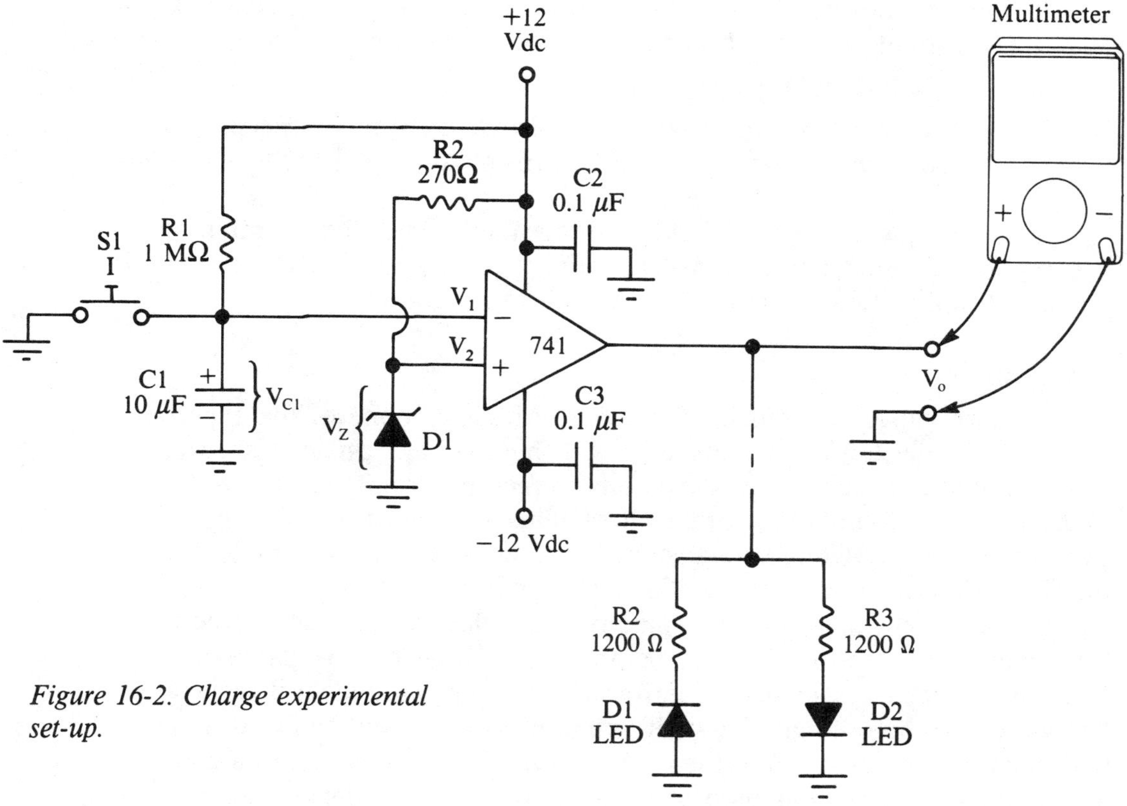

Figure 16-2. Charge experimental set-up.

issues an output to indicate which relationship is true: $V_1 = V_2$, $V_1 < V_2$, or $V_1 > V_2$. Figure 16-2 shows a voltage comparator that is connected as a trip-point circuit. In this circuit, the noninverting input is biased to V_2 by a zener-diode voltage regulator. The zener-diode potential, V_z, is V_2.

Voltage V_1 is voltage V_{C1} across capacitor C1. When the circuit is first turned on, V_2 rapidly goes to V_z, but V_1 is zero. In this case, $V_1 < V_2$, so output voltage V_o is high (positive).

Capacitor C1 charges through R1 immediately, however, so V_{C1} begins rising. At some point, $V_1 = V_2$, so output V_o drops to zero, but does not remain there long. Almost immediately, voltage $V_1 > V_2$, so the output snaps from zero to low (negative). From the observer's point of view, the switch from high to low (i.e., positive to negative) is almost immediate.

Experiment 16-3

Connect the circuit of Fig. 16-2. Monitor the output, either with a voltage meter or an LED circuit, shown in Fig. 16-2. In either case, monitor the voltage across C1 with a voltmeter.

1. Turn on the circuit and let the output stabilize to $V_1 > V_2$. One LED (D1) should be turned on and the other turned off. Measure V_2.
2. Press S1 and hold it down until V_{C1} is zero.
3. Release S1, while watching both the LEDs and the voltmeter. Notice the voltage at which one LED turns off and the other turns on.

Note that this time is predictable and repeatable. Therefore, combining a trip-point circuit (such as a voltage comparator) and an RC network makes a timing circuit possible for a waveform generator.

Safety note

High-voltage capacitors can be dangerous. Although none of the experiments and projects in this book deal with high-voltage capacitors, some readers might acquire them at some point in their electronics careers. Some of the capacitors found in the ripple filters of high-voltage dc power supplies can contain a high-voltage charge with sufficient stored current to kill humans. *Do not rely on the 5RC rule in safety situations.* When high voltages are used, there may be some residual electrical energy stored in the capacitor dielectric after discharge. This charge can be dangerous. Repetitive discharges should be performed in order to reduce this residual charge to zero before handling it. If you are unsure of yourself in dealing with high-voltage capacitors, then refer servicing and handling to a qualified person. This phenomenon is less a factor in low-voltage capacitors, but is a hazard in the high-voltage types. *Be careful*!

Now that the basic RC networks and trip-point circuits are covered, let's look at oscillators, multivibrators, and other waveform-generator circuits.

Summary

Charging and discharging capacitors in an RC circuit is a logarithmic function. The product RC is called the *time constant,* and a capacitor is considered fully charged or discharged after 5 time constants. Because the capacitor voltage changes with time, it can be used with a fixed voltage reference to form a trip-point circuit. In this circuit, when the voltage reaches the fixed reference voltage, the output of an op amp will change state between positive and negative saturation values.

Questions

1. In an RC circuit with a dc power source, how does capacitor voltage vary with time during charging? during discharging?
2. What is the time constant for an RC circuit? How many time constants does it take for the capacitor to be considered fully charged or discharged?
3. Explain the operation of the circuit in Fig. 16-2.

Problems

1. For the circuit in Fig. 16-1A, $V = 12$ volts, $R = 10$ kohms, and $C = 0.22\ \mu\text{F}$. Calculate the voltage across the capacitor 5 ms after the switch is moved to position B.
2. For Problem 1, determine the voltage across the capacitor 15 ms after the switch is moved to position B.
3. In Fig. 16-1A, the capacitor is fully charged to 15 volts when the switch is moved to position C. If $R = 100$ kohms and $C = 10\ \mu\text{F}$, what is the voltage after 3 seconds?
4. For the circuit in Fig. 16-1A, the capacitor is fully charged to 12 volts when the switch is moved to position C. If $R = 22$ kohms and $C = 1.2\ \mu\text{F}$, what is the voltage after 30 ms?
5. For Problem 1, how long does it take to charge from 2 volts to 4 volts?
6. For Problem 2, how long does it take to charge from 8 volts to 10 volts?
7. For the circuit in Fig. 16-2, if the zener voltage is 4.7 volts, at what time does the output change state?
8. For the circuit in Fig. 16-2, when does the output change state if the zener voltage is 8.2 volts?

17 OSCILLATOR AND MULTIVIBRATOR CIRCUITS

AFTER COMPLETING THIS chapter, you will be able to:

- □ Describe the two major types of oscillator circuits
- □ Explain the criteria for oscillation for feedback oscillator circuits

Signal- and waveform-generator circuits produce sine-wave (square wave, triangle wave, or other waveform) output signals for purposes of operation, testing, troubleshooting, and alignment of electronic circuits. Although the subject of designing top-notch signal-generator circuits is beyond the scope of this chapter, it is possible to greatly reduce the possibilities for people whose needs are less severe. This chapter looks at some signal generators that you can build, and which can be used in a wide variety of applications for electronic and radio work. But first, look at the generic types of oscillator circuits that you'll encounter.

Types of oscillator circuits

There are two major categories of oscillator circuits are found in electronics textbooks: relaxation oscillators and feedback oscillators. The *relaxation oscillator* uses some sort of voltage breakdown device, such as a neon glow-lamp, tunnel diode, or unijunction transistor. The *feedback oscillator* (Fig. 17-1) uses an amplifier circuit and a frequency-sensitive feedback network to start and sustain oscillations on a particular frequency. Most of the oscillators that are useful for sine-wave signal-generator circuits are of the feedback-oscillator class.

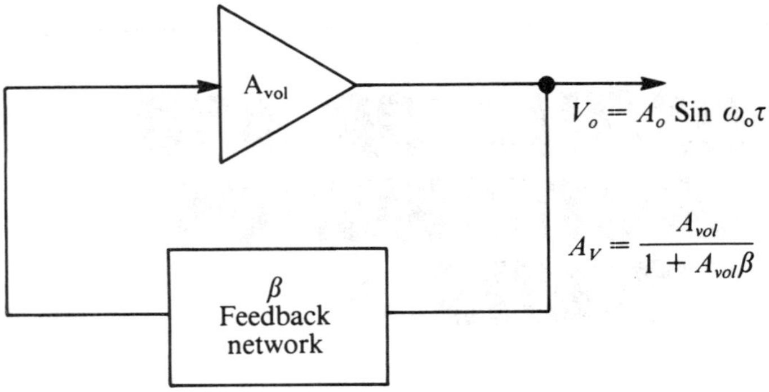

Figure 17-1. Block diagram of the feedback oscillator.

The requirements for sustained oscillation, called *Nyquist's criteria*, are: the loop gain, between feedback network losses and amplifier gain, must be greater than or equal to one at the frequency of oscillation; and the feedback signal must be in-phase with the input signal at the frequency of oscillation. The second of these criteria means that the feedback signal must be phase-shifted 360 degrees—180 degrees is usually obtained from the inversion of the amplifier and 180 degrees from the frequency-selective feedback network.

Feedback oscillators can also be classified according to the nature of the feedback network. Three basic types of RF oscillators exist (and many others are variations on the themes): Armstrong, Colpitts or Clapp, and Hartley. Other feedback networks are used at audio frequencies, but these are the principal forms of RF oscillator. You can tell which is which by looking at the feedback network (Fig. 17-2). The *Armstrong oscillator* (Fig. 17-2A) uses a separate *tickler coil* (L2) to provide feedback to the main tuning coil (L1). These coils are usually all wound on the same coil form. The Colpitts oscillator (Fig. 17-2B) uses a parallel resonant-tuned circuit and (this is the key) a tapped-capacitor voltage divider (C1 and C2) to provide feedback. The voltage divider might or might not be a direct part of the resonant circuit. The Clapp oscillator is a variant of the Colpitts circuit, in which a series resonant-tuning circuit is used. Finally, the Hartley oscillator uses a tapped-inductor voltage divider (Fig. 17-2C) for the feedback network.

The resonator of the oscillator might be either an inductor-capacitor (LC) tuned circuit, as shown in the examples of Fig. 17-2, or a piezoelectric crystal resonator. The latter is a single-channel resonator, but it is more stable than an LC-tuned circuit.

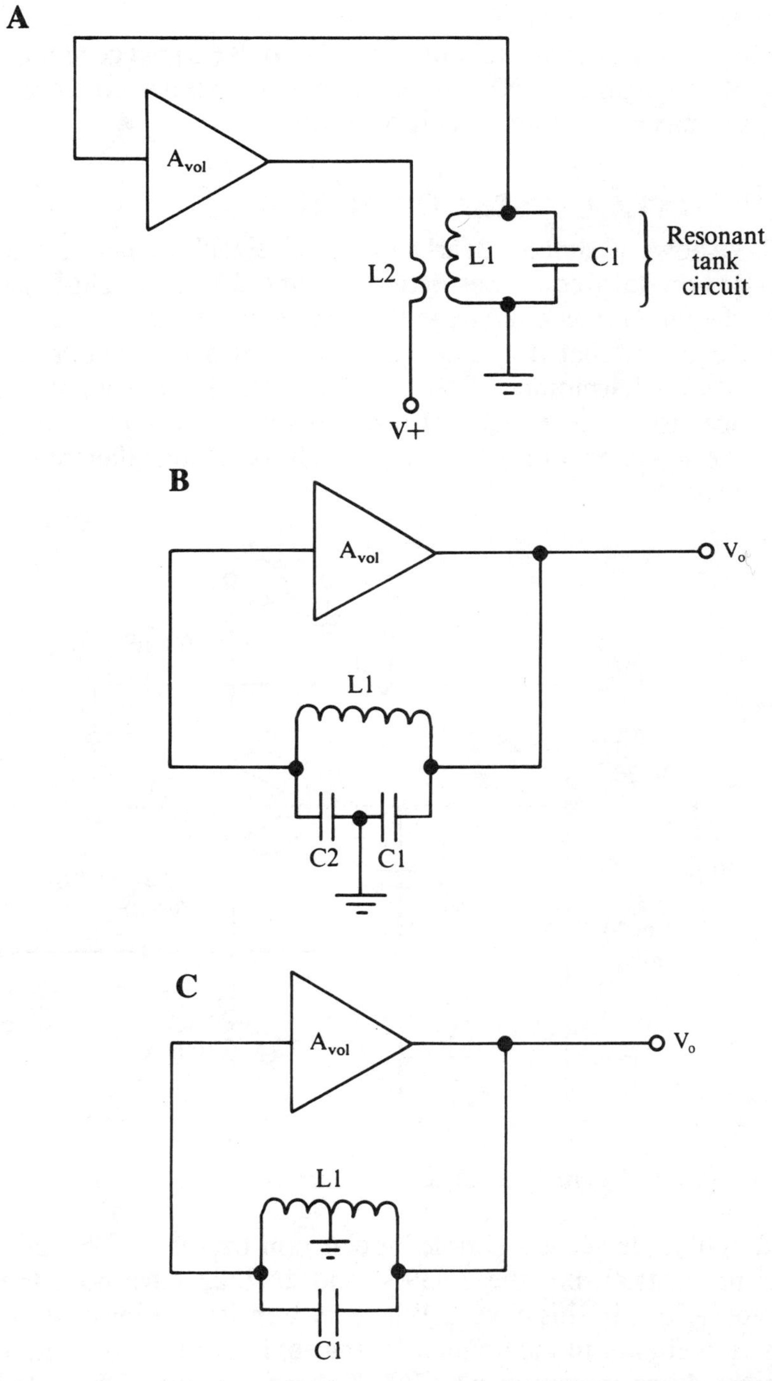

Figure 17-2. A) Armstrong oscillator, B) Colpitts oscillator, C) Hartley oscillator.

Circuit examples and projects

In the sections to follow, you will find a number of RF signal generators that can be used to align and test RF circuits. These projects are simple to build, yet they quite nicely illustrate circuit operation.

1-to-20 MHz crystal oscillator

Figure 17-3 shows a simple, nearly universal signal generator that can accommodate crystal frequencies between 1 and 20 MHz. This oscillator operates in the *fundamental mode*, so the marked frequency of the crystal is the actual frequency that it operates on (as opposed to *overtone crystals*, which operate on harmonics of the fundamental frequency). If you are given a choice, ask for a crystal that operates into a capacitance of 32 pF. Otherwise, the actual operating frequency might be a little different than the specified frequency.

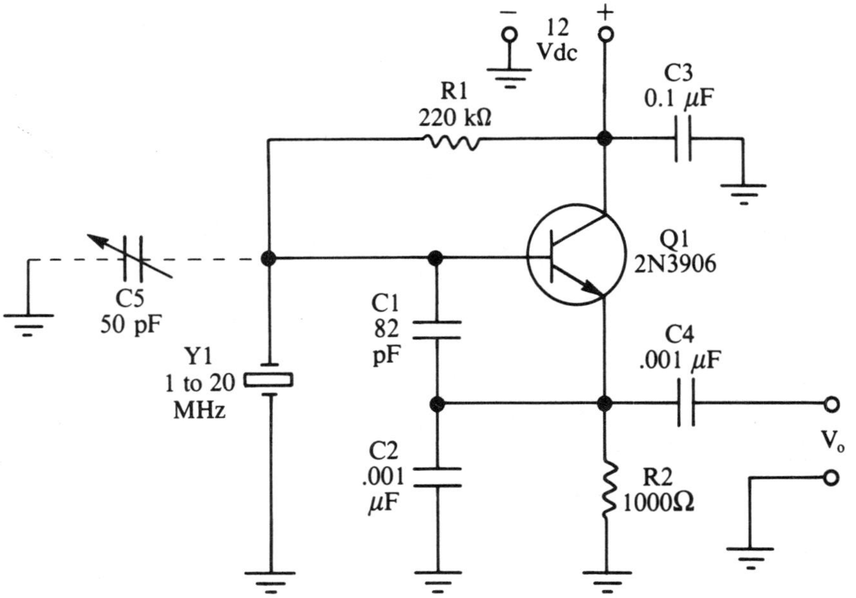

Figure 17-3. Crystal Colpitts oscillator.

The amplifier device is a simple bipolar npn transistor. The transistor selected is not critical, but the 2N3904 and 2N2222 have been tried on numerous occasions in this circuit. Whatever transistor is selected, it must operate as an oscillator in the frequency range of interest. If you want to use a similarly rated pnp transistor (e.g., 2N3906), simply reverse the polarity of the dc power supply. The transistor uses the simplest form of dc bias network: R1 is directly connected between the transistor's collector and its base. The output signal is taken through a dc blocking capacitor (C4) and across the emitter resistor (R2). A bypass capacitor (value 0.01 to 0.1 μF) is connected between the collector and ground. This capacitor sets the collector at ground potential for ac potentials, while it keeps the rated dc power-supply potential at 9 to 12 Vdc.

The feedback network consists of capacitors C1 and C2. The ratio of C_1/C_2 is selected to achieve a reasonable tradeoff of output level and stability. These capacitors should be either silvered mica or (preferably) NPO disk-ceramic capacitors. Use of these capacitor types will prevent frequency drift of the oscillator as a result of temperature-dependent changes in C_1 and C_2. For the same reason, capacitor C3 should also be a silvered mica or NPO disk-ceramic type.

The circuit, as shown in Fig. 17-3, operates at a single frequency because it is crystal controlled. Although you might think that the crystal always operates on the specified frequency, the differences in circuit capacitance will probably make it oscillate at a different frequency. If the exact frequency is important, replace C3 with a variable capacitor. Alternatively, connect a 50-pF variable capacitor in parallel across the crystal (C5 in Fig. 17-3).

The circuit of Fig. 17-3 is popular with people who troubleshoot FM radio receivers. Figure 17-4 shows a signal generator that can be used to align and test common FM radio receivers. The basic circuit is the same as Fig. 17-3, but the crystal is replaced with a SPDT switch and a pair of crystals. The 10.7-MHz crystal is used to test the FM IF-amplifier stages, so it is on the standard FM IF frequency.

The 9-MHz crystal is used to test the front-end and tuning-dial accuracy of the receiver. A 9-MHz crystal will output sufficient harmonics at 90 MHz, 99 MHz, and 108 MHz to check the calibration of the low-, middle-, and high-end of the dial, respectively. Notice that this signal generator does not produce an audio tone in the output of the receiver, but rather it is indicated by "quieting" of the receiver and an increase in the strength-meter reading.

HF/VHF buffer amplifier

A *buffer amplifier* isolates the output of the oscillator from the load or circuits that follow. Variations in the load of an oscillator can "pull" the frequency incorrectly, so the buffer is used to prevent that problem. Figure 17-5 shows the circuit for a buffer amplifier than can be used in the low-frequency (LF), high-frequency (HF), and in the lower end of the VHF ranges.

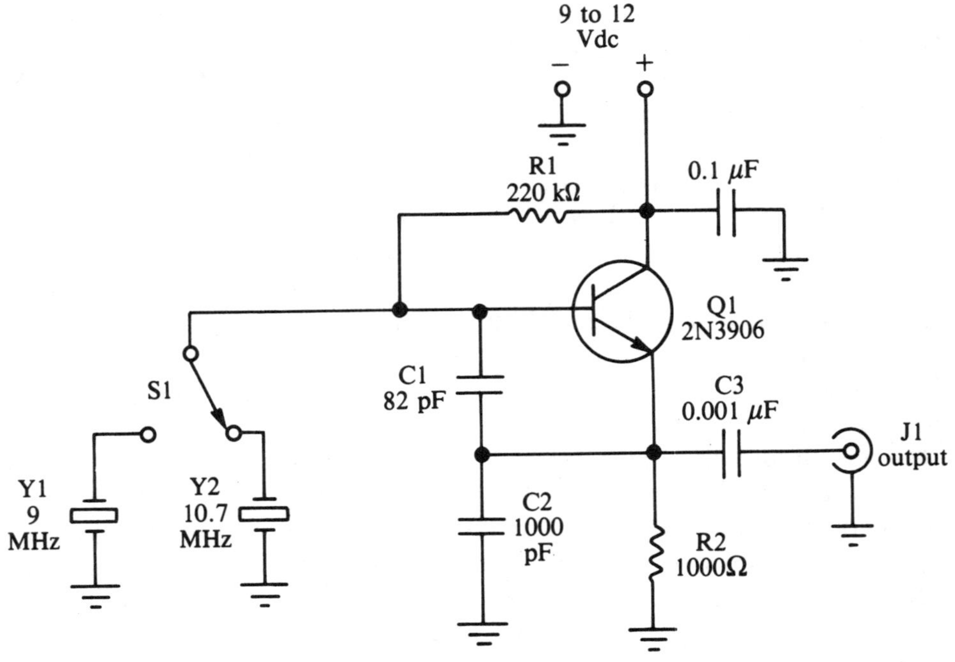

Figure 17-4. FM aligner oscillator.

The amplifier device in the buffer amplifier is a 40673 dual-gate MOSFET transistor (or NTE-222, which is a replacement). The MOSFET is connected in the standard grounded-source (S) configuration. A small source bias is provided by the 100-ohm resistor (R2) from source to ground. A 0.1-μF capacitor that is shunted across R2 keeps the source resistor at ground potential for RF signals, and it also keeps the circuit at the small positive dc potential that is caused by the voltage drop across the resistance.

The positive-operating dc potential on the drain (D) is supplied through a 270-ohm collector load resistor (R3) and a 1000-μH RF choke (RFC1). The choke has an inductive reactance that increases linearly with frequency. So, the combination of R3 and RFC1 provides a load impedance that increases with frequency, which overcomes the tendency of the transistor to offer less gain at higher frequencies. The dc drain-load network (R3/RFC1) is decoupled by a pair of 0.1-μF bypass capacitors. Use either disk ceramic, polyester, or other capacitor forms for this application, but make sure when ordering that they will work as bypass capacitors at the low-VHF region (see the write-up in the catalog at the beginning of each capacitor section).

Input is applied through a dc-blocking capacitor (C1), across a gate resistor (R1), to MOSFET gate G1. The output signal is taken from the drain (D) through a 0.001-μF dc blocking capacitor (C3).

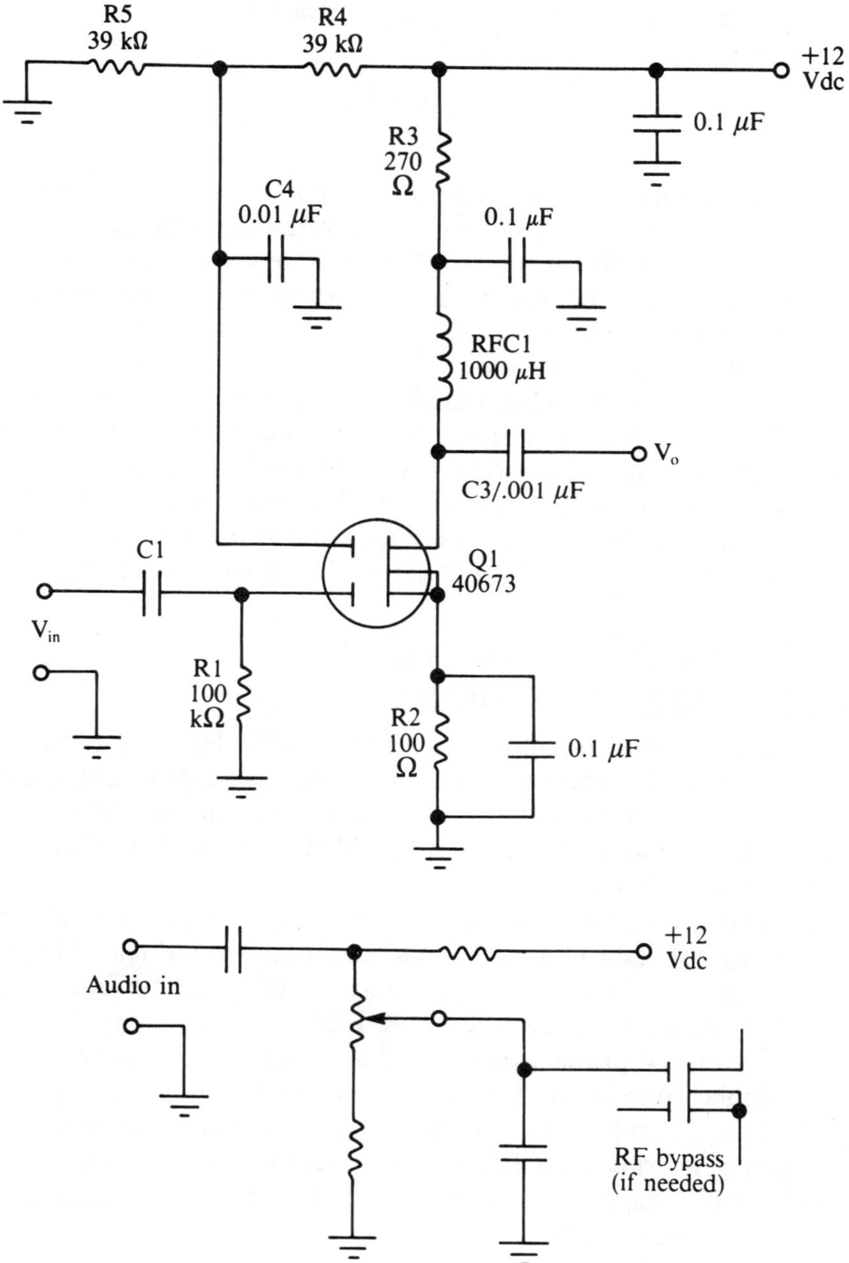

Figure 17-5. Buffer amplifier (wideband).

The second gate (G2) of the 40673 MOSFET is biased to a potential of about 10 Vdc, which is set from the +12-Vdc power supply by voltage divider R4/R5. The G2 terminal is set to ac ground by a bypass capacitor (C4) that is selected to have a low capacitive reactance at the minimum operating frequency, relative to R5. In practical terms:

$$X_{C4} < \frac{R_5}{10}$$

A variation on the theme can be built by connecting the voltage-divider network R_4/R_5 to a potentiometer or other variable voltage source instead of the V+ power supply (as shown). The variable voltage can then be used as an output-level control. In some signal generators, the G2 terminal of the MOSFET buffer amplifier is connected to an *automatic gain control* (AGC) to stabilize the output signal level.

Another use for the G2 terminal is to amplitude modulate (AM) the output signal level. A potentiometer and a capacitor (see inset to Fig. 17-5) is used to connect the G2 circuit to an audio sine-wave source. Make sure that the sine-wave amplitude is high enough to make the amplifier nonlinear (otherwise modulation will not occur), but not so much that the output waveform peaks and valleys do not "flat top" (as shown on an oscilloscope).

455 kHz AM IF-amplifier test/alignment oscillator

The modern superheterodyne AM radio uses a 455-kHz IF amplifier (262.5 kHz is used on AM car radios). Figure 17-6 shows a simple signal-generator circuit that can be used to test, troubleshoot, or align the AM IF stage in common radios. The active element is a MPF-102 (or NTE-312) *junction field-effect transistor (JFET)*.

As you can see by comparing Fig. 17-6 to Fig. 17-2, this circuit is of the basic Hartley-oscillator class because it uses a tapped inductor (T1) in the feedback network. The transformer used for this purpose is a standard 455-kHz IF transformer (Digi-Key TK-1301, or equivalent). This transformer has a tapped primary that is used to match the low-collector impedance of bipolar (pnp or npn) transistor amplifiers, but you can press that tapped LC tank circuit into service as a Hartley oscillator. The secondary of the transformer forms a handy output port to pass a 455-kHz signal to the following buffer stage (see Fig. 17-5 and the previous discussion for a description of this circuit).

Signal generator for the AM and shortwave bands

The same general type of circuit can be used for the AM band (530 to 1610 kHz) or for the shortwave bands (1610 kHz to 30 MHz). The version shown

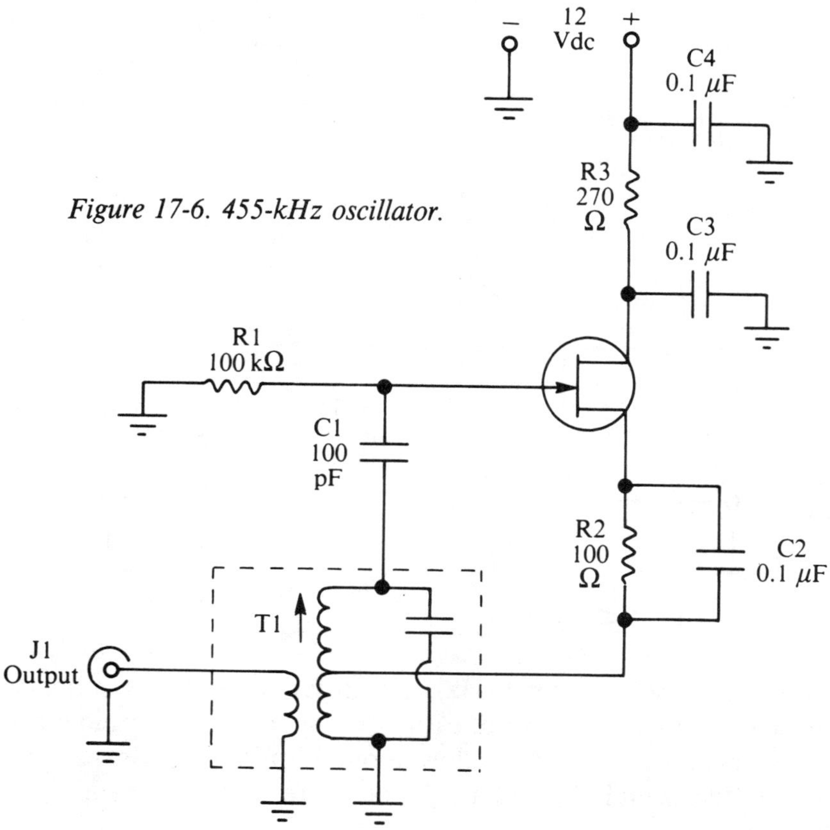

Figure 17-6. 455-kHz oscillator.

in Fig. 17-7 uses a standard transformer that has a 217-μH inductance in the primary winding. This particular example uses a Toko coil from Digi-Key (Part TK-1903) in conjunction with a 365-pF broadcast variable capacitor to cover the AM broadcast band. For other frequencies:

$$F = \frac{1}{2\pi\sqrt{LC}} \tag{17-1}$$

or, if F is known, but C is unknown:

$$C = \frac{1}{39.5F^2L} \tag{17-2}$$

or, if F is known, but L is unknown:

$$L = \frac{1}{39.5F^2C} \tag{17-3}$$

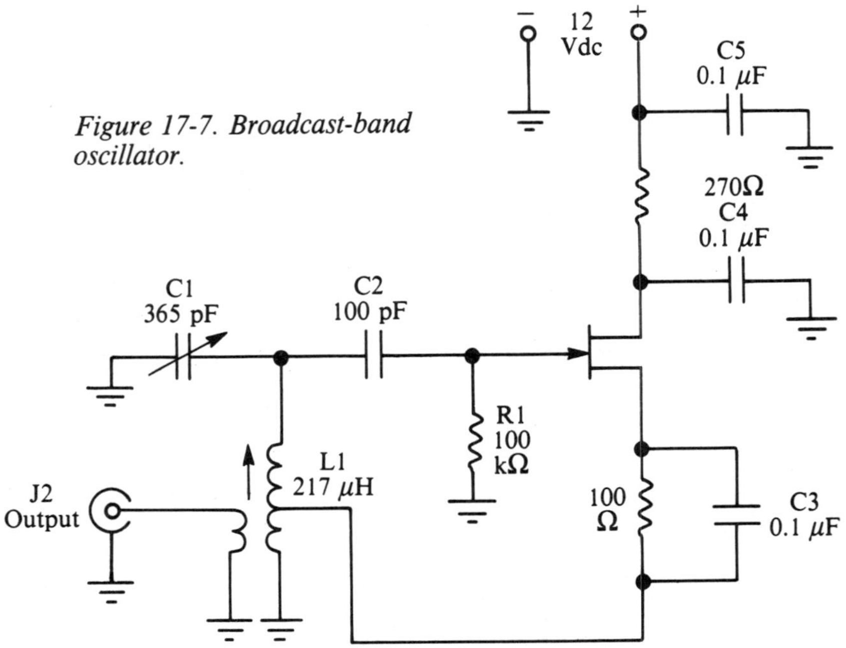

Figure 17-7. Broadcast-band oscillator.

In all three cases, *F* is in hertz (Hz), *C* is in farads, and *L* is in henrys. For frequencies at which a coil is not easily available, select a toroidal core and wind it yourself. In either case, allow about 20 pF for stray capacitances in the circuit (the actual amount might be more or less, depending on the layout).

A VHF voltage-tuned oscillator circuit is shown in Fig. 17-8A. I've used this circuit at frequencies from 20 to 150 MHz. The circuit uses a feedback capacitor (C2) across the collector-emitter circuit of the transistor. This capacitor is critical, and the circuit probably won't oscillate without it. The tuning network consists of inductor L1, dc blocking capacitor C1, and varactor D1. *Varactors* are voltage-variable capacitor diodes. The inductor is wound on either a VHF toroidal coil or on an air form that is about ⅜ inch in diameter. If #20 or so solid wire is used for L1, the coil will be self-supporting after it is removed from the coil form.

The VHF oscillator of Fig. 17-8A is tuned with a dc voltage (V_t) that is applied through a 150-kohm resistor (R4). This voltage is positive in order to reverse-bias D1. A problem with this circuit is that the frequency is a nonlinear function because of the characteristics of D1. Figure 17-8B shows a somewhat more linear version of this same circuit. In this case, the dc-blocking capacitor is replaced with a second varactor diode that is identical to the first. The total capacitance of the combination is half the capacitance of one diode, but the voltage-vs.-frequency characteristic is more linear.

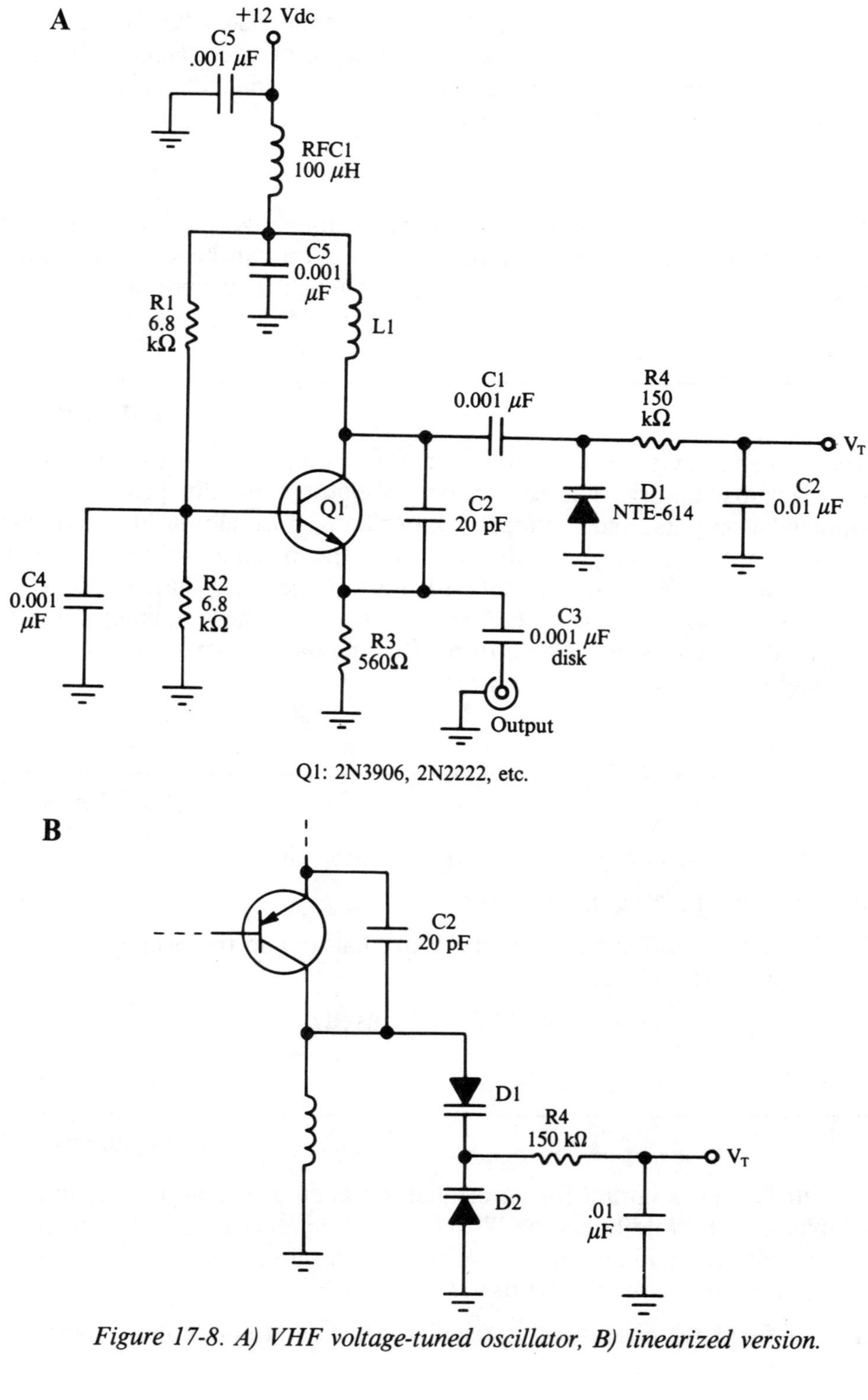

Figure 17-8. A) VHF voltage-tuned oscillator, B) linearized version.

Both of these circuit variations can be either fixed-tuned with a dc voltage, variable-tuned through a potentiometer, or swept across a band of frequencies with a sine-wave (FM) or sawtooth signal. In the case of the sawtooth, use a 10- to 60-Hz signal.

Conclusion

The RF signal-generator circuit is generally well-behaved and thus is easy to build and align, if proper technique is used. You can be quite successful making these circuits perform the tasks for which they were designed.

Summary

The two primary types of oscillator circuits are relaxation oscillators and feedback oscillators. The relaxation oscillator generally produces non-sinusoidal outputs, and the feedback oscillator normally produces a sine-wave output. In order to sustain oscillation, the product of the gain and the attenuation of the feedback network must be greater than 1. Most RF sine-wave oscillators use an LC network to select the frequency of oscillation. To keep the output load from affecting the frequency, the oscillator is usually isolated by a buffer amplifier.

Questions

1. What are the two major types of oscillator circuits?
2. What are the Nyquist criteria?
3. Name three different types of RF oscillators and the primary characteristic of each.
4. What is a buffer amplifier? Why is it used?

Problems

1. The feedback circuit for an oscillator has an attenuation of 30 and a phase shift of 180 degrees. What is the minimum gain of the amplifier in order to maintain oscillation? Should an inverting or a noninverting amplifier be used?
2. The feedback circuit for an oscillator has an attenuation of 3 and a

phase shift of 0 degrees. What is the minimum gain of the amplifier in order to maintain oscillation? Should an inverting or a noninverting amplifier be used?

3. The circuit in Fig. 17-6 is used to provide a frequency of 455 kHz. If the inductor is 217 μH, what size capacitor should be used?
4. If the inductor in Fig. 17-6 is changed to 100 μH, what size capacitor should be used in order to obtain a frequency of 455 kHz?

18 SQUARE-WAVE GENERATORS

AFTER COMPLETING THIS chapter, you will be able to:

- ☐ Explain astable multivibrator circuits
- ☐ Understand the Fourier series as it applies to nonsinusoidal waveforms
- ☐ Discuss waveform symmetry

The square-wave oscillator is, perhaps, the most common form of waveform generator, other than the sine-wave oscillator. These circuits are easily designed from RC networks and a variety of active devices (we will use operational amplifiers). This chapter shows how these circuits work, how to design them, and how to perform some experiments that are based on these circuits.

Astable (free running) circuits

An *astable multivibrator* (*AMV*) is free-running; that is, it can output a waveform that repeats itself without being either triggered or retriggered. The output of the AMV is a periodic (i.e., repeats itself regularly) pulse or wave train. In a periodic signal, the wave repeats itself indefinitely until the circuit is either turned off or is otherwise inhibited.

Astable multivibrators are oscillators. Waveforms that are available from the AMV include square waves, triangle waves, and sawtooth waves. Sine waves are also available from oscillator circuits, but most of those circuits operate differently from the others.

Nonsinusoidal waveform generators

The nonsinusoidal AMV circuits will produce square, triangle, or sawtooth waves, depending on the design. When combined with a monostable multivibrator (MMV), in order to produce variable-width pulses (of a duration less than the period of the square wave), a variable pulse-generator circuit results. Because the square-wave generator is the most basic form, the discussion of AMV circuits begins with square waves, the subject of this chapter.

Square waves

Figure 18-1 shows the classic square wave. Each time interval of the wave is quasistable, so you might conclude that the square wave generator has no stable states (hence it is astable). The waveform snaps back and forth between $-V$ and $+V$, dwelling on each level of a time duration (t_a or t_b). The period, T, is the sum of these dwell times:

$$T = t_a + t_b \tag{18-1}$$

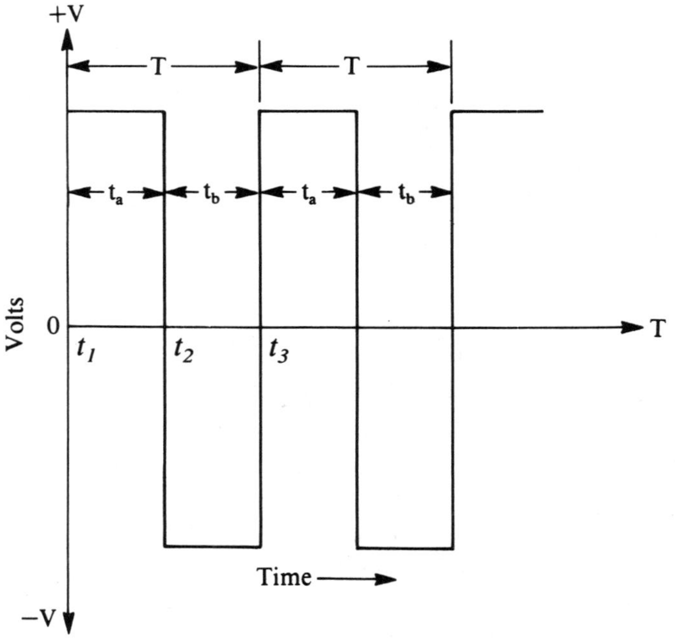

Figure 18-1. Square-wave oscillation.

Where:

T is the period of the square wave (t_1 to t_3)
t_a is the interval t_1 to t_2
t_b is the interval t_2 to t_3

The frequency of oscillation (F) is the reciprocal of T:

$$F = \frac{1}{T} \tag{18-2}$$

Where: F is in Hertz and T is in seconds.

The ideal square wave is both base-line and time-line symmetrical, $|+V| = |-V|$ and $t_a = t_b$. Under time-line symmetry $t_a = t_b = t$, so $T = 2t$ and $f = 1/2t$.

Fourier series

All continuous periodic signals can be represented by a fundamental-frequency sine wave and a collection of harmonics of that fundamental sine wave that are summed together linearly. These frequencies comprise the Fourier series of the waveform. The elementary sine wave is described by:

$$V = V_m \text{Sin } (2\omega t) \tag{18-3}$$

Where:

v is the instantaneous amplitude of the sine wave at time t
V_m is the peak amplitude of the sine wave
ω is the angular frequency ($2\pi F$) of the sine wave
t is the time (in *seconds*)

The *period* of the sine wave is the time between reoccurrence of identical events, or $T = 2\pi/\omega = 1/F$ (where F is the frequency in cycles per second).

Figure 18-2 shows three complex waveforms and the sine and cosine waves that they consist of. Figure 18-2A shows the basic symmetrical square wave. In Fig. 18-2B, the harmonic (curve A) sine wave of frequency F is added to its third harmonic (curve B) at frequency $3F$ to produce a distorted square wave, shown as curve C. For comparison, the idealized square wave is shown superimposed on the group of waves. In Figs. 18-2C and 18-2D, the process continues with harmonics up to the seventh harmonic ($7F$); notice that the square wave is becoming more like the underlying ideal square wave. The construction of the sawtooth and triangle waveforms are shown in Figs. 18-3 and 18-4, respectively.

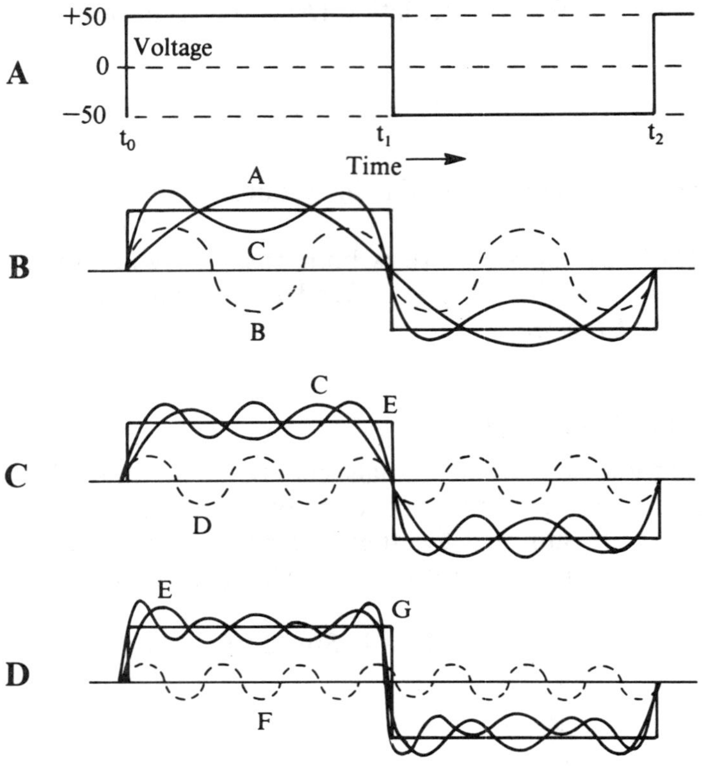

A Fundamental
B 3rd harmonic
C Fundamental plus 3rd harmonic
D 5th harmonic
E Fundamental plus 3rd and 5th harmonic
F 7th harmonic
G Fundamental plus 3rd, 5th, and 7th harmonic

Figure 18-2. Constructing a square wave from a sine wave and its harmonics.

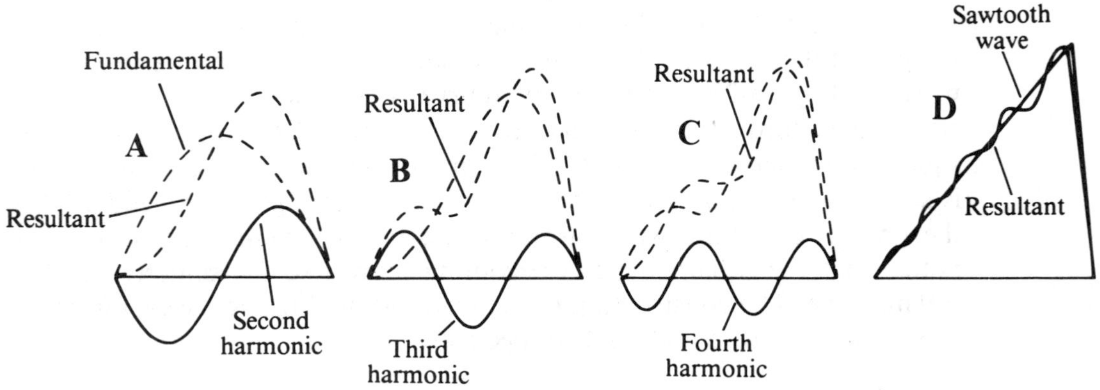

Figure 18-3. Constructing a sawtooth from a sine wave and its harmonics.

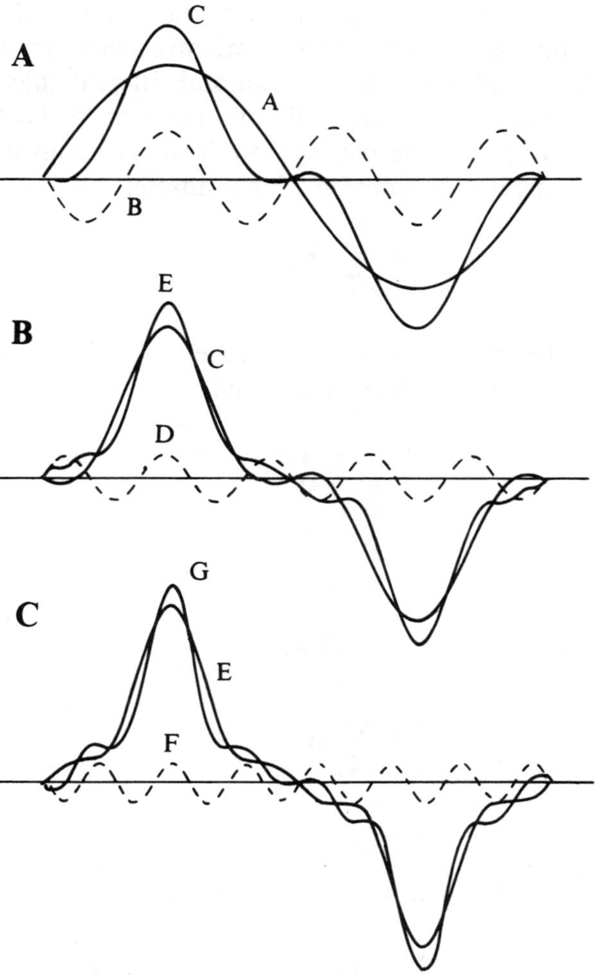

A Fundamental
B 3rd harmonic
C Fundamental plus 3rd harmonic
D 5th harmonic
E Fundamental plus 3rd and 5th harmonic
F 7th harmonic
G Fundamental plus 3rd, 5th, and 7th harmonics

Figure 18-4. Constructing a peaked or triangle wave from a sine wave and its harmonics.

The Fourier series that forms a waveform can be found if a given waveform is decomposed into its constituent frequencies—either by a bank of harmonically related frequency-selective filters or in a computer that uses uses a digital signal-processing algorithm, called the *Fast Fourier Transform (FFT)*.

The Fourier series that is needed to construct a particular waveform from the ground up can be calculated mathematically. In some of the following math, the notation of summation and integral calculus is used. You don't need to know advanced mathematics to understand these concepts. In fact, after you read the passage, don't worry about it again.

The symbol Σ means summation. For example:

$$Y = \sum_{n=1}^{5} X_n \tag{18-4}$$

means to add together all values of X_n together over the range: $n = 1$ to $n = 5$. For example, suppose the data set is:

$$X_1 = 2.5$$
$$X_2 = 0$$
$$X_3 = 1.7$$
$$X_4 = 2.8$$
$$X_5 = 3.4$$

The operation of Eq. (18-4) is as follows:

$$Y = \sum_{n=1}^{5} X_n \tag{18-5}$$

$$Y = X_1 + X_2 + X_3 + X_4 + X_5 \tag{18-6}$$

$$Y = 2.5 + 0 + 1.7 + 2.8 + 3.4 \tag{18-7}$$

$$Y = 10.4 \tag{18-8}$$

Remember these calculus concepts.

1. Integration, symbolized by $\int$, is the math process of finding the area under a curve. A range is specified on some integral symbols; these symbols define the points between which area is taken.

2. For sine waves, the action of integration causes the phase of the sine wave to shift 90 degrees.

The Fourier series for any waveform can be expressed in the form:

$$f(t) = \frac{a_o}{2} + \sum_{n=1}^{\infty} [a_n \mathrm{Cos}(n\omega t) + b_n \mathrm{Sin}(n\omega t)] \tag{18-9}$$

Where:

$a_o/2$ is the dc offset component (if any)
a_n and b_n represent the amplitudes of the harmonics (see below)
n is an integer
$f(t)$ means that the series is a function of time, t

The sine and cosine wave amplitude coefficients in Eq. (18-9), a_n and b_n, are expressed by:

$$a_n = \frac{2}{T}\int_0^t f(t)\,\mathrm{Cos}\,(n\omega t)\,dt \tag{18-10}$$

and,

$$b_n = \frac{2}{T}\int_0^t f(t)\,\mathrm{Sin}(n\omega t)\,dt \tag{18-11}$$

The amplitude terms are nonzero at the specific frequencies determined by the Fourier series. Because only certain frequencies, determined by integer n, are allowable, the spectrum of the periodic signal is said to be *discrete.*

The term $a_o/2$ in the Fourier series expression is the average value of f(t) over one complete cycle (one period) of the waveform. In practical circuit terms, it is also the *dc component* of the waveform. When the waveform possesses *halfwave symmetry* (i.e., the peak amplitude above zero is equal to the peak amplitude below zero at every point in t, or $+V_m = |-V_m|$), no dc component exists, so $a_o/2 = 0$.

An alternative Fourier series expression replaces the $a_n\mathrm{Cos}(nwt) + b_n\mathrm{Sin}(nwt)$ with an equivalent expression of another form:

$$f(t) = \frac{2}{T}\sum_{n=1}^{\infty} C_n\,[n\omega t - \Phi_n] \tag{18-12}$$

Where:

$$C_n = \sqrt{(a_n)^2 + (b_n)^2} \tag{18-13}$$

$$\Phi = \arctan\left(\frac{a_n}{b_n}\right) \tag{18-14}$$

All other terms are as previously defined.

You can infer certain things about the harmonic content of a waveform by examining its symmetries. From the previous equations, you can conclude that the harmonics extend to infinity on all waveforms. Clearly, in practical systems a much less than infinite bandwidth is found, so some of those harmonics will be removed by the normal action of the electronic circuits. Also, the higher harmonics might not be truly significant, so they can sometimes be ignored. As n becomes larger, amplitude coefficients a_n and b_n become smaller. At some point, the amplitude coefficients are reduced so that their contribution to the shape of the wave is either negligible or is totally unobservable in practical terms. The value of n, at which this action occurs, depends partially on the rise time of the waveform. *Rise time* is usually defined as the time required for the waveform to rise from 10 percent to 90 percent of its final amplitude.

The square wave represents another case altogether because it has a very fast rise time. Theoretically, the square wave contains an infinite number of harmonics, but not all of the possible harmonics are present. For example, in the case of the square wave, only the odd harmonics are typically found (e.g., 3, 5, 7). According to some standards, accurately reproducing the square wave requires 100 harmonics, and others claim that 1000 harmonics are needed. Choosing the standard depends on the specifics of the application.

Another factor that determines the profile of the Fourier series of a specific waveform is whether the function is odd or even. The even function is: $f(t) = f(-t)$ and the odd function is: $-f(t) = f(-t)$. In the even function, only cosine harmonics are present, so sine amplitude coefficient b_n is zero. Similarly, in the odd function only sine harmonics are present, so cosine amplitude coefficient a_n is zero.

Waveform symmetry

Both symmetry and asymmetry can occur in several ways in a waveform, and those factors can affect the nature of the Fourier series of the waveform. In Fig. 18-5, you see the case of a waveform with a dc component. Or, in terms of the Fourier-series equation, the term a_o is nonzero. The dc component represents a case of asymmetry in a signal because the upper half (above the zero-volts baseline) is not an exact image of the lower half. The amplitude of the pure square wave is V_a, and $|+V_a| = |-V_a|$. This dc offset can seriously affect instrumentation electronic circuits that are dc-coupled, and can thereby cause serious errors in the output circuit.

Two different forms of symmetry are shown in Fig. 18-6. *Zero-axis symmetry* occurs when, on a point-for-point basis, the waveshape and amplitude above the zero baseline is equal to the amplitude below the baseline: $|+V_m| = |-V_m|$. Both square waves (Fig. 18-6A) and triangle

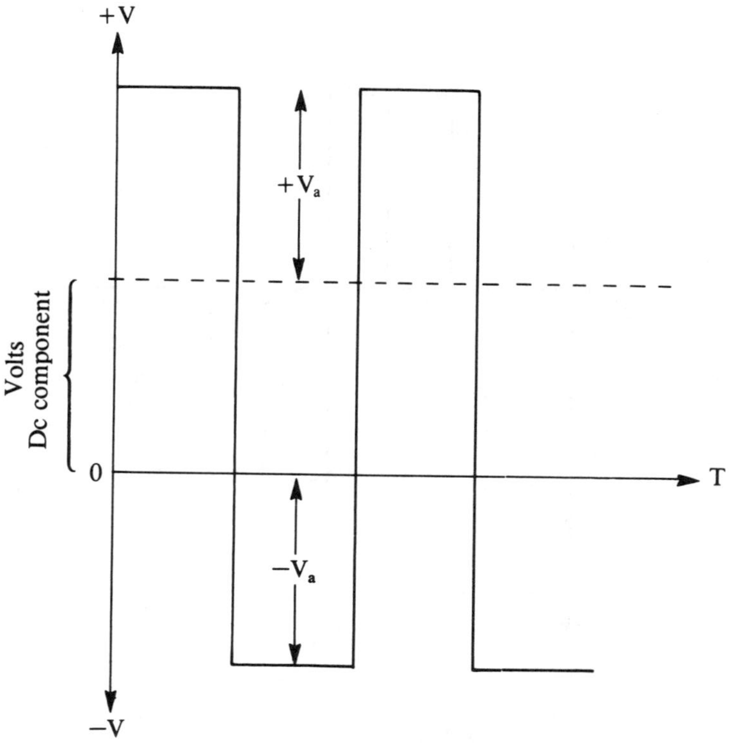

Figure 18-5. A square-wave with a dc component is asymmetrical.

waves (Fig. 18-6B) are shown. When a waveform possesses zero-axis symmetry, it will usually only contain odd harmonics, not even harmonics. This situation is found in square waves and triangle waves (for example, the Fourier spectrum for a square wave in Fig. 18-7).

Note in Fig. 18-6 that the positive and negative amplitudes are equal. This fact and the shape makes the waveform zero-baseline symmetrical, called *halfwave symmetry*. In this type of symmetry, the shape of the wave above the zero baseline is a mirror image of the shape of the waveform below the baseline. Half-wave symmetry also implies a lack of even harmonics.

An exception to the no-even-harmonics rule is that even harmonics are present in the zero-axis symmetrical waveform if the even harmonics are in-phase with the fundamental sine wave. This condition will neither produce a dc component, nor disturb the zero-axis symmetry.

Quarter-wave symmetry exists when the left- and right-half sides of the waveforms are mirror images of each other on the same side of the zero-axis. The ideal square wave meets this requirement (Fig. 18-8A). The vertical dotted line divides the two halves on either side of the zero-baseline. Notice that in these cases, the left and right halves are mirror images of each other.

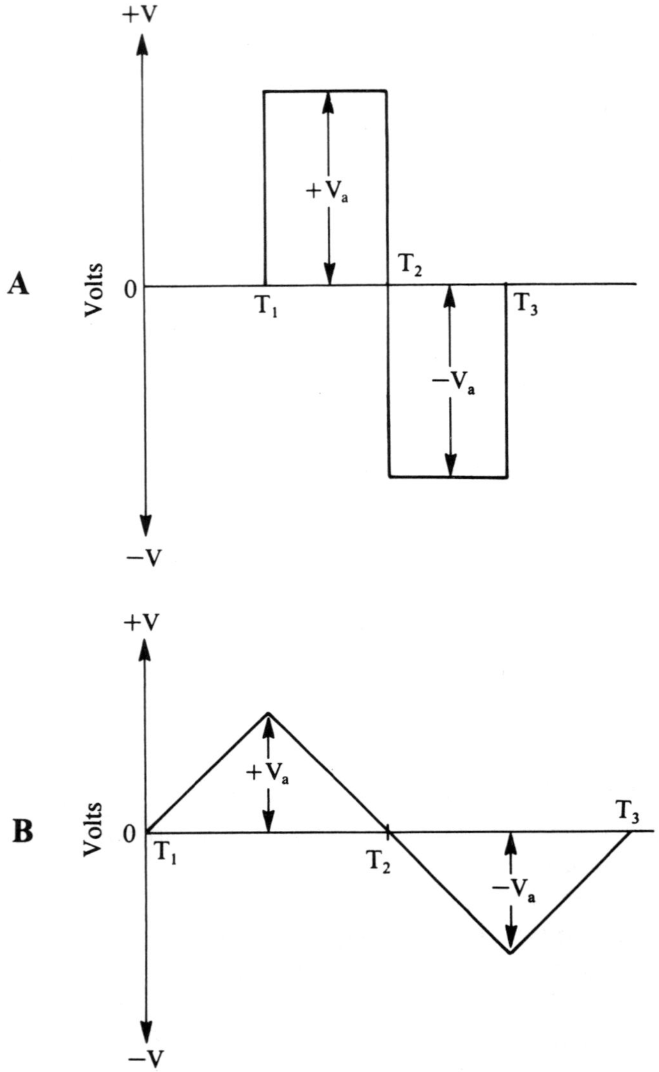

Figure 18-6. Symmetrical waveforms: A) square wave, B) triangle wave.

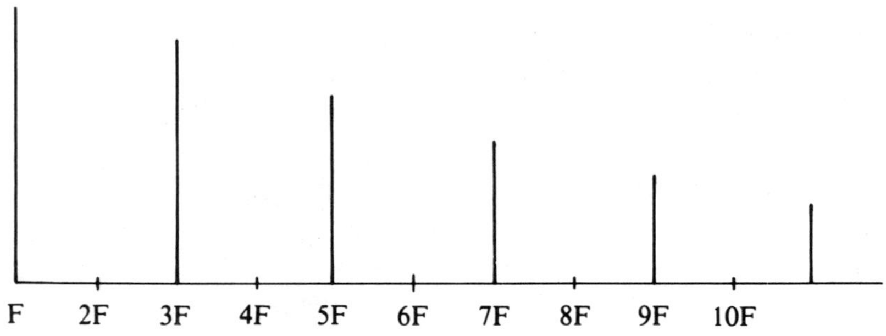

Figure 18-7. Harmonics of a square wave (no even-order harmonics).

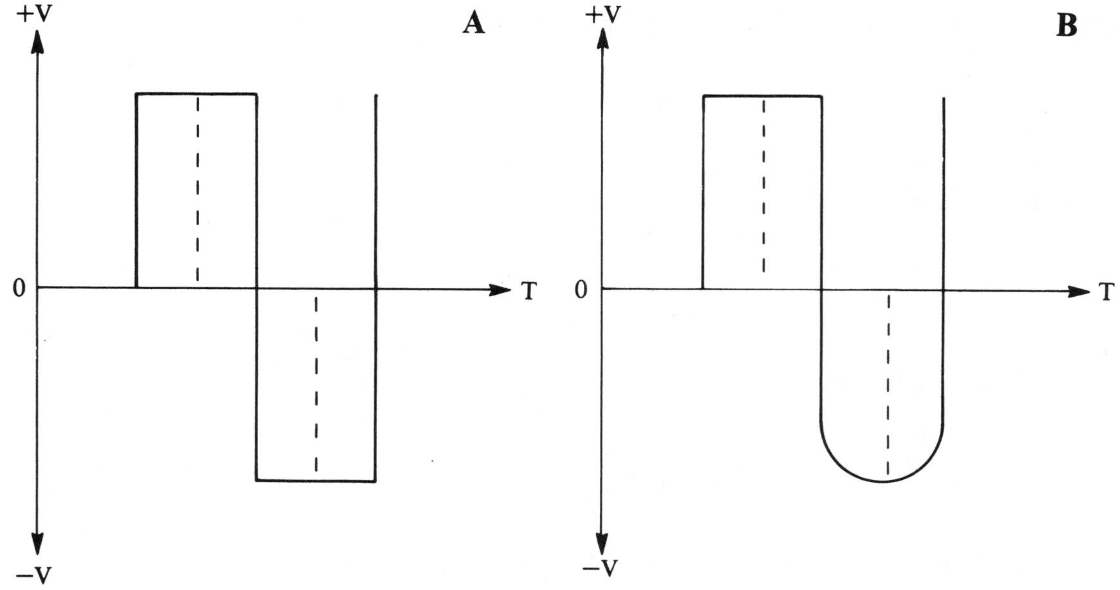

Figure 18-8. A) Square wave with both half-wave and quarter-wave symmetry, B) waveform with quarter-wave symmetry, and half-wave asymmetry.

Note in Fig. 18-8B, however, that waveforms that are not half-wave symmetrical can be quarter-wave symmetrical. In this example, the portion above the zero-axis is like a square wave, and indeed the left- and right-hand sides are mirror images of each other. Similarly, below the zero-axis, the rounded waveform has a mirror-image relationship between the left and right sides. In this case, a full set of even harmonics exists; any odd harmonics that are present are in-phase with the fundamental sine wave.

In the ideal symmetrical square wave, the Fourier spectrum consists of the fundamental frequency (F), plus the odd order harmonics ($3F$, $5F$, $7F$, etc). Furthermore, the harmonics are in-phase with the fundamental. Theoretically, an infinite number of odd-number harmonics are present in the ideal square wave. However, in practical square waves, the "ideal" is considered to be satisfied with harmonics to about $999F$. That ideal is almost never reached, however, as a result of the normal bandwidth limitations of the circuit. An indicator of harmonic content is the rise time of the square waves: the faster the rise time, the larger the number of harmonics.

The circuit for an op-amp square-wave generator is shown in Fig. 18-9A. The operation of this circuit depends on the relationship between $V_{(-In)}$ and $V_{(+In)}$. In the circuit of Fig. 18-9A, the voltage that is applied to the noninverting input ($V_{(+In)}$ is determined by a resistor voltage divider, R2 and R3). This voltage is called V_1 in Fig. 18-9A and is:

$$V_1 = \frac{V_o R_3}{R_2 + R_3} \tag{18-15}$$

or, when V_o is saturated:

$$V_1 = \frac{V_{sat} R_3}{R_2 + R_3} \tag{18-16}$$

Once again, the factor $R_3/(R_2 + R_3)$ is often designated β:

$$\beta = \frac{R_3}{R_2 + R_3} \tag{18-17}$$

Because Eq. (18-17) is always a fraction: $V_1 < V_{sat}$ and V_1 is of the same polarity as V_{sat}.

The voltage applied to the inverting input ($V_{(-In)}$) is the voltage across capacitor C1, i.e., V_{C1}. This voltage is created when C1 charges under the influence of current I, which in turn is a function of V_o and the time constant of R_1C_1. Timing operation of the circuit is shown in Fig. 18-9B.

At turn-on, $V_{C1} = 0$ volts and $V_o = +V_{sat}$, so $V_1 = +V_1 = \beta(+V_{sat})$. Because $V_{C1} < V_1$, the op amp sees a negative differential-input voltage, so the output remains at $+V_{sat}$. During this time, however, V_{C1} is charging towards $+V_{sat}$ at a rate of:

$$V_{C1} = V_{sat}(1 - \epsilon^{-t2/R1C1}) \tag{18-18}$$

When V_{C1} reaches $+V_1$, however, the op amp sees $V_{C1} = V_1$, so $V_{id} = 0$. The output now snaps from $+V_{sat}$ to $-V_{sat}$ (time T_2 in Fig. 18-9B). The capacitor now begins to discharge from $+V_1$ toward zero, and then re-charges toward $-V_{sat}$. When it reaches $-V_1$, the inputs once again are zero, so the output again snaps to $+V_{sat}$. The output continuously snaps back and forth between $-V_{sat}$ and $+V_{sat}$, thereby producing a square-wave output signal.

The time-constant that is required to charge from initial voltage V_{C1} to end voltage V_{C2} in time T is:

$$RC = \frac{-T}{Ln\left(\frac{V - V_{c2}}{V - V_{c1}}\right)} \tag{18-19}$$

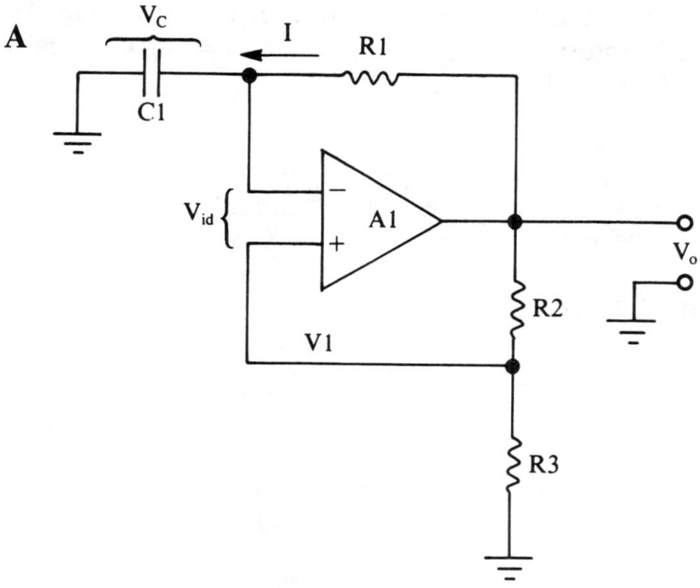

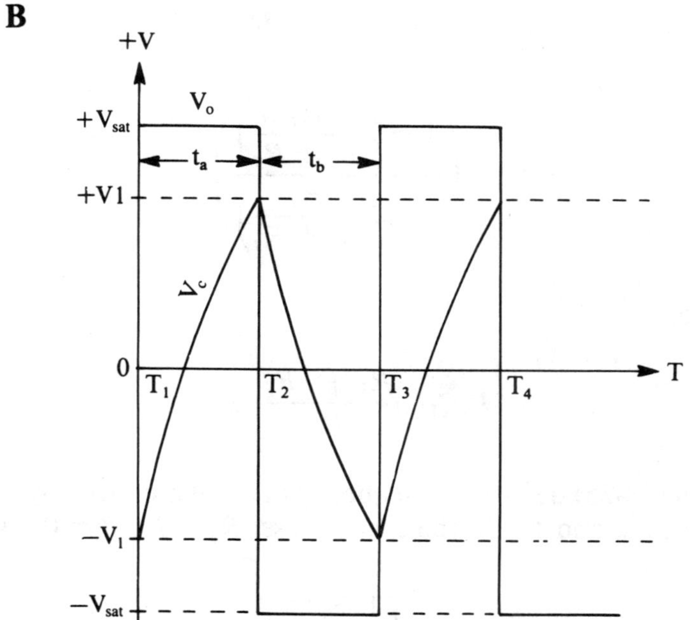

Figure 18-9. A) Operational amplifier square-wave oscillator, B) timing waveforms.

In Fig. 18-9A, the RC time constant is R_1C_1. From Fig. 18-9B, it is apparent that for interval T_a: $V_{C1} = -\beta V_{sat}$, $V_{C2} = +\beta V_{sat}$, and $V = V_{sat}$. To calculate the period T, then:

$$2R_1\,C_1 = \frac{-T}{Ln\left(\dfrac{V_{sat} - \beta\,V_{sat}}{V_{sat} - (-\beta\,V_{sat})}\right)} \tag{18-20}$$

or, rearranging Eq. (18-20):

$$-T = 2R_1C_1Ln\left(\frac{V_{sat} - \beta\,V_{sat}}{V_{sat} + \beta\,V_{sat}}\right) \tag{18-21}$$

$$-T = 2R_1C_1\,Ln\left(\frac{1-\beta}{1+\beta}\right) \tag{18-22}$$

$$T = 2R_1C_1Ln\left(\frac{1+\beta}{1-\beta}\right) \tag{18-23}$$

Because $\beta = R_3/(R_2 + R_3)$:

$$T = 2R_1C_1Ln\left(\frac{1+\left(\dfrac{R_3}{R_2+R_3}\right)}{1-\left(\dfrac{R_3}{R_2+R_3}\right)}\right) \tag{18-24}$$

which reduces to:

$$T = 2R_1C_1Ln\frac{R_2 + 2R_3}{R_2} \tag{18-25}$$

Equation (18-25) defines the frequency of oscillation for any combination of R_1, R_2, R_3, and C_1. In the special case: $R_2 = R_3$, $\beta = 0.5$ so:

$$T = 2R_1C_1Ln\left(\frac{1+0.5}{1-0.5}\right) \tag{18-26}$$

$$T = 2R_1C_1Ln\left(\frac{1.5}{0.5}\right) \tag{18-27}$$

$$T = 2R_1C_1Ln(3) = 2R_1C_1\ (1\ 1) = 2.2R_1C_1 \tag{18-28}$$

$$T = 2.2R_1C_1 \tag{18-29}$$

and, since $F = 1/T$:

$$F = \frac{0.4545}{R_1C_1} \tag{18-30}$$

Experiment 18-1

This experiment shows the basic form of square-wave generator circuit. It will produce a symmetrical square wave, in which the amplitudes of positive and negative swings are equal. This experiment does not require an oscilloscope because it uses a pair of LEDs as output indicators (see the circuit in Experiment 18-1).

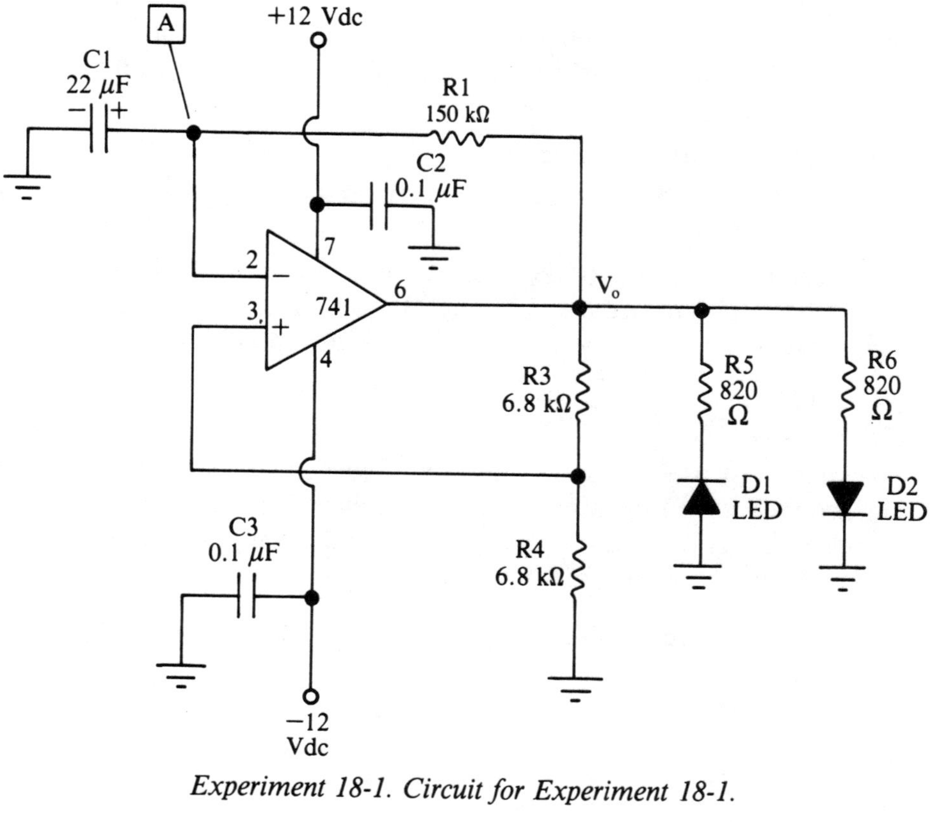

Experiment 18-1. Circuit for Experiment 18-1.

A LED is a special pn-junction diode that emits light when it is forward-biased. In this experiment, we will use two red LEDs wired back to back. In Experiment 18-1, the two LEDs are shown connected across the output of the operational amplifier. Each LED has a resistor (R5 and R6) in series in order to limit the current to less than 20 mA. When the output voltage V_o is low (i.e., at $-V_{sat}$), diode D1 is forward-biased and it emits light; when the output is high (i.e., at $+V_{sat}$), the other diode (D2) is forward-biased (on).

1. Connect the circuit of Experiment 18-1. Use a capacitor for C1 that is rated for between 16 and 30 μF and a resistance of 120 to 220 kohms.
2. Calculate the time: $T = 2.2R_1C_1$.
3. Turn on the circuit by applying ± 12 Vdc.
4. If the circuit works properly, the LEDs will alternate on and off at a rate equal to time T; measure the time with the second hand of a watch. If only one LED turns on, or if neither LED turns on, assume that either some wiring is incorrect or an op amp is bad.
5. Shunt a second timing resistor across R1 so that its value is approximately 50 percent of the first value (i.e., parallel to another resistor of the same value across R1.
6. Calculate the new time, T.
7. Measure T with a sweep second hand and compare it to the predicted value.
8. Change the values of the capacitor and resistor to various combinations and repeat the experiment. If you select an RC time constant that is too short, the LEDs will blink too rapidly.

Experiment 18-2

This experiment uses the same circuit as Experiment 18-1, except that an oscilloscope is used to examine the output waveform. If you have a two-channel oscilloscope, connect one probe across the output of the op-amp, and the other across the timing capacitor.

1. Build the circuit of Experiment 18-1 with the following values for R_1 and C_1:

$$R_1 = 47 \text{ kohms}$$
$$C_1 = 0.01\ \mu\text{F}$$

2. Calculate the value of T.

3. Examine the output waveform on an oscilloscope.
4. Measure the period of the waveform and compare it with step 2.
5. If you have a two-channel oscilloscope, then connect the second input across the timing capacitor, C1.
6. Compare the capacitor waveform with the output waveform, paying attention to the amplitude and timing aspects.
7. What can you tell about the feedback voltage from this comparison? Does this tell you the ratio of R_3/R_4?

Time asymmetrical square waves

The circuit of Fig. 18-9A produces time-line symmetrical square waves (i.e., $t_a = t_b$). If time-line asymmetrical square waves are required, then a circuit, such as either Fig. 18-10 or Fig. 18-12A is required. The circuit in Fig. 18-10 uses a potentiometer (R4) and a fixed resistor (R5) to establish a variable

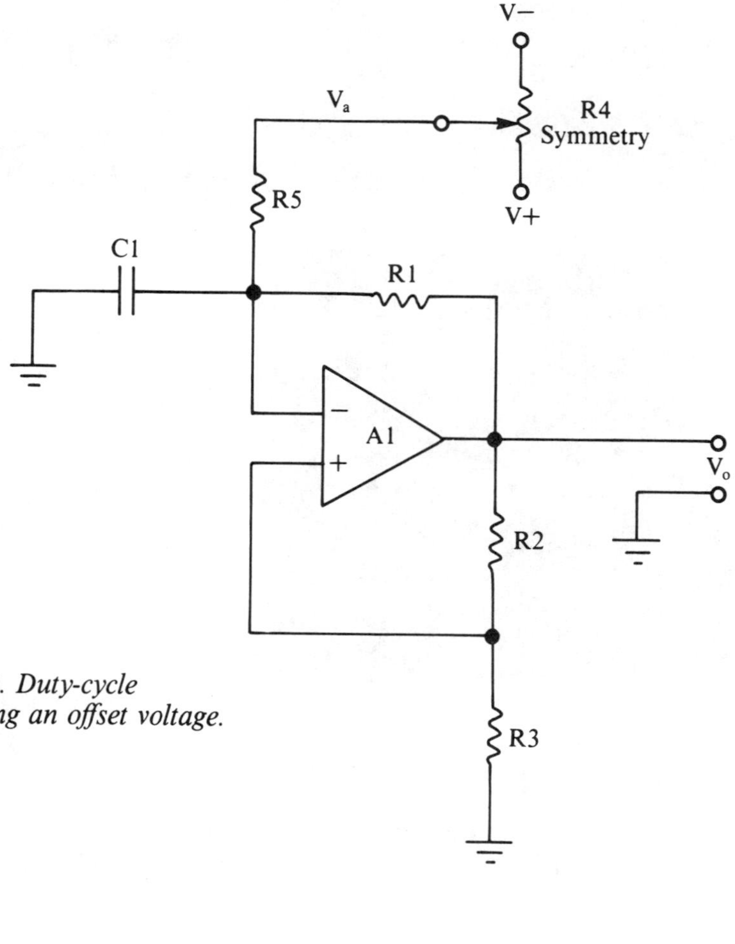

Figure 18-10. Duty-cycle variation using an offset voltage.

duty-cycle asymmetry. The circuit is similar to Fig. 18-9A, but with an offset circuit (R_4/R_5) added. The assumptions are $R_5 = R_1$, and $R_4 << R_1$. If V_a is the potentiometer output voltage, C1 charges at a rate of $(R_1/2)C_1$ toward a potential of $(V_a + V_{sat})$. After output transition, however, the capacitor discharges at the same $(R_1/2)C_1$ rate toward $(V_a - V_{sat})$. The two interval times are therefore different; t_a and t_b are no longer equal.

Figure 18-11 shows three extremes of V_a: $V_a = +V$ (Fig. 18-11A), $V_a = 0$ (Fig. 18-11B) and $V_a = -V$ (Fig. 18-11C). These traces represent very long, equal, and very short-duty cycles, respectively.

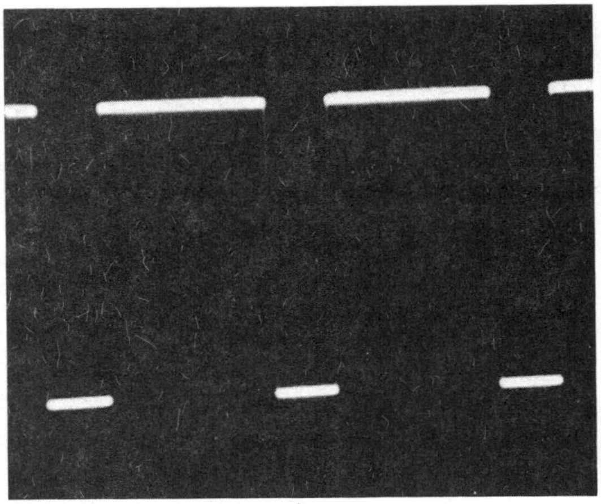

A

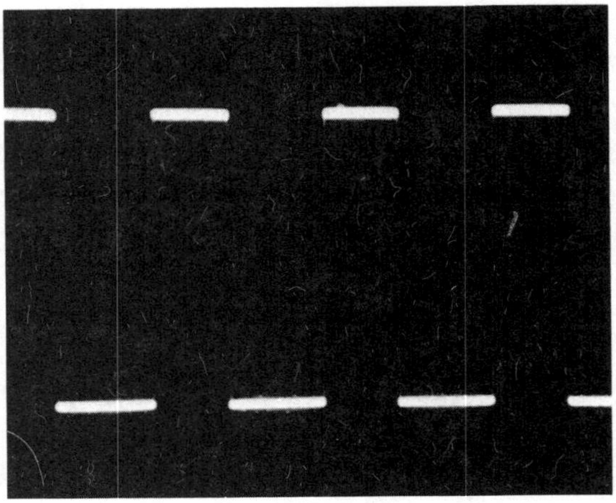

B

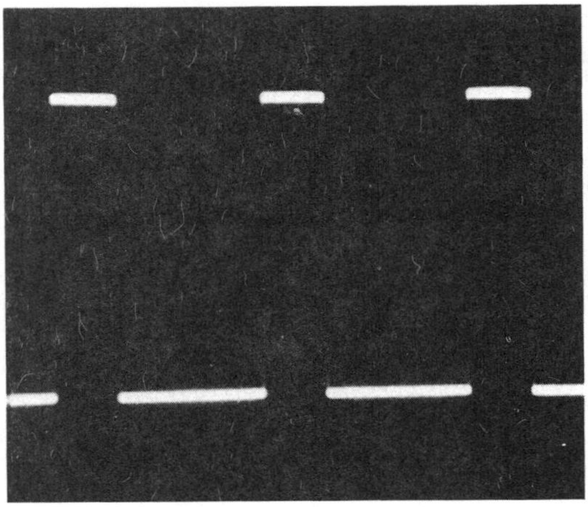

C

Figure 18-11. Three settings of potentiometer (Fig. 18-10), voltage output is: A) maximum negative, B) zero, C) maximum positive.

Experiment 18-3

This circuit examines the asymmetrical square-wave generator circuit. The symmetry of the waveform is set by a potentiometer and fixed resistor added to the circuit of Experiment 18-1. These components (Experiment 18-2) are connected into the original circuit at point A. All other connections remain the same.

1. Build and test the circuit used in Experiment 18-1; use the values recommended and record the timing. Turn the circuit off when the testing is finished.
2. Add the circuit of Experiment 18-2.

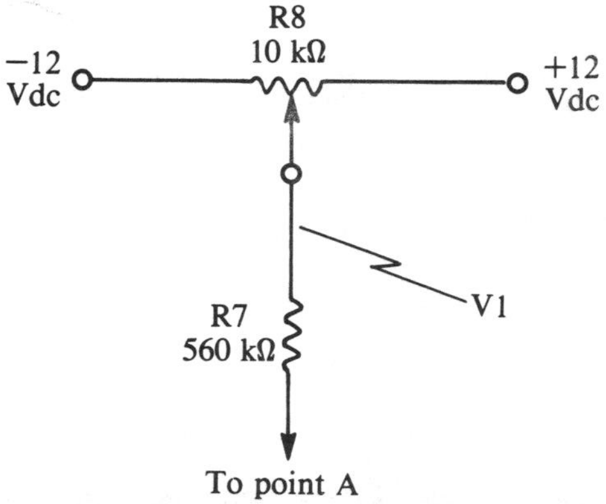

Experiment 18-2. Circuit for Experiment 18-2.

3. Turn the circuit on.
4. Use a voltmeter to measure the voltage at the wiper of the potentiometer (V_1 in Experiment 18-2). Set the voltage $V_1 = 0$.
5. Observe the output LEDs and compare the timing with the original configuration.
6. Adjust the potentiometer R8 to set V_1 to its maximum positive value.
7. Repeat Step 5.
8. Readjust the potentiometer for maximum negative voltage at V_1.
9. Repeat Step 5.
10. Compare the timing data for all three conditions.
11. Experiment with various settings of R8, and with different values of R7.

Readers who own an oscilloscope can perform this experiment with the component values used in Experiment 18-2.

The circuit of Fig. 18-12A also produces asymmetrical square waves, but the duty cycle is fixed, rather than variable. Once again the basic circuit is like Fig. 18-9A, but with added components. In Fig. 18-12A, the RC

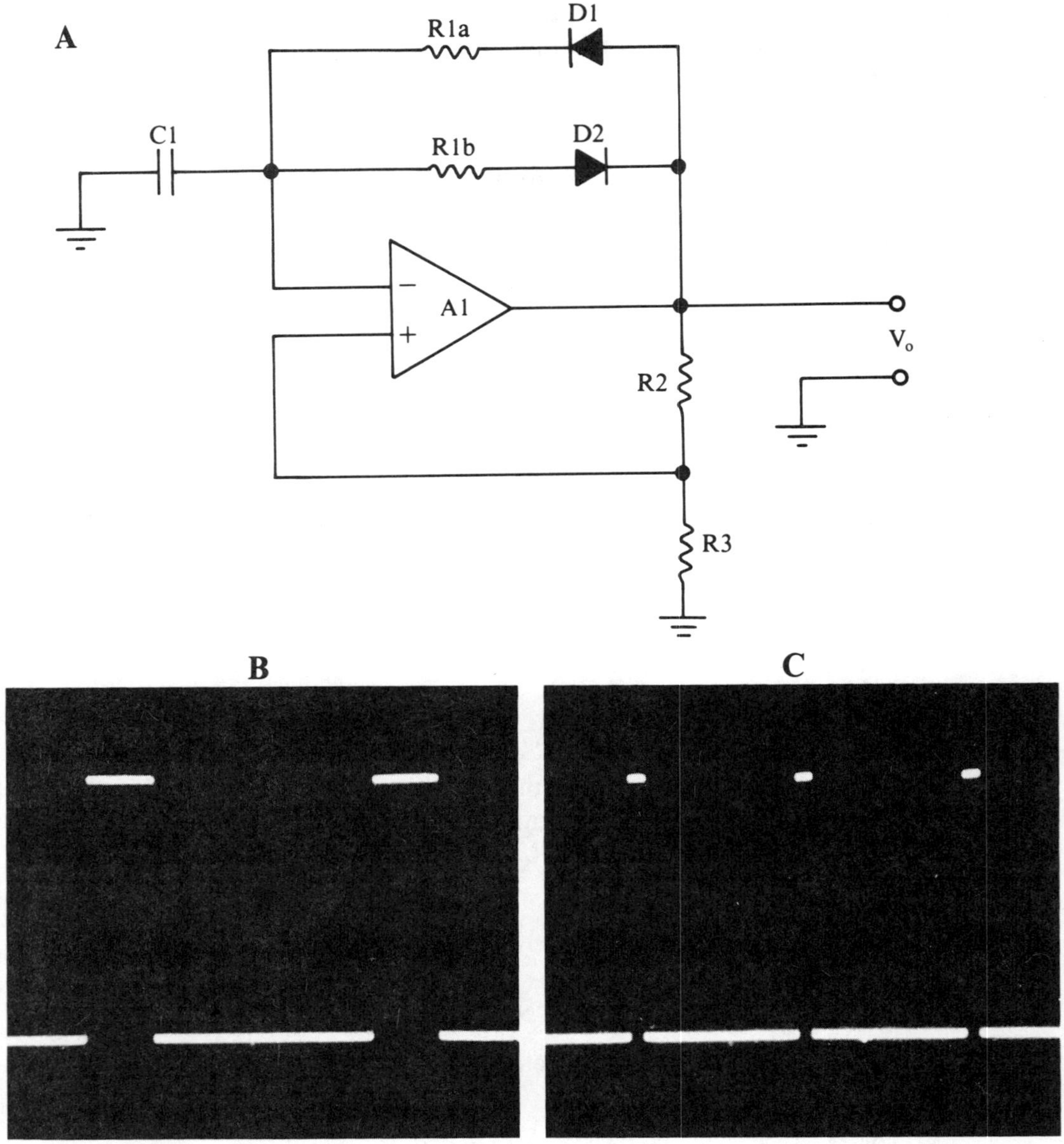

Figure 18-12. A) Fixed asymmetrical duty cycle: B) 3:1 ratio, C) 10:1 ratio.

timing network is altered so that the resistors are different on each swing of the output signal. During t_a, $V_a = +V_{sat}$, so diode D1 is forward-biased and D2 is reverse-biased. For this interval:

$$t_a = (R_{1A})\,(C_1)\,Ln\left(1 + \frac{2\,R_2}{R_3}\right) \qquad (18\text{-}31)$$

During the alternate half-cycle (t_b), the output voltage V_o is at $-V_{sat}$, so D1 is reverse-biased and D2 is forward-biased. During this interval, R1B is the timing resistor, and R1A is effectively out of the circuit. The timing equation is:

$$t_b = (R_{1B})\,(C_1)\,\mathrm{Ln}\left(1 + \left(\frac{2\,R_2}{R_3}\right)\right) \tag{18-32}$$

The total period, T, is $t_a + t_b$, so:

$$T = (R_{1A})\,(C_1)\,Ln(+\left(\frac{2\,R_2}{R_3}\right)) + (R_{1B})\,(C_1)\,Ln\left(1 + \left(\frac{2\,R_2}{R_3}\right)\right) \tag{18-33}$$

Collecting terms:

$$T = (R_{1A} + R_{1B})\,C_1\,Ln\left(1 + \frac{2\,R_2}{R_3}\right) \tag{18-34}$$

Equation (18-32) defines the oscillation frequency of the circuit in Fig. 18-12A. Figures 18-12B and 18-12C show the effects of two values of the R_{1A}/R_{1B} ratio. In Fig. 18-12B, the ratio $R_{1A}/R_{1B} = 3{:}1$; in Fig. 18-12C, the ratio $R_{1A}/R_{1B} = 10{:}1$.

The effect of this circuit on capacitor charging can be seen in Fig. 18-13. A relatively low R_{1A}/R_{1B} ratio is seen in Fig. 18-13A. Notice in the lower trace that the capacitor charge time is long compared with the discharge time. An even greater effect is seen in the case of a high R_{1A}/R_{1B} ratio (Fig. 18-13B).

A **B**

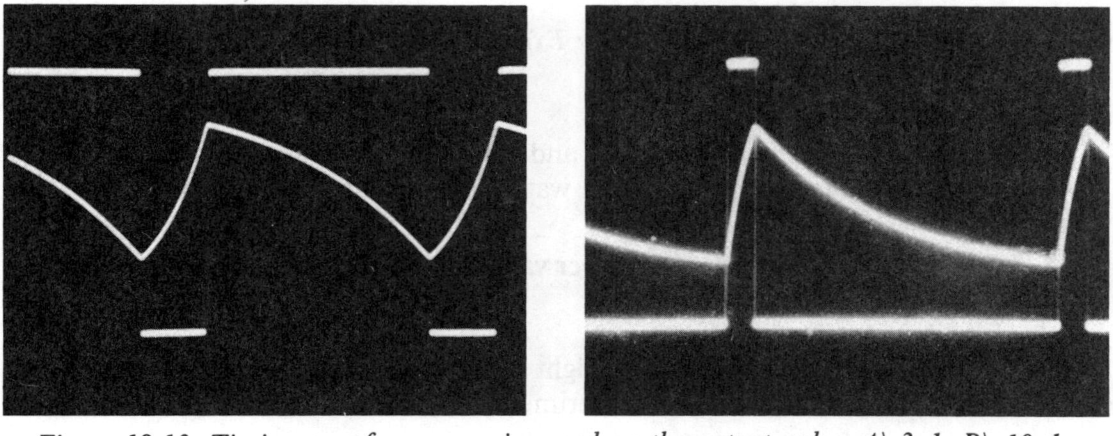

Figure 18-13. Timing waveforms superimposed on the output pulse: A) 3:1, B) 10:1.

Experiment 18-4

This experiment deals with the other method for producing asymmetrical waveforms: the diode-controlled switch. This experiment is based on Experiment 18-1.

1. Wire and test the circuit of Experiment 18-1. Turn the circuit off when testing is completed.
2. Modify the circuit of Experiment 18-1 in the manner shown in Experiment 18-3. First, place a diode in series with R1 (which is now relabelled "R1a"). Next, add a second feedback circuit that consists of R1b and D2. Notice that the diodes are connected backwards, with respect to each other. The direction of the arrow, with respect to the actual diode, is shown in the inset to Experiment 18-3.

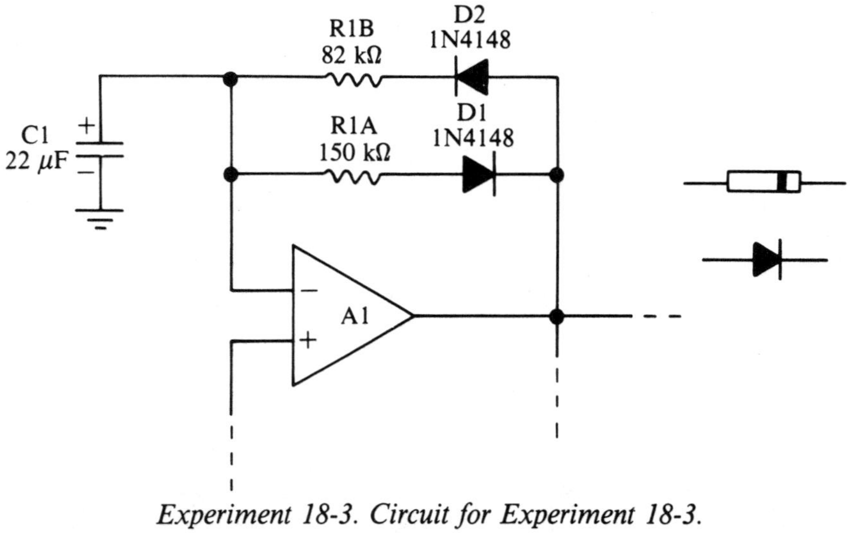

Experiment 18-3. Circuit for Experiment 18-3.

3. Turn on the circuit.
4. Observe the timing of both LEDs and compare their respective on periods. Use the second hand of a watch to make timing measurements.
5. Change R_{1b} to both higher and lower values and compare the results to the original measurement.

Readers who own an oscilloscope might wish to perform this experiment with the component values in Experiment 18-2.

Output voltage limiting

The standard op-amp MMV or AMV circuit sometimes produces a relatively sloppy square output wave. By adding a pair of back-to-back zener diodes (Fig. 18-14) across the output, however, the signal can be cleaned up (at the expense of amplitude). For each polarity, the output signal sees one forward-biased and one reverse-biased zener diode. On the positive swing, the output voltage is clamped at $[V_{Z1} + 0.7]$ volts. The 0.7-volts factor represents the normal junction potential across the forward-biased diode (D2). On negative swings of the output signal, the situation reverses. The output signal is clamped to $[-(V_{Z2} + 0.7)]$ volts.

Experiment 18-5

Perform Experiment 18-2, but modify the circuit as shown previously in Fig. 18-14. The series resistor (R4) should have a value of 1000 ohms, while the zener diodes should be 5.6 to 6.8 volt units (make them identical). Compare the waveforms (as to rise time, rounding, and amplitude) at the output of the op amp (pin 6) and across the diodes.

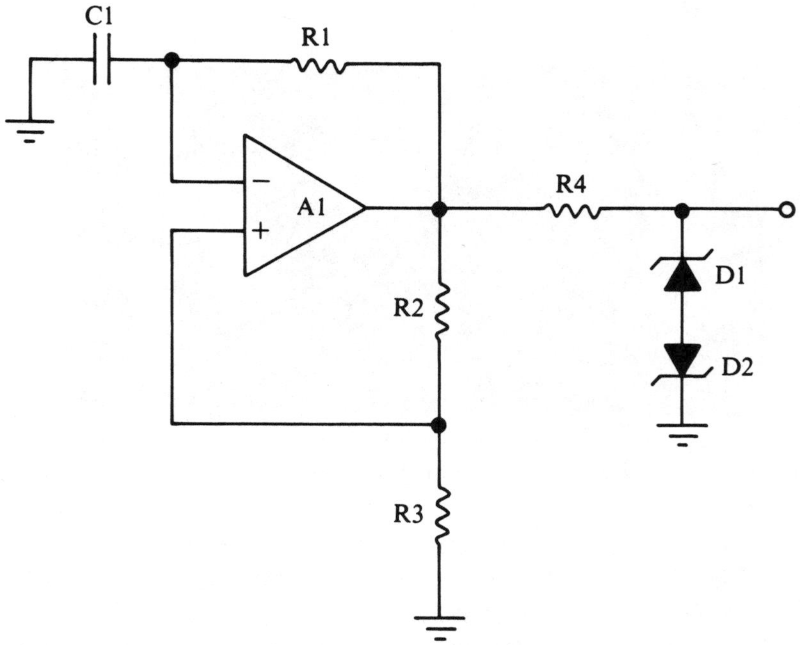

Figure 18-14. Square-wave circuit with output limiting.

Square waves from sine waves

Figure 18-15 shows a method for converting sine waves to square waves. This circuit is shown in Fig. 18-15A and the waveforms are shown in Fig. 18-15B. The circuit is an op amp that is connected as a comparator. Because the op amp has no negative-feedback path, the gain is very high (i.e., A_{vol}); in op amps, gains of 20,000 to 2,000,000 (250,000 typical) are found. Thus, a voltage difference across the input terminals of only a few millivolts will saturate the output. From this behavior, the operation of the circuit and the waveform in Fig. 18-8B can be understood.

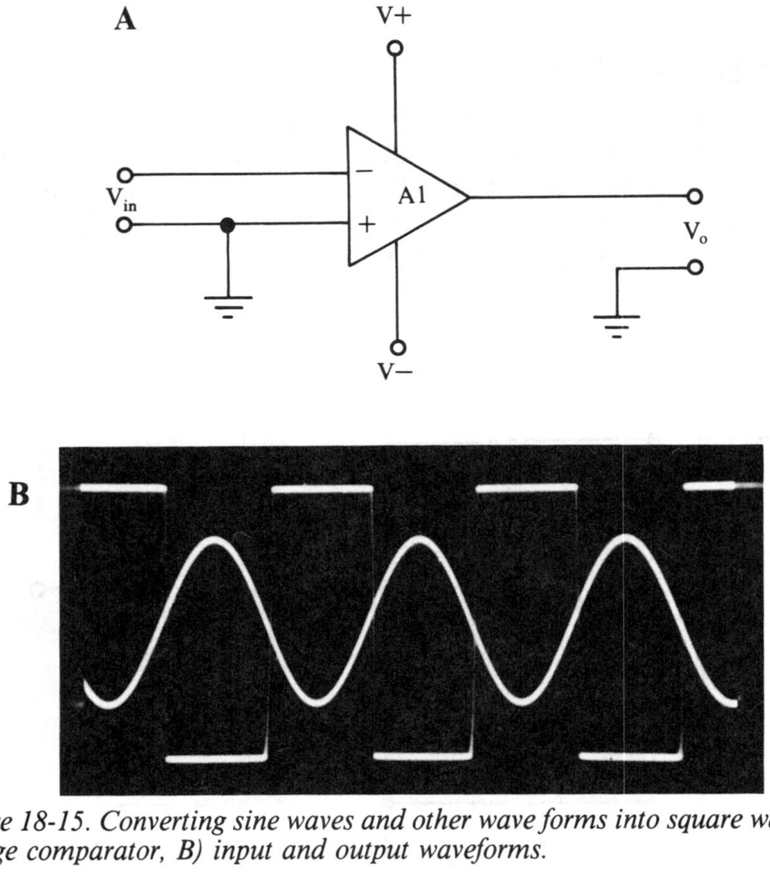

Figure 18-15. Converting sine waves and other wave forms into square waves: A) basic circuit is a voltage comparator, B) input and output waveforms.

The input waveform is a sine wave. Because the noninverting input is grounded (Fig. 18-15A), the output of the op amp is zero only when the input signal voltage is also zero. When the sine wave is positive, the output signal will be at $-V_o$; when the sine wave is negative, the output signal will be at $+V_o$. The output signal will be a square wave at the sine-wave frequency, with a peak-to-peak amplitude of $[(+V_o) - (-V_o)]$.

Project 18-1
Variable square-wave generator

This circuit puts your knowledge square-wave generator circuits into practice. The circuit of Project 18-1 is a two-stage oscillator-amplifier that is based on the RCA/GE CA-3240 BiMOS op amp. The CA-3240 is a dual version of the famous CA-3140 device; it is similar in form (but not capability) to the lower-grade 1456 devices (which are 741-family op amps).

The frequency of the square-wave generator is set by a complex network that consists of R1, R2 (the adjust-frequency control), switch S1 (which selects decade differences in capacitance), and timing capacitors C1 through C6. Switch S1 selects the frequency range, while potentiometer R2 selects the exact frequency within that range. If a scale that is connected to R2 is calibrated for one range, it will also serve for the other ranges, because a 10 : 1 ratio exists between the ranges.

The output circuit consists of the output-limiting circuit for A1a (zener diodes D1 and D2), a level-set control and a unity-gain noninverting follower that is used as a buffer amplifier.

The ±-Vdc power supplies are connected to the op amp at pins 4 (for −12 volts) and 8 (for +12 volts). Each power-supply terminal is bypassed by a 0.1-μF capacitor. This capacitor must be placed as close as possible to the body of the op amp. If the circuit rings or oscillates at a high frequency (>100 kHz), increase the value of the bypass capacitors.

Summary

An astable multivibrator circuit is a form of oscillator that generally creates nonsinusoidal signals. These signals may be square, triangle, or sawtooth waves. Each of these signals may be considered the sum of a series of sine waves; this is known as a Fourier series. Because these waves are at different frequencies, amplifiers must be capable of wideband operation in order to accurately reproduce the shape of the input signal. Square waves can be produced by using an RC network and an op amp. Placing an additional voltage on the capacitor converts the square wave from symmetrical to asymmetrical.

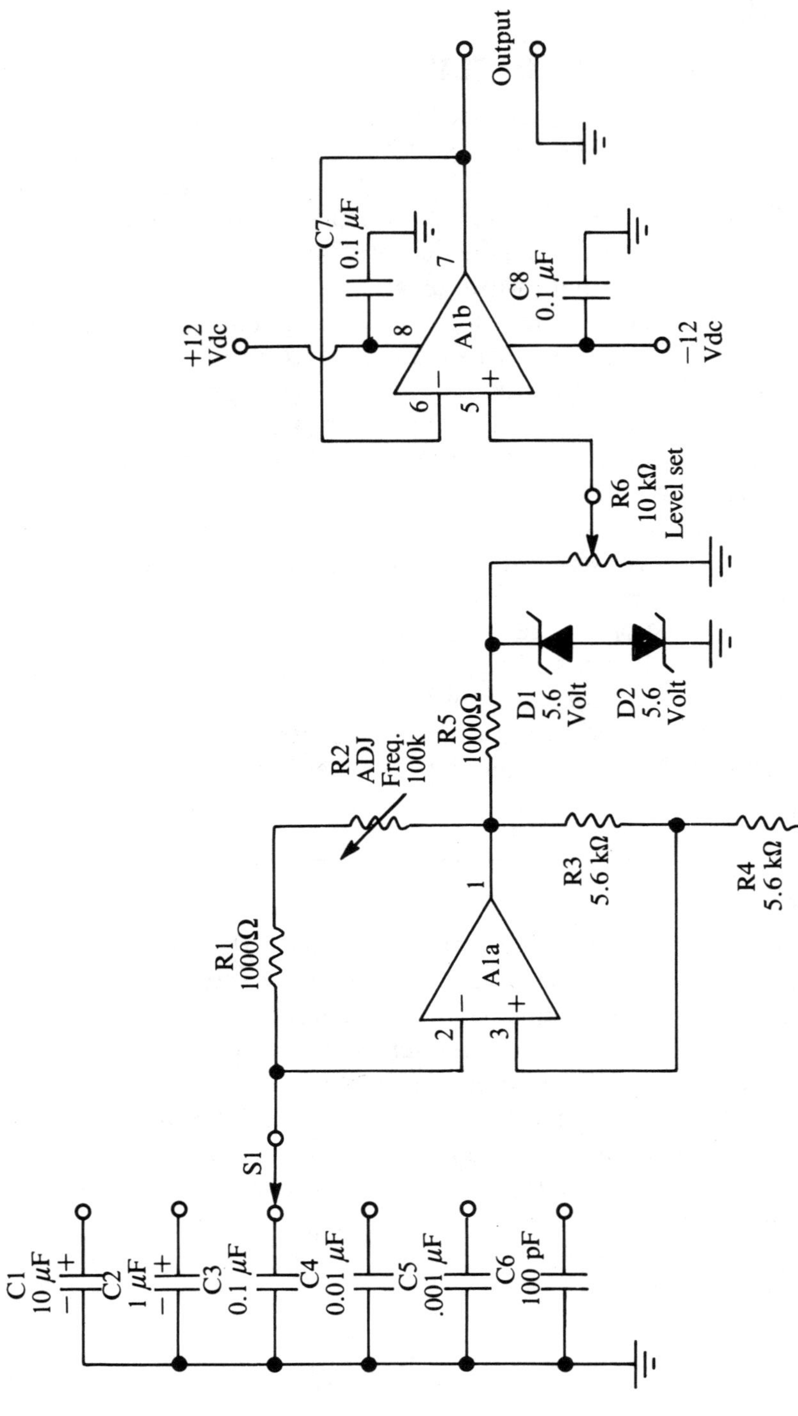

Project 18-1. Multi-frequency square-wave generator.

Questions

1. What is an astable multivibrator?
2. What is meant by waveform symmetry?
3. What are the characteristics of a totally symmetrical square wave?
4. What is a Fourier series?
5. Write the Fourier series for any periodic waveform.
6. What are zero-axis symmetry, half-wave symmetry, and quarter-wave symmetry?

Problems

1. For the circuit in Fig. 18-16, R_1 = 10 kohms, R_2 = 20 kohms, R_3 = 100 kohms, C = 0.1 μF. Calculate **a.** maximum and minimum output voltages **b.** maximum and minimum voltages at point A **c.** frequency

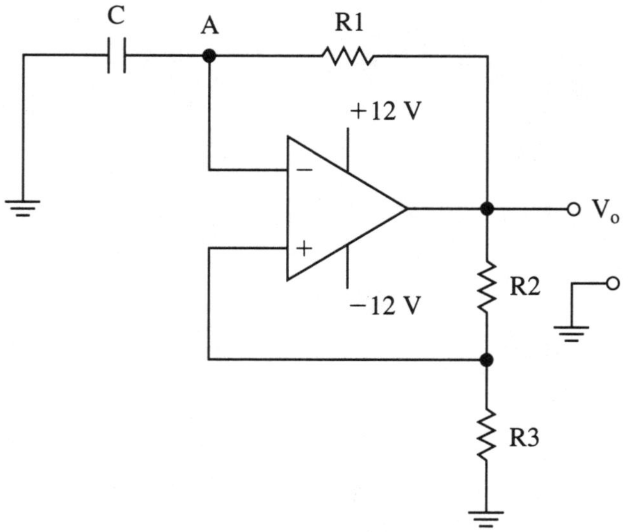

Figure 18-16.

2. Repeat Problem 1 for R_1 = 20 kohms, R_2 = 450 kohms, R_3 = 50 kohms, and C = 2.2 μF.
3. For Problem 1, sketch the capacitor and output waveforms.
4. For Problem 2, sketch the capacitor and output waveforms.

19 TRIANGLE- AND SAWTOOTH-WAVE GENERATORS

AFTER COMPLETING THIS chapter, you will be able to:

- ☐ Explain the generation and characteristics of triangle waves
- ☐ Explain the generation and characteristics of sawtooth waves

The triangle and sawtooth waveforms (Fig. 19-1) are examples of *periodic ramp functions*. The sawtooth (Fig. 19-1A) is a single-ramp waveform. The voltage begins to rise linearly at time t_1 until it reaches a maximum. At time t_2, the waveform abruptly drops back to zero, where it again starts to ramp up linearly. The sawtooth is usually periodic, although single-sweep and triggered-sweep variants are sometimes seen. The period is designated T (see Fig. 19-1A), the frequency (F) is $1/T$.

The triangle waveform (Fig. 19-1B) is a double ramp. The waveform begins to ramp up linearly at time t_1. It reverses direction at time t_2, then ramps downward linearly until time t_3. At time t_3, the waveform again reverses direction and begins ramping upwards. The period of the triangle waveform (T) is $T_3 - T_1$, and again frequency (F) is $1/T$, or $1/(T_3 - T_1)$.

Ramp generators are derived from capacitor-charging circuits. The familiar RC charging curve was discussed in chapter 16, and it is reproduced in simplified form in Fig. 19-1C. The RC-charging waveform has an exponential shape, so it is not well-suited to generate a linear-ramp function. Two approaches force the capacitor-charging waveform time to be more linear. The first is to limit the charging time to the short quasilinear segment, shown in Fig. 19-1C. The obtained ramp is not very linear, is limited in amplitude to a small fraction of V_1, and has a relatively steep

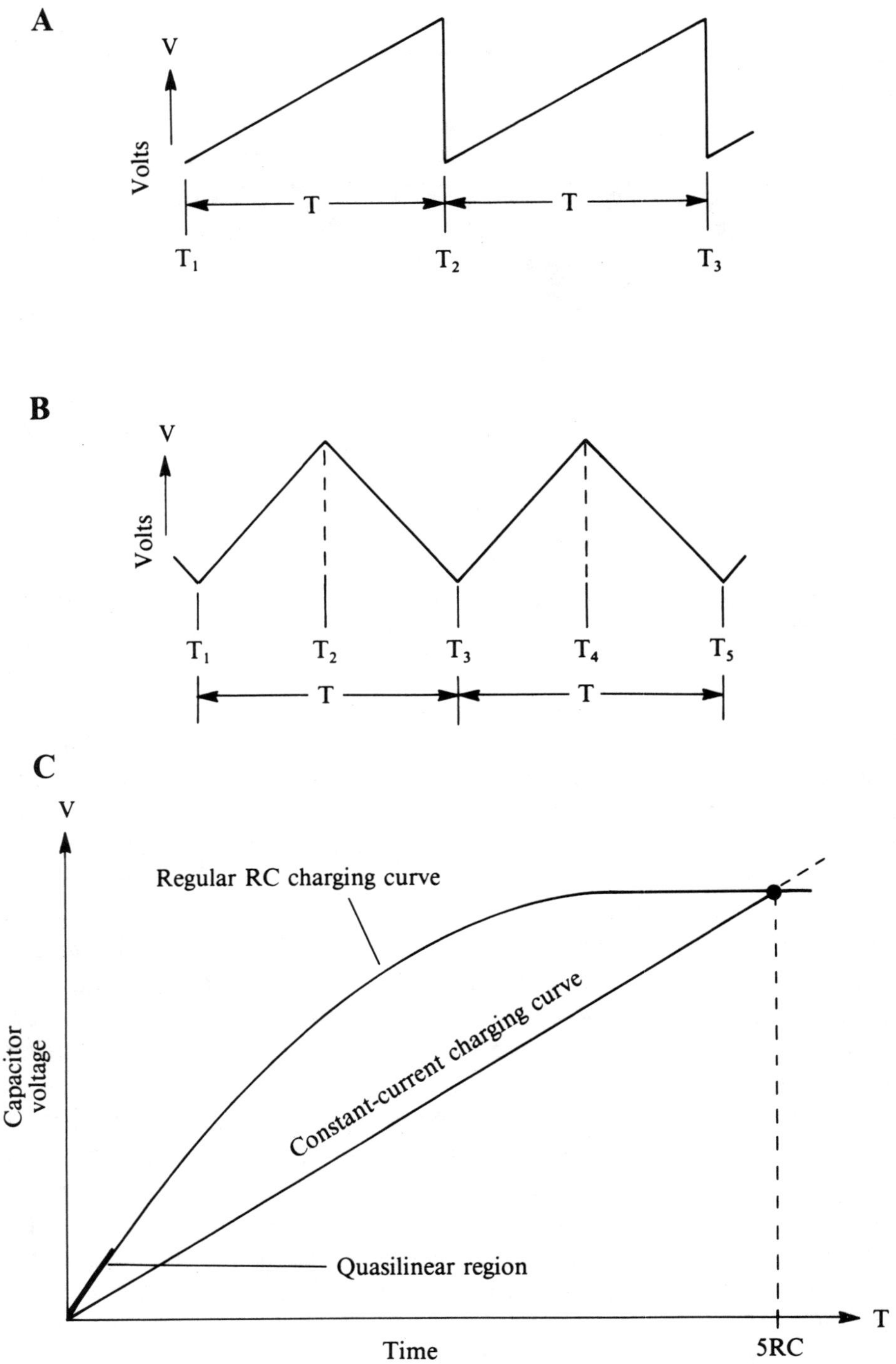

Figure 19-1. A) Sawtooth wave form, B) triangle wave form, C) generating a ramp from capacitor charging waveforms.

slope that might not be useful for any given application. A superior method is to charge the capacitor through a constant-current source (CCS). The linear ramp that is shown in Fig. 19-1C was made by using the CCS to charge the capacitor.

Triangle and sawtooth waveform oscillators create the constant-current form of ramp generator by using a *Miller-integrator circuit* to charge the capacitor (Fig. 19-2A). When a Miller integrator is driven by a stable reference voltage, the output is a linearly rising ramp. The ramp voltage (V_o) is:

$$V_o = \frac{V_{ref}}{T} \qquad (19\text{-}1)$$

or, because $T = RC$:

$$V_o = \frac{V_{ref}}{RC} \qquad (19\text{-}2)$$

If $V_{ref} = +10$ Vdc and the RC time-constant is $T = RC = 0.001$ seconds, the ramp slope is:

$$V_o = \frac{10 \text{ volts}}{0.001 \text{ sec}} \qquad (19\text{-}3)$$

$$V_o = 10 \text{ volts/millisecond} \qquad (19\text{-}4)$$

Triangle generator circuits

Figure 19-2A shows a simplified circuit model of a triangle waveform generator and Fig. 19-2B shows the expected waveforms. This circuit consists of a Miller integrator as the ramp generator and an SPDT switch (S1) that can select either positive ($+V_{ref}$) or negative ($-V_{ref}$) reference voltages. For purposes of this discussion, S1 is an electronic switch that is toggled between positions A and B by a square wave that is applied to the control terminal (CT). Assume an initial condition at time t_2 (at which point $V_o = -V_1$) and the input of the integrator is connected to $-V_{ref}$. At time t_2, the square-wave switch driver changes to the opposite state, so S1 toggles to connect $+V_{ref}$ to the integrator input. The ramp output will rise linearly at a rate of $+V_{ref}$/RC, until the switch again toggles at time t_3. At this point, the ramp is under the influence of $-V_{ref}$, so drops linearly from $+V_1$ to $-V_1$. The switch continuously toggles back and forth between $-V_{ref}$ and $+V_{ref}$, so the output (V_o) continuously ramps back and forth between $-V_1$ and $+V_1$.

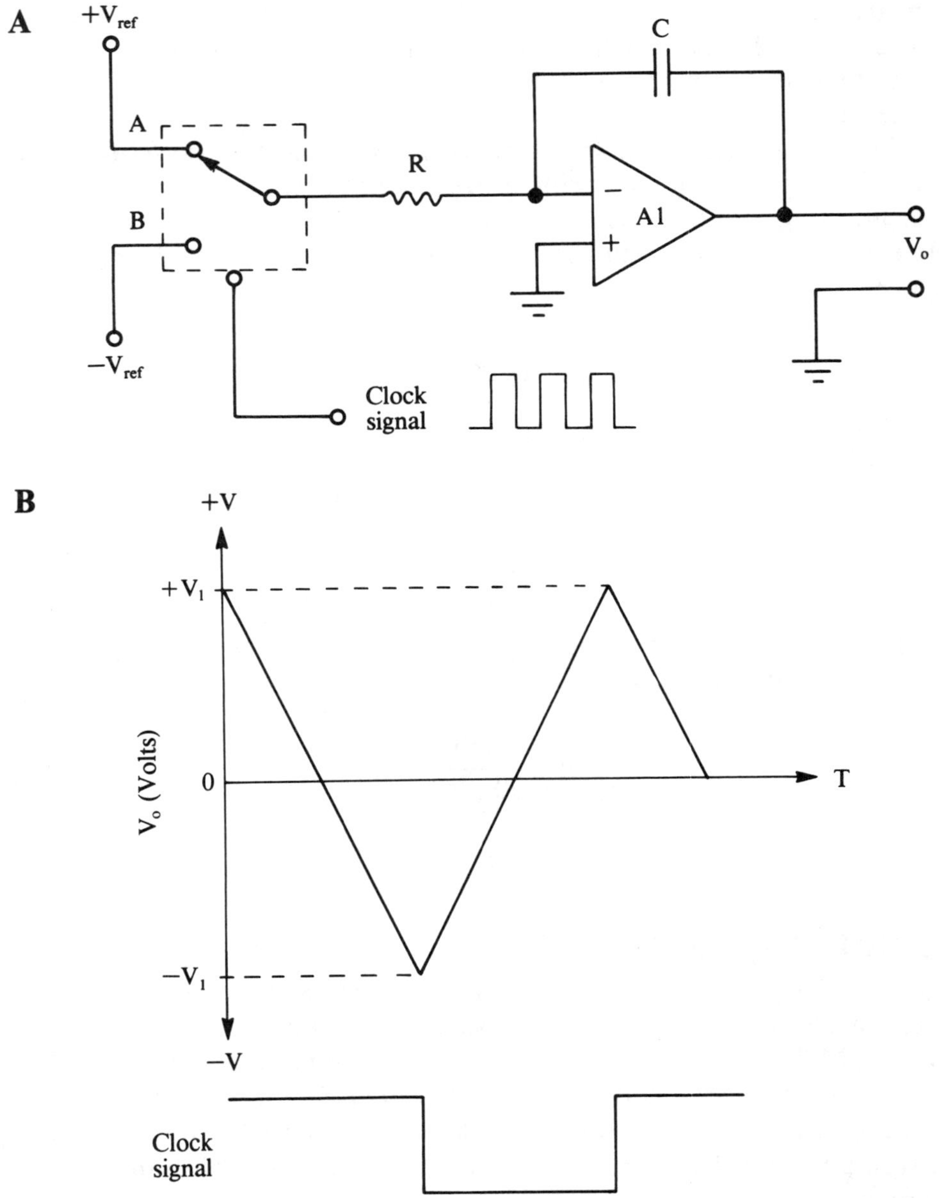

Figure 19-2. A) Notional triangle wave form generator, B) timing diagram.

The circuit of Fig. 19-2A is not practical, but it serves as an analogy for an actual circuit. Figure 19-3A shows the circuit for a practical triangle waveform generator in which a Miller integrator forms the ramp generator and a voltage comparator serves as the switch (see also the waveforms in Fig. 19-3B). The comparator uses the positive-feedback configuration, so it

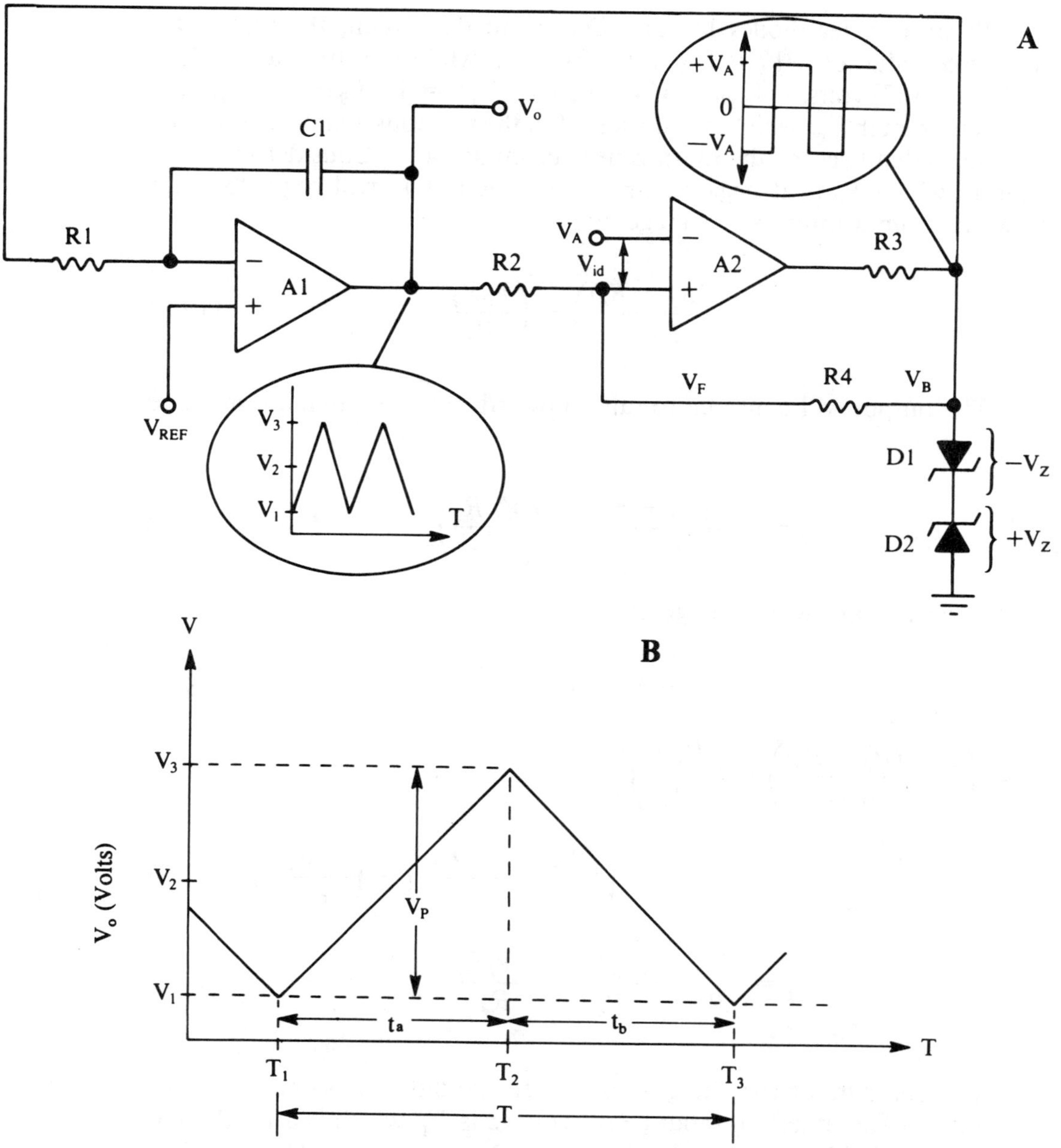

Figure 19-3. A) Triangle waveform generator, B) timing waveform.

operates as a noninverting Schmitt trigger. Such a circuit snaps high ($V_B = +V_{sat}$) when the input signal crosses a certain threshold voltage in the positive-going direction. It will snap low again ($V_B = -V_{sat}$) when the input signal crosses a second threshold in a negative-going direction. The two thresholds are not always at the same potential.

Because zener diodes D1 and D2 are in the circuit, the maximum allowable value of $+V_B$ is $[V_{ZD1} + 0.7]$ volts, while the limit for $-V_B$ is $-[V_{ZD2} + 0.7]$ volts. If $V_{ZD1} = V_{ZD2}$, then $|+V_B| = |-V_B|$. These potentials represent $\pm V_{ref}$, which was discussed in the previous analogy, as are the potentials that become the ramp-generator input signal. Consider an initial state in which V_B is at negative limit $-V_B$. Output V_o will begin to ramp upwards from a minimum voltage of:

$$V_1 = \left(\frac{V_A\,(R_2 + R_4)}{R_4}\right) - \left(\frac{V_B R_2}{R_4}\right) \tag{19-5}$$

The output will continue to ramp upwards toward a maximum value of:

$$V_3 = \left(\frac{V_A(R_2 + R_4}{R_4}\right) + \left(\frac{V_B\,R_2}{R_4}\right) \tag{19-6}$$

and cause a peak swing voltage of:

$$V_p = V_3 - V_1 \tag{19-7}$$

$$V_p = \left(\left(\frac{V_A\,(R_2 + R_4)}{R_4}\right) + \left(\frac{V_B R_2}{R_4}\right)\right) - \left(\left(\frac{V_A\,(R_2 + R_4)}{R_4}\right) - \left(\frac{V_B R_2}{R_4}\right)\right) \tag{19-8}$$

$$V_p = \frac{V_B R_2}{R_4} + \frac{V_B R_2}{R_4} = \frac{2\,V_B R_2}{R_4} \tag{19-9}$$

The comparator switching occurs when the differential-input voltage, V_{id}, is zero. The inverting-input (−In) voltage is V_A, which is a fixed reference potential. The noninverting input (+In) is at a voltage (V_F), which is the superposition of two voltages, V_o and V_B:

$$V_F = \left(\frac{V_o R_4}{R_2 + R_4}\right) + \left(\frac{\pm V_B R_2}{R_2 + R_3}\right) \tag{19-10}$$

If $+V_B = -V_B$, then the positive and negative thresholds are equal. The duration of each ramp (t_a and t_b) can be found from:

$$t_{a,b} = \frac{V_p}{\left(\dfrac{V_B}{R_1C_1}\right)} \tag{19-11}$$

The value of V_p is selected from $-V_B$ or $+V_B$, as is needed. In Eq. (19-9), it was found that $V_p = 2V_BR_2/R_4$, so:

$$t_{a,b} = \frac{\left(\dfrac{2\ V_BR_2}{R_4}\right)}{\left(\dfrac{V_B}{R_1C_1}\right)} \tag{19-12}$$

$$t_{a,b} = \left(\frac{R_1C_1}{V_B}\right)\left(\frac{V_BR_2}{R_4}\right) \tag{19-13}$$

$$t_{a,b} = R_1C_1\left(\frac{2\ R_2}{R_4}\right) \tag{19-14}$$

or, in the less general (but more common) case of $t_a = t_b$:

$$T = 2\ R_1C_1\left(\frac{2\ R_2}{R_4}\right) \tag{19-15}$$

The frequency (F) of the triangle wave is the reciprocal of the period (1/T), so:

$$F = \frac{1}{T} \tag{19-16}$$

$$F = \frac{1}{\left(\dfrac{4\ R_1C_1R_2}{R_4}\right)} \tag{19-17}$$

$$F = \frac{R_4}{4\ R_1C_1R_2} \tag{19-18}$$

Experiment 19-1

This experiment looks at the use of a Miller integrator circuit to generate a triangle waveform. The Miller integrator is an op-amp circuit, in which a capacitor is used as the negative-feedback element. When a dc voltage is applied to the input of the integrator circuit, the output voltage, V_o, rises in a ramp-function manner.

1. Connect the circuit of Experiment 19-1. Use either a CA-3140, which is a BiMOS op amp, or an equivalent BiFET op amp (a device with a minimal input bias current is needed). Use a pair of 1.5-volt AA cells for the dc signal source. Connect either a dc-coupled oscilloscope or an voltmeter to the output. If the voltmeter is an analog type, it should be set to display zero volts at the center of the scale.
2. The gain of the Miller integrator of Experiment 19-1 is $-1/R_1C_1$, where R_1 is in ohms and C_1 is in farads. Calculate the gain for the circuit shown. You should get an answer of 1.55.

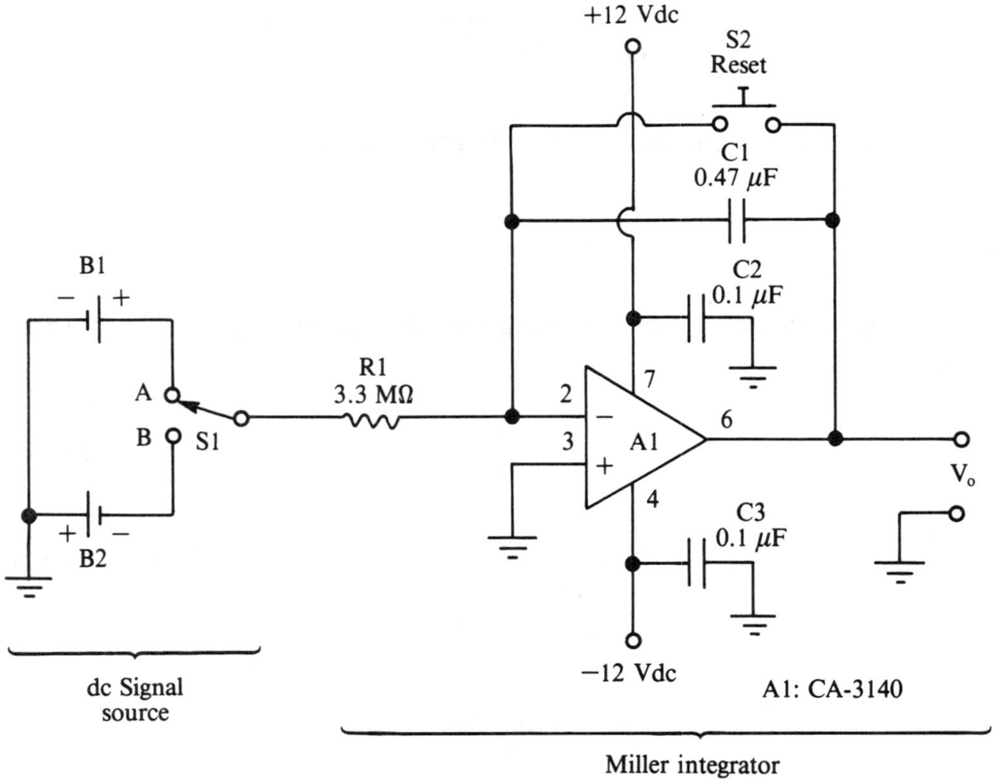

Experiment 19-1. Circuit for Experiment 19-1.

3. Set switch S1 to position A, then turn on the power to the circuit.
4. Press switch S2 to discharge the capacitor. Immediately after releasing S2, watch the output indicator ('scope or voltmeter). It should begin rising.
5. When V_o gets to about 10 volts, change switch S1 to position B and observe what happens to the output voltage.

Sawtooth generators

The *sawtooth wave* (Fig. 19-1A) is a single-slope ramp function. The wave ramps linearly upwards (or downwards), then abruptly snaps back to the initial baseline condition. Figure 19-4A shows a simple model of a sawtooth generator circuit. A constant-current source charges a capacitor in a manner

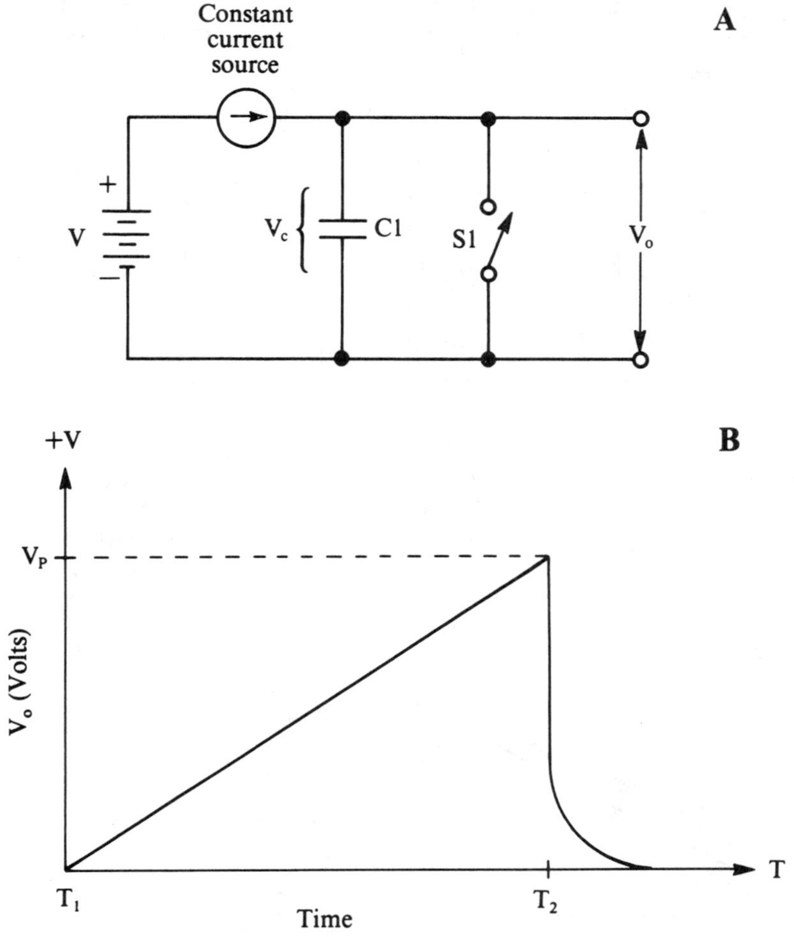

Figure 19-4. A) Notional sawtooth generator, B) timing waveform.

that generates the linear ramp function (Fig. 19-4B). When the ramp voltage (V_c) reaches the maximum point (V_p), switch S1 is closed, which forces V_c back to zero by discharging the capacitor. If switch S1 remains closed, the sawtooth is terminated. If S1 reopens, however, a second sawtooth is created as the capacitor recharges.

Figure 19-5A shows the circuit for a periodic sawtooth oscillator. It is similar to Fig. 19-4A, except that a JFET, Q1, is used as the discharge switch. When Q1 is turned off, the output voltage ramps upward (see Fig. 19-5B). When the gate is pulsed on hard, the drain-source channel resistance drops from a very high value to a very low value, forcing C1 to discharge rapidly. In the absence of a gate pulse, however, the channel resistance remains very high. At time t_1, the gate is turned off, so V_c begins to ramp upwards. At t_2, the JFET gate is pulsed, so C1 rapidly discharges to zero. When the pulse ($t_2 - t_3$) ends, however, Q1 turns off again and the ramp restarts. The same circuit can be used for single-sweep operation by replacing the pulse train that is applied to the gate of Q1 with the output of a monostable multivibrator.

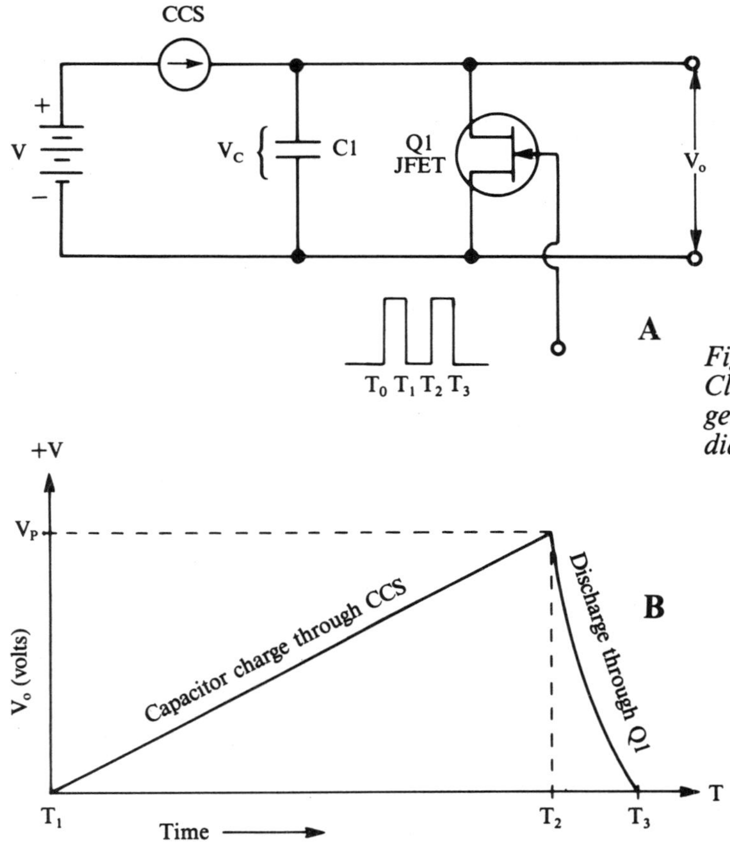

Figure 19-5. A) Clock-driven sawtooth generator, B) timing diagram.

The circuit of Fig. 19-6A shows a sawtooth generator that uses a Miller integrator (A1) as a ramp generator, and replaces the discharge switch with an electronic switch, which is driven by a voltage comparator and a one-shot circuit. The timing diagram for this circuit is shown in Fig. 19-6B.

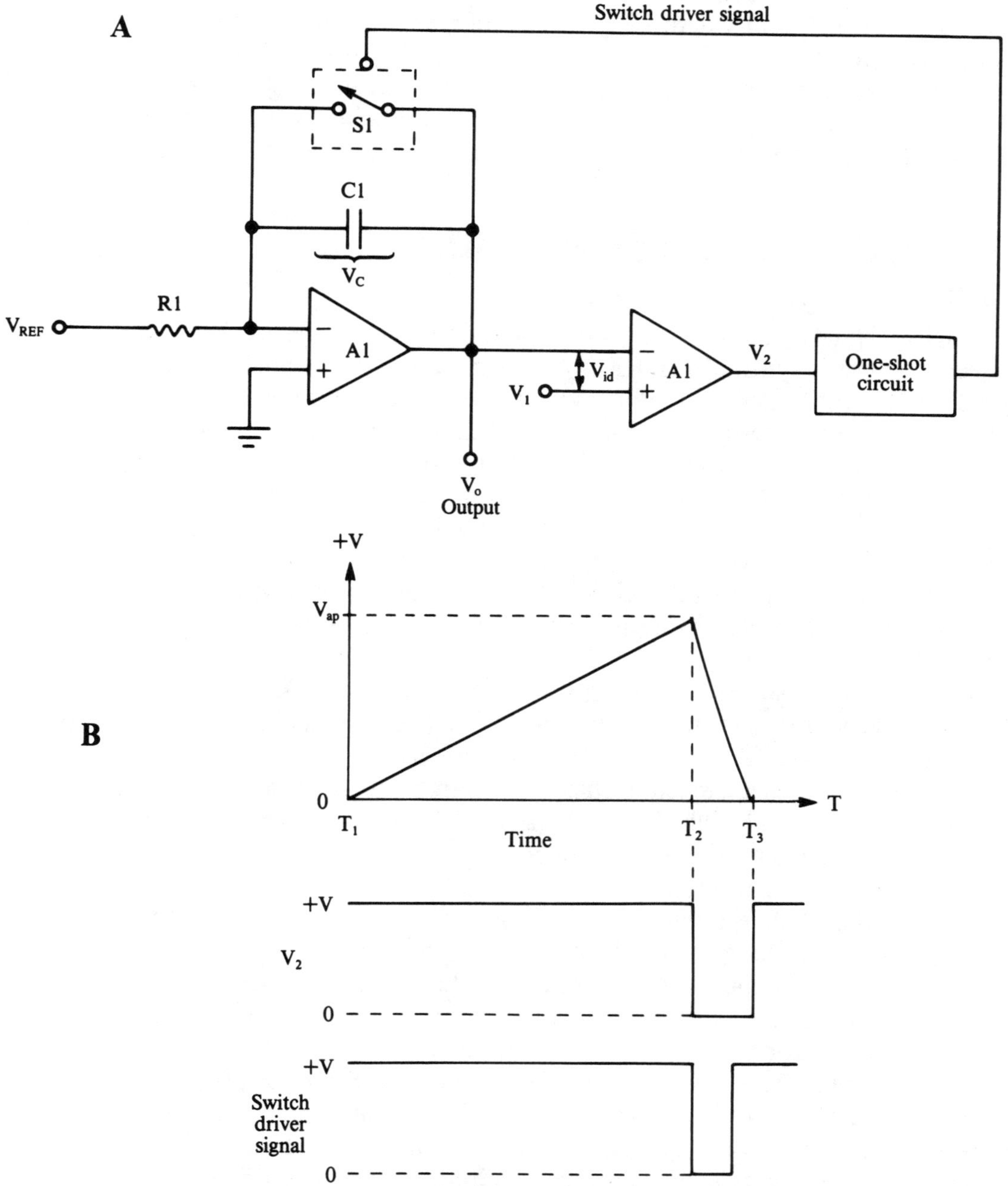

Figure 19-6. A) Sawtooth generator, B) timing diagram.

Under the initial conditions, at time t_1, the output voltage (V_o) ramps upward at a rate of $[-(-V_{ref})/R_1C_1]$. The voltage comparator (A2) is biased with the noninverting input (+In) set to V_1, and the inverting input at V_o. The comparator differential input voltage is: $V_{id} = (V_1 + V_o)$. As long as $V_1 > V_o$, the comparator sees a negative input, so it produces a high output of $+V_{sat}$. At the point where $V_1 = V_o$, the differential input voltage is zero, so the output of A2 (voltage V_2) drops low (i.e., $-V_{sat}$). The negative-going edge of V_2 at time t_2 triggers the one-shot circuit. The output of the one-shot briefly closes electronic switch S1 and causes the capacitor to discharge. The one-shot pulse ends at time t_3, so S1 reopens and allows V_o to again ramp upwards.

Digitally generated sawtooth and triangle waveforms

Sawtooth signal generators are used for a variety of purposes in electronics: electronic music synthesizers, RF sweeping circuits, audio signal generators, and voltage-controlled oscillators (VCOs), certain bench tests, oscilloscope calibrators, and other applications. In addition, many circuit applications exist for embedded sawtooth generators. In many situations, the sawtooth is used to precisely calibrate an oscilloscope time base. If the sawtooth that is used to sweep the oscilloscope horizontally is controlled from a stable crystal oscillator, a very precise sweep rate is possible. The "standard" solid-state sawtooth generator circuit consists of an op-amp Miller integrator circuit that is excited by a square wave. But problems do exist with such circuits.

Figure 19-7 shows part of the problem with the standard op-amp sawtooth generator circuit. The waveform has two principal defects. First, the ramp-up edge ($T_1 - T_2$) of the sawtooth is not linear. Because the original design uses a capacitor charge/discharge circuit in the Miller integrator, the ramp naturally has a shape like the normal capacitor-charge waveform. What is required of a proper sawtooth is a linear ramp (i.e., one that rises as a straight line).

The second defect with the waveform in Fig. 19-7 is the fall-time ($T_2 - T_3$): it's too long. Although proper design will make the Miller integrator sawtooth generator closer to the ideal, the use of a few low-cost digital components produces a better sawtooth generator without headaches.

The circuit for a digitally synthesized sawtooth generator is shown in Fig. 19-8. The heart of this circuit is IC1, a DAC0806 8-bit digital-to-analog converter (DAC). This DAC is based on the MC-1408 family of DACs and it was selected because it is well-behaved in simple, easy-to-use circuitry. The DAC is easily available through mailorder sources, such as Jameco Electronics, or through local distributors in blister packs.

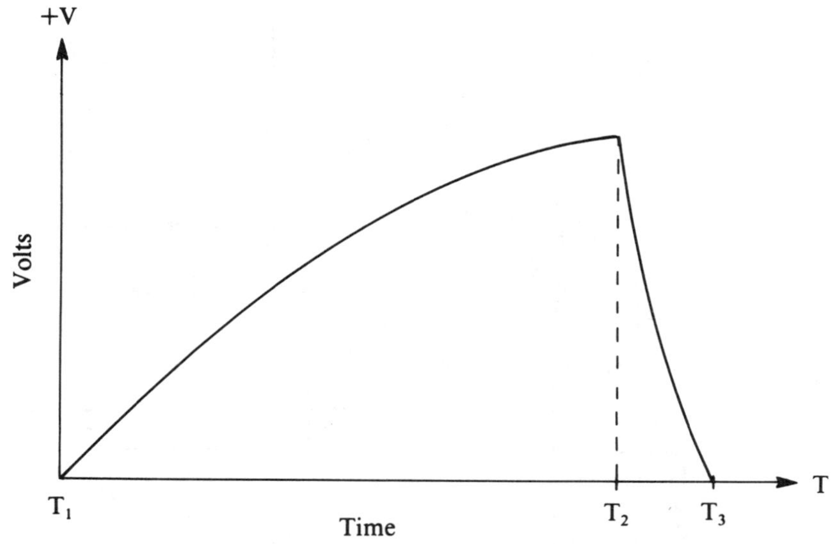

Figure 19-7. Typical capacitor-charge sawtooth waveform.

A DAC produces an output current that is proportional to the reference voltage or current, and the binary word applied to the digital inputs. The controlling function for the DAC selected that was for this section is:

$$I_o = I_{ref}\left(\frac{A}{256}\right) \tag{19-19}$$

Where:

I_o is the output current from pin 4
I_{ref} is the reference current applied to pin 14
A is the decimal value of the binary word applied to the eight binary inputs (pins 5 through 12)

The reference current is found from Ohm's law and it is the quotient of the reference voltage and the series resistor at pin 14 (R4). In data acquisition systems (the main use for the DAC), the reference voltage is a precision regulated potential. But in this case, the precision is not necessary, so use the V+ power supply as the reference voltage. Therefore, the reference current is (+12 Vdc)/R_4. With the value of R4 shown (6,800 ohms), I_{ref} is 0.00018 amperes (1.8 mA). Values from 500 μA to 2 mA are permissible with this device. If you elect to change the reference current, keep $R_4 = R_5$.

The reference current sets the maximum value of output current, I_o. When a full-scale binary word (11111111) is applied to the binary inputs,

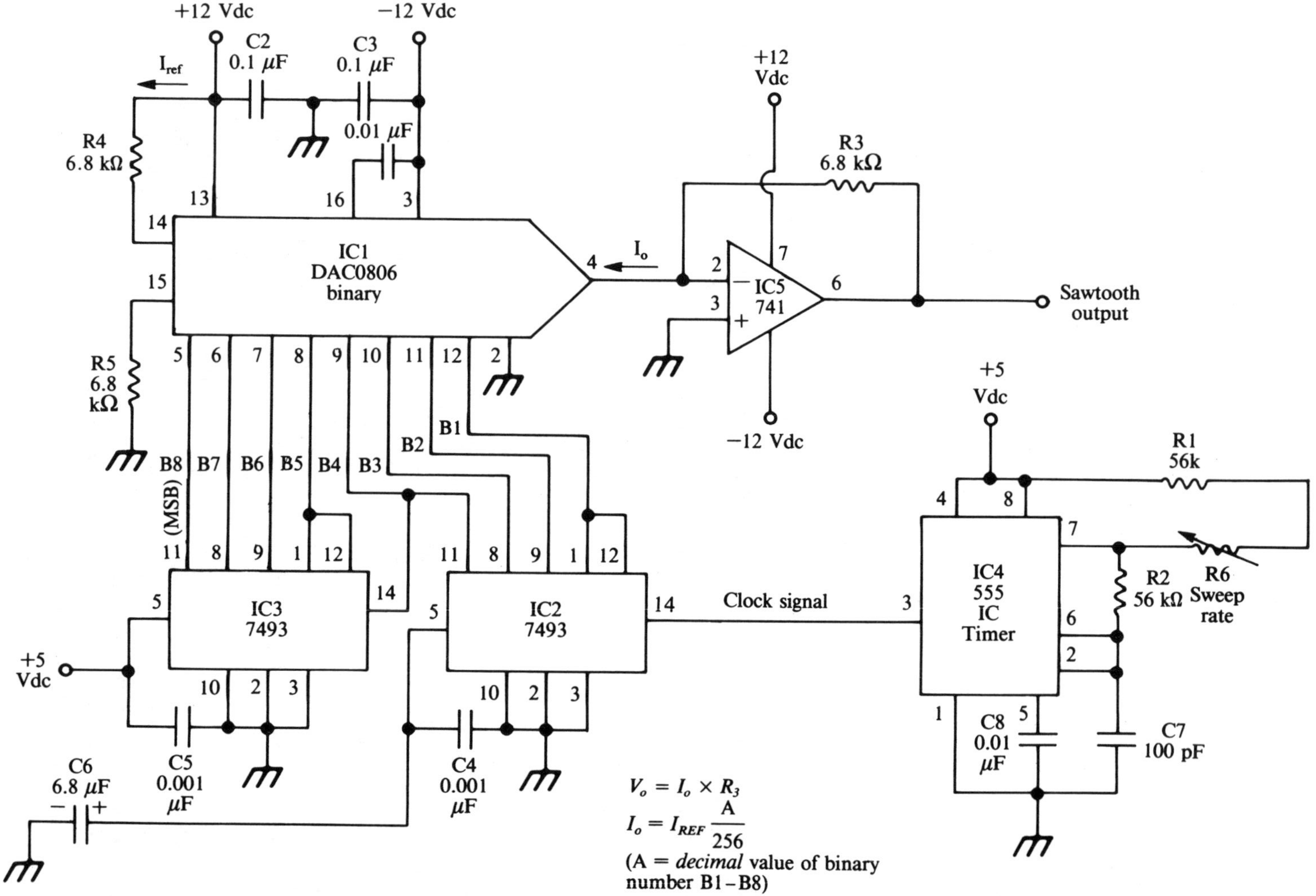

Figure 19-8. Digital sawtooth generator.

the output current I_o is:

$$I_o = (1.8 \text{ mA}) \left(\frac{255}{256}\right) \tag{19-20}$$

$$I_o = (1.8 \text{ mA}) \, (0.996) \tag{19-21}$$

$$I_o = 1.78 \text{ mA} \tag{19-22}$$

The DAC0806 is a current-output DAC, so you must use an op-amp current-to-voltage converter in order to make a sawtooth voltage function. Such a circuit is an ordinary inverting follower without an input resistor. The output voltage (V_o) will rise to $I_o R_3$.

The actual output waveform is "staircased" in binary steps that are equal to the 1-LSB current of IC1 (the 1-LSB voltage of V_o). The 1-LSB voltage is the smallest step change in output-potential cause by flipping the least-significant bit (B1) either from 0 to 1 or 1 to 0. The reason why you don't see the steps is that the frequency response of the 741 operational amplifier (used for the current to voltage converter) acts as a low-pass filter to smooth the waveform. If a higher frequency op amp is used, a capacitor that shunts R3 will low-pass filter the waveform. A -3-dB frequency (F) of 1 or 2 kHz will smooth the waveform. The value of the capacitor is calculated from:

$$C_{\mu F} = \frac{1{,}000{,}000}{2\pi R_3 F} \tag{19-23}$$

Where:

$C_{\mu F}$ is the capacitance (in *microfarads*)
F is the -3-dB cut-off frequency (in *hertz*)
R_3 is expressed (in *ohms*)

This circuit is synchronized by a clock oscillator that consists of a single 555 IC timer (see chapters 22 and 23). Although not a TTL device, the 555 is TTL-compatible when the V+ potential that is applied to pins 4 and 8 is limited to +5 Vdc. The 555 is connected in the astable-multivibrator configuration, so it outputs a chain of pulses with +4-volt amplitude. The operating frequency is set by three resistors (R1, R2, and R6) and a capacitor (C7). The actual clock frequency is:

$$F = \frac{1.44}{((R3 + R12) + 2\,R4)\,C} \tag{19-24}$$

Where:

F is the frequency (in *hertz*)
C is the value of C7 (in *farads*)
R_1, R_2, and R_6 are the resistances (in *ohms*)

Select a clock frequency that is 256 times the desired sawtooth fundamental frequency.

Other waveform selections

As electronic music buffs will testify, almost any waveform can be supplied if the right binary words are supplied to the digital inputs of the DAC0806. Because I wanted a sawtooth waveform, I connected the DAC digital inputs to the outputs of an 8-bit binary counter that was built from a pair of 7493 TTL base-16 counter chips. Each chip is a four-bit counter, so they are cascaded to produce the 8-bit binary word that is needed to drive the DAC. If you want a detailed description of this chip, I recommend Don Lancaster's *TTL Cookbook*.

The function of this counter is to increment in steps from 00000000 to 11111111 under the control of a clock signal that is applied to the input (pin 14) of IC2. You could use any 8-bit counter that outputs a TTL-compatible signal in place of the 7493 devices that I selected. The 7493 was selected for the best of all engineering reasons: I had a pair of them in my junk box.

If you want a triangle waveform, it is possible to replace the 7493 devices with a base-16 up/down counter chip. Arrange the digital control logic to reverse the direction of the count when the maximum state (11111111) is sensed.

Two ways exist to generate waveforms other than a sawtooth or triangle, and both of them involve using a computer memory. The binary-bit pattern that represents the waveform is stored in memory, then it is output in the correct sequence. One method uses a Read-Only Memory (ROM), which you preprogram with the bit pattern that represents the waveform. A binary counter circuit that is connected as an address generator selects the bit-pattern sequence.

The second method is to store the bit pattern in a computer, then output it under program control through an 8-bit parallel output port. This method is usable for both generating special waveforms and for linearizing the tuning characteristic of circuits, such as VCOs, swept oscillators, etc. The digital solution to the linearization problem involves storing a look-up table in either a ROM or computer memory. I learned this system in a laboratory where it was once used to linearize low-level pressure transducer measurements. Interfacing a computer output (assuming that you have an 8-bit parallel port available) is simple. Connect the output of the computer directly to the input of the DAC.

The digitally synthesized sawtooth generator is easy to build and it is well-behaved, so its construction shouldn't haunt you. In addition, it is easy to generate any waveform. You can expect to see more direct digitally synthesized test circuits and communications equipment in the future.

Summary

Two common nonsinusoidal signals are the triangle wave and the sawtooth wave. Both involve a ramp circuit, which creates a linearly increasing voltage as a function of time. The ramp circuit generally involves the charging or discharging of a capacitor by a constant-current source, or through a resistor. In the sawtooth wave, a single ramp circuit is used for charging, and discharge takes place extremely rapidly, whereas in the triangle wave, charging and discharging normally take place at the same rate. These waveforms can also be generated through the use of digital circuits and through the use of a digital-to-analog converter.

Questions

1. What are two examples of periodic ramp functions? What characteristics do they share? How are they different?
2. What component is generally present in both of these circuits?
3. How can a triangle wave be made asymmetrical?
4. What is one application of a sawtooth-wave generator?
5. For the circuit in Fig. 19-6, what is the effect of increasing the value of V_I?
6. How can a digital signal be used to create a sawtooth wave? a triangular wave?
7. When a digital counter is used to create a sawtooth wave, what is the relationship between the digital clock frequency and the frequency for the sawtooth wave?

Problems

1. For the circuit in Fig. 19-3A, R_1 = 10 kohms, C_1 = 2.2 μF, R_2 = 20 kohms, R_3 = 50 kohms, R_4 = 100 kohms. Calculate the period and frequency of the output waveform.
2. Repeat Problem 1 for R_1 = 20 kohms, C_1 = 0.022 μF, R_2 = 50 kohms, R_3 = 50 kohms, and R_4 = 1 Mohm.
3. For the circuit in Fig. 19-6A, the reference voltage is 12 volts and V_1 is 8 volts. If R_1 = 10 kohms and C = 0.01 μF, calculate the time it takes for the output to go from 0 to its maximum value.
4. Repeat Problem 3 for V_1 = 6 volts, R_1 = 100 kohms and C = 2.2 μF.
5. For the circuit in Fig. 19-8, what is the output current when the input word is 1101 1001?
6. For the circuit in Fig. 19-8, what is the output current when the input word is 0010 1101?

20 SINE-WAVE OSCILLATORS

AFTER COMPLETING THIS chapter, you will be able to:

- ☐ Understand the requirements for oscillation
- ☐ Describe the different types of sine-wave oscillators
- ☐ Know the ways in which stability of oscillators can be improved

A feedback oscillator (Fig. 20-1) consists of an amplifier with an open-loop gain of A_{vol} and a feedback network with a gain or transfer function β. It is called a *feedback oscillator* because the output signal of the amplifier is fed back to the amplifier's own input, by way of the feedback network. Figure 20-1 is a block-diagram model of the feedback oscillator. Its resemblance to a feedback amplifier is no coincidence. Indeed, as anyone who has misdesigned or misconstructed an amplifier knows all too well, a feedback oscillator is an amplifier in which special conditions, called *Barkhausen's criteria for oscillation*, prevail:

1. Feedback voltage V_F must be in-phase (360 degrees) with the input voltage
2. The loop gain βA_{vol} must be unity (1).

The first of these criteria means that the total phase shift from the input of the amplifier, to the output of the amplifier, around the loop, and back to the input, must be 360 degrees (2π radians) or an integer (N) multiple of 360 degrees (i.e., $N2\pi$ radians).

The amplifier can be one of many different devices. In some circuits, it is a common-emitter bipolar transistor (either npn or pnp devices). In others, it is a JFET or a MOSFET. In older equipment, it is a vacuum tube. In modern circuits, the active device is probably either an IC op amp or some other form of linear IC amplifier.

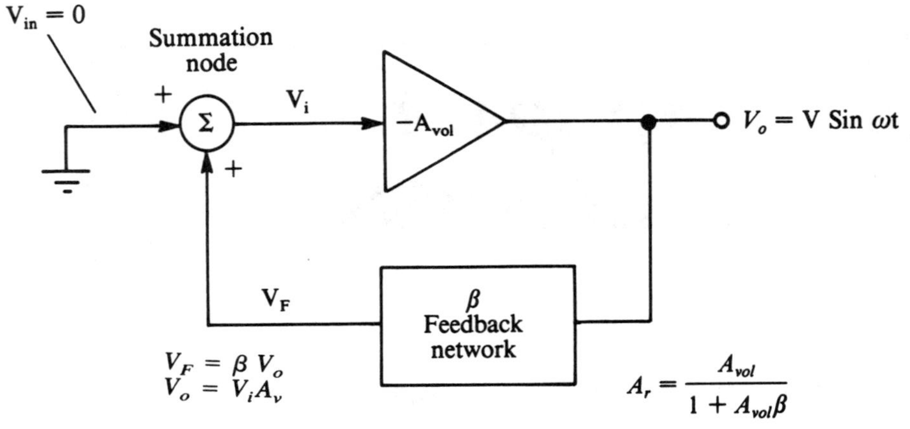

Figure 20-1. Feedback oscillator block diagram.

The amplifier is usually an inverting type, so the output is out of phase with the input by 180 degrees. As a result, in order to obtain the required 360 degrees phase shift, an additional phase shift of 180 degrees must be provided in the feedback network at the frequency of oscillation only. If the network is designed to produce this phase shift at only one frequency, the oscillator will produce a sine-wave output on that frequency.

Before considering specific sine-wave oscillator circuits, examine Fig. 20-1 more closely. Several things can be determined about the circuit; first:

$$V_i = V_{in} + V_F \tag{20-1}$$

so,

$$V_{in} = V_i - V_F \tag{20-2}$$

and also,

$$V_F = \beta V_o \tag{20-3}$$

$$V_o = V_i A_{vol} \tag{20-4}$$

The transfer function (or gain) A_v is:

$$A_v = \frac{V_o}{V_{in}} \tag{20-5}$$

substituting Eqs. (20-2) and (20-4) into Eq. (20-5):

$$A_v = \frac{V_i A_{vol}}{V_i - V_F} \tag{20-6}$$

from Eq. (20-3), $V_F = \beta V_o$, so:

$$A_v = \frac{V_i A_{vol}}{V_i - \beta\, V_o} \tag{20-7}$$

but Eq. (20-4) shows $V_o = V_i A_{vol}$, so Eq. (20-7) can be written:

$$A_v = \frac{V_i A_{vol}}{V_i - \beta\, V_i A_{vol}} \tag{20-8}$$

and, dividing both numerator and denominator by V_i:

$$A_v = \frac{A_{vol}}{1 - \beta A_{vol}} \tag{20-9}$$

Equation (20-9) serves for both feedback amplifiers and oscillators. But in the special case of an oscillator $V_{in} \rightarrow 0$, so $V_o \rightarrow \infty$. Implied, therefore, is that the denominator of Eq. (20-9) must also be zero:

$$1 - \beta A_{vol} \tag{20-10}$$

so, for the case of the feedback oscillator:

$$\beta A_{vol} = 1 \tag{20-11}$$

βA_{vol} is the loop gain of the amplifier and feedback network, so Eq. (20-11) meets Barkhausen's second criterion.

Sine-wave oscillators

Sine-wave oscillators produce an output signal that is sinusoidal. Such a signal is ideally very pure, and if indeed it is perfect, then its Fourier spectrum will contain only the fundamental frequency and no harmonics. The harmonics in a nonsinusoidal waveform give it a characteristic shape.

The active element in this circuit is the op amp. However, any linear amplifier will also work in place of the op amp. The one circuit that shows the principles most clearly is the *RC phase-shift oscillator.*

Stability in oscillator circuits can refer to several different phenomena. First, is *frequency stability,* which refers to the ability of the oscillator to remain on the design frequency over time. Several different factors affect frequency stability, but the most important are *temperature* and *power-supply voltage variations.* Another form of stability is *amplitude stability.* Because sine-wave oscillators do not operate in the saturated mode, minor variations in overall circuit gain can affect the amplitude of the output signal. Again, the factors most often cited for this problem include temperature and dc power-supply variations. The latter is overcome with regulated dc power supplies for the oscillator. The former is overcome with either a temperature-compensated design or by maintaining a constant operating temperature. Some variable sine-wave oscillators will exhibit amplitude variation of the output signal when the operating frequency is changed. In these circuits, either a self-compensation element or an automatic level control amplifier stage is used.

Still another form of stability regards the purity of the output signal. If the circuit exhibits spurious oscillations, they will be superimposed on the output signal. As with any circuit that contains an op amp or any other high gain linear amplifier, you must properly decouple the dc power-supply lines. It might also be necessary to frequency-compensate the circuit.

RC phase-shift oscillator circuits

The RC shift oscillator is based on a three-stage cascade resistor-capacitor network, such as is shown in Fig. 20-2A. An RC network will exhibit a phase shift (ϕ) (Fig. 20-2B) that is a function of resistance (R) and capacitive reactance (X_c). Because X_c is inversely proportional to frequency ($1/2\pi FC$), the phase angle shift across the network is therefore a function of frequency.

The goal in designing the RC phase-shift oscillator is to create a phase shift of 180 degrees between the input and output of the network at the desired frequency of oscillation. It is conventional to make the three stages of the network identical so that each provides a 60-degree phase shift. However, this technique is not strictly necessary, provided that the total phase shift is 180 degrees. One reason for using identical stages, however, is that it is possible for the nonidentical designs to have more than one frequency for which the total phase shift is 180 degrees. This phenomenon can lead to undesirable multi-modal oscillation.

Figure 20-3 shows the circuit for an op-amp RC phase-shift oscillator.

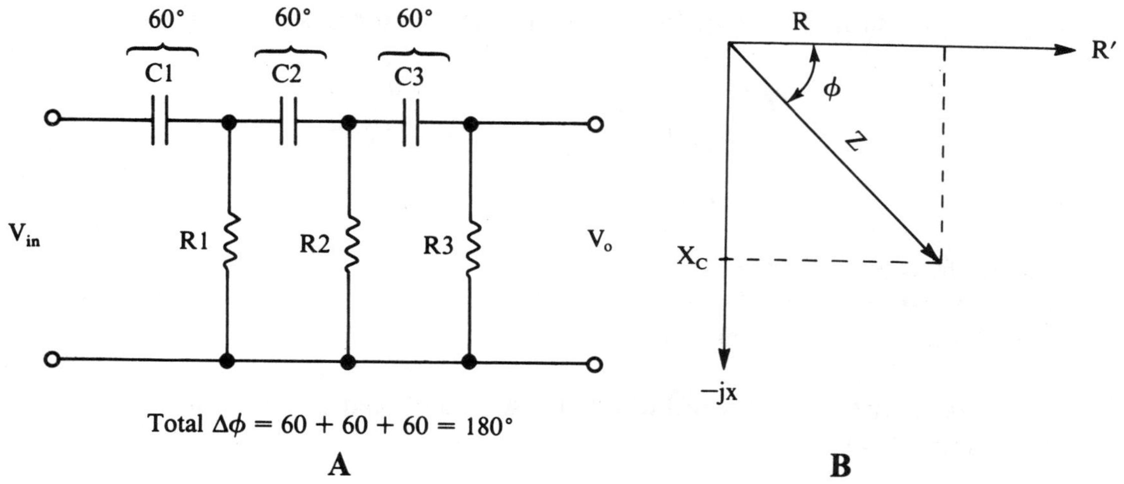

Figure 20-2. A) Phase-shift network, B) vector-phase relationships.

Figure 20-3. Op-amp RC phase-shift oscillator.

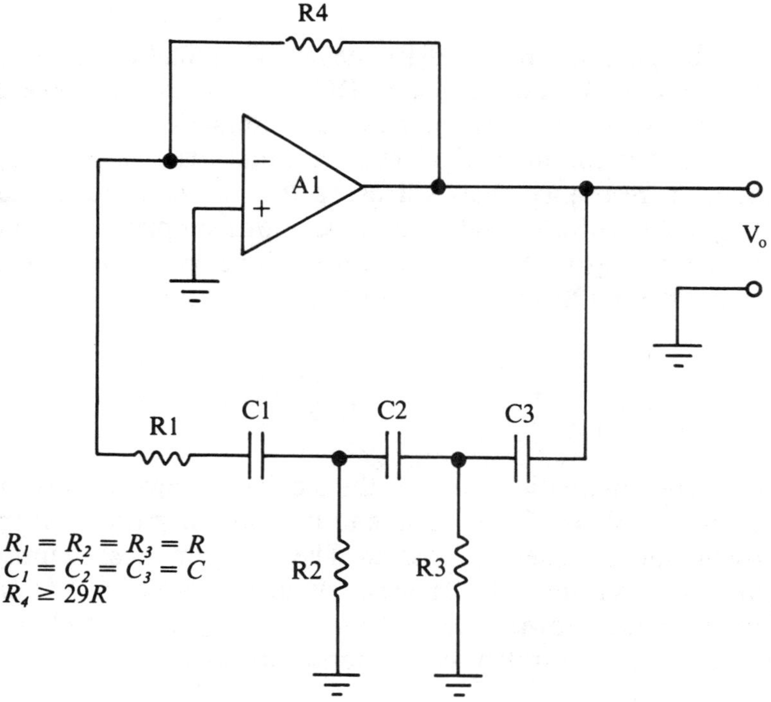

The cascade phase-shift network $R_1R_2R_3/C_1C_2C_3$ provides 180 degrees of phase shift at a specific frequency, and the amplifier provides another 180 degrees (because it is an inverting follower). The total phase shift is therefore 360 degrees at the frequency for which the RC network provides a

180-degree phase shift. The frequency of oscillation (F) for this circuit is:

$$F = \frac{1}{2\pi\sqrt{6}\,RC} \qquad (20\text{-}12)$$

Where:

f is the frequency (in *hertz*)
R is the resistance (in *ohms*)
C is the capacitance (in *farads*)

Commonly, the constants in Eq. (20-12) are combined to arrive at a simplified expression:

$$F = \frac{1}{15.39RC} \qquad (20\text{-}13)$$

Because the required frequency of oscillation is usually determined from the application, select an RC time constant to force the oscillator to operate as needed. Also because capacitors are available in fewer standard values, it is common to select an arbitrary trial value of capacitance, then select a resistance that will cause the oscillator to produce the correct frequency. Also, to make the calculations simpler, it is prudent to express the equation so that the capacitance can be specified (C) in microfarads. As a result, Eq. (20-13) is sometimes rewritten as:

$$R = \frac{1{,}000{,}000}{15.39C_{\mu F}F} \qquad (20\text{-}14)$$

The attenuation through the feedback network must be compensated by the amplifier if loop gain is to be unity or greater. At the frequency of oscillation, the attenuation is $1/29$. The loop gain must be unity, so the gain of amplifier A1 must be at least 29 to satisfy $A\beta = 1$. For the inverting follower (as shown), $R_1 = R$ and $A_v = R_4/R_1$. Therefore, $R_4 \geq 29R$, in order to meet Barkhausen's criterion for loop gain.

Experiment 20-1

Design a 1000-Hz sine-wave oscillator that is based on the RC phase-shift circuit of Fig. 20-3. Select values for R, C, and the feedback resistor.

1. Select trial value for C: 0.01 μF is a good starting value.
2. Solve Eq. (20-13) for R as a function of F:

$$R = \frac{1}{15.39FC}$$

$$R = \frac{1}{(15.39)\ (1000\ \text{Hz})\ C}$$

$$R_{(@1000\ Hz)} = \frac{1}{15{,}390\ C}$$

3. Substitute the first trial for C:

$$R_{(@1000\ Hz)} = \frac{1}{(15{,}390)\ (0.01 \times 10^{-6}\ \text{farads})}$$

$$R = \frac{1}{1.539 \times 10^{-4}} = 6{,}498\ \text{ohms}$$

4. The value of R is 6,498 ohms, which can be obtained as a 1% precision resistor.
5. Select a minimum value of R_4:

$$R_4 \geq 29R$$

$$R_4 \geq (29)\ (6{,}498) \geq 188{,}434\ \text{ohms}$$

6. A practical value of R_4 is greater than the calculated value, such as 200 or 220 kohms. It should not be much higher or stability problems might arise.

Experiment 20-2

Build the sine-wave oscillator discussed in the previous experiment. Use a two-channel oscilloscope to examine the following:

1. The output voltage V_o.
2. The signal that is applied to the inverting input of the op amp. This step may require disconnecting the −In end of R1.

Wein-bridge oscillator circuits

The Wein-bridge circuit is shown in Fig. 20-4. Like several other well-known bridge circuits, the Wein bridge consists of four impedance arms. Two of the arms (R_1, R_2) form a resistive voltage divider that produces a voltage V_1 of:

$$V_1 = \frac{V_{ac} R_2}{R_1 + R_2} \tag{20-15}$$

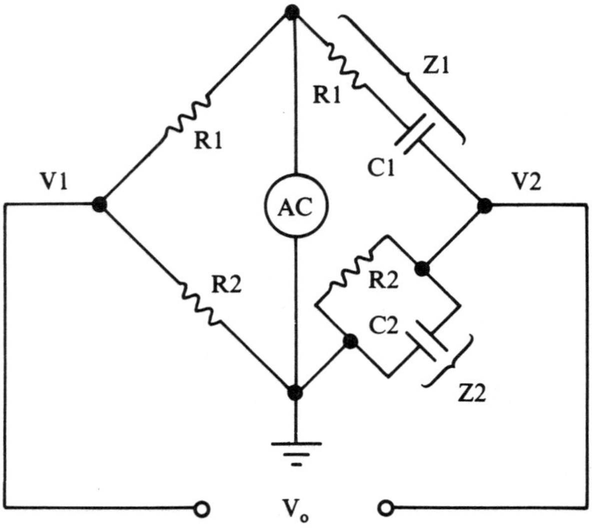

$R_1 = R_2 = R$
$C_1 = C_2 = C$

Figure 20-4. Wein bridge.

The remaining two arms (Z_1, Z_2) are complex RC networks that each consist of one capacitor and one resistor each. Impedance Z_1 is a series RC network and Z_2 is a parallel RC network. The voltage and phase shift that is produced by the Z_1/Z_2 voltage divider are functions of the RC values and of the applied frequency. Notice that $V_2 = V_{ac}/3$ and that V_2 and V_{ac} are in-phase with each other.

Figure 20-5 shows the circuit for a Wein-bridge oscillator. The resistive voltage divider supplies V_1 to the inverting input (−In) and V_2 is applied to the noninverting input (+In). In Fig. 20-5, the bridge signal source is the output of the amplifier (A1). The ac signal is applied to +In, so the gain it sees is found from:

$$A_v = \frac{R_3}{R_4} + 1 \tag{20-16}$$

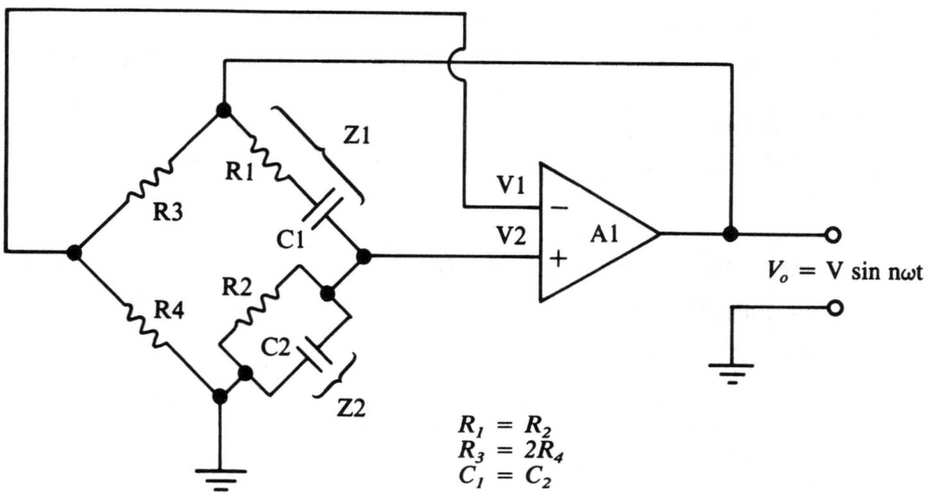

Figure 20-5. Wein-bridge oscillator.

The ac feedback applied to +In is:

$$\beta = \frac{Z_2}{Z_1 + Z_2} \tag{20-17}$$

At resonance, $\beta = 1/3$, so (as shown in Fig. 20-4B):

$$V_2 = \frac{V_o}{3} \tag{20-18}$$

Because $A_v = V_o/V_2$ by definition, satisfying Barkhausen's loop gain criterion ($-A_v\beta = 1$) requires that $A_v = V_o/V_2 = 3$. Using this result:

$$A_v = \frac{R_3}{R_4} + 1 \tag{20-19}$$

or,

$$R_3 = 2R_4 \tag{20-20}$$

If $R_1 = R_2 = R$ and $C_1 = C_2 = C$, the resonant frequency of the Wein bridge is:

$$F = \frac{1}{2\pi RC} \tag{20-21}$$

For the standard Wein-bridge oscillator, in which $R_1 = R_2 = R$, $C_1 = C_2 = C$, and $R_3 = 2R_4$, a sine-wave output will result on frequency F.

Experiment 20-3

Design a 2,000-Hz sine-wave oscillator with the Wein-bridge circuit of Fig. 20-5.

1. Select a trial capacitance of 0.01 μF.
2. Calculate the value of R by solving Eq. (20-21) for R when F is 2,000 Hz:

$$R = \frac{1}{2\pi FC}$$

$$R = \frac{1}{(2)\,(3.14)\,(2{,}000\ \text{Hz})\,(0.01 \times 10^{-6}\ \text{farads})}$$

$$R = \frac{1}{1.256 \times 10^{-4}} = 7{,}961\ \text{ohms}$$

3. The nearest standard value for R is 8,200 ohms. If the exact frequency is not important, select one of these units for the resistor. Otherwise, use a precision 1% resistor for R. Also, the precision resistor offers better frequency stability with temperature.

Amplitude stability

The oscillations in the Wein-bridge circuit build up without limit when the gain of the amplifier is high. Figure 20-6 shows the result of when the gain is only slightly above that which is required for stable oscillation. Note that some clipping begins to appear on the sine-wave peaks. At even higher gains, the clipping becomes more severe, until it eventually looks like a square wave. Figure 20-7 shows several methods to stabilize the waveform

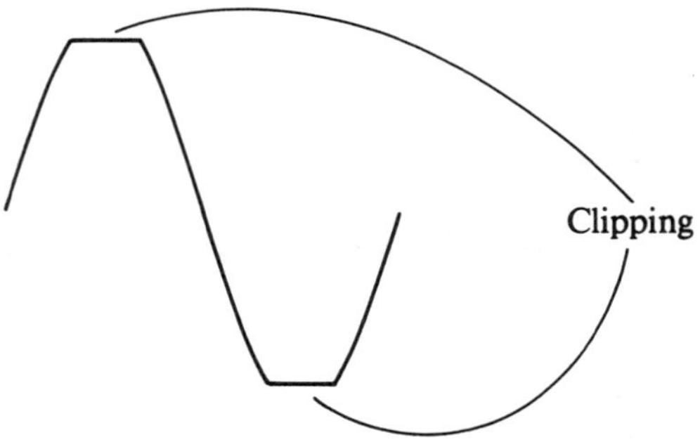

Figure 20-6. Clipping in the Wein-bridge oscillator.

amplitude. Figure 20-7A shows the use of small-signal diodes, such as the 1N914 and 1N4148 devices. At low-signal amplitudes, the diodes are not sufficiently biased, so the gain of the circuit is:

$$A_v = \frac{R_1 + R_3}{R_2} \tag{20-22}$$

As the output signal voltage increases, however, the diodes become forward-biased. D1 is forward-biased on negative peaks of the signal, while D2 is forward-biased on the positive peaks. Because D1 and D2 are shunted across R3, the total resistance R_3 is less than R3. By inspecting Eq. (20-22), you can determine that the change from R3 to R_3 reduces the gain of the circuit. The circuit is thus self-limiting.

Another variant of the gain-stabilized Wein-bridge oscillator is shown in Fig. 20-7B. In this circuit, a pair of back-to-back zener diodes provide the gain-limitation function. With the resistor ratios shown, the overall gain is limited to slightly more than unity, so the circuit will oscillate. The output peak voltage of this circuit is set by the zener voltages of D1 and D2, which should be equal for low-distortion operation.

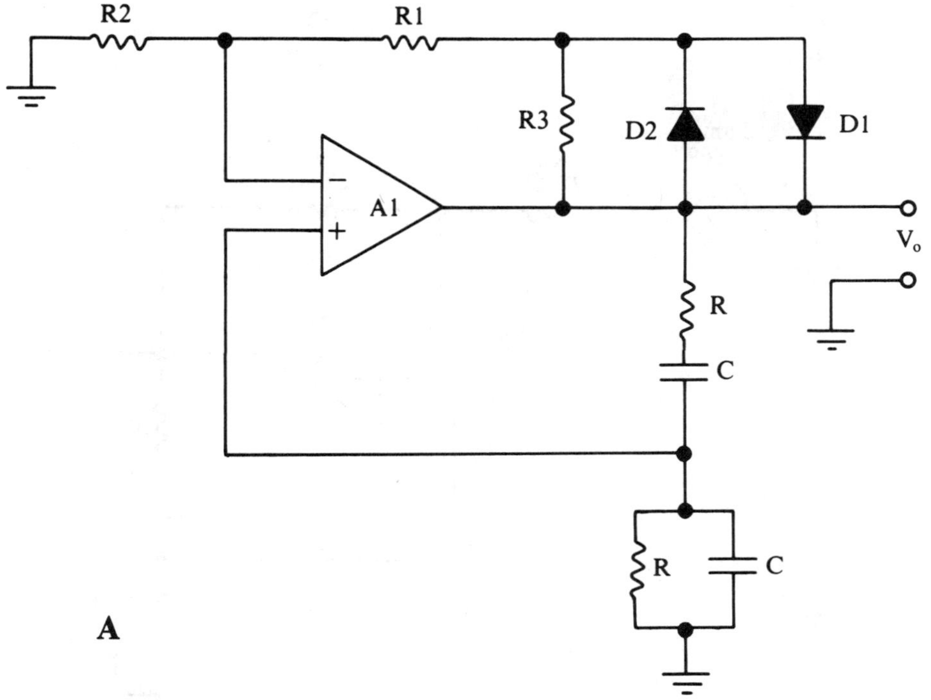

Figure 20-7. A) Improved circuit for limiting clipping, B) zener-diode version, C) incandescent-lamp version.

Figure 20-7. Continued.

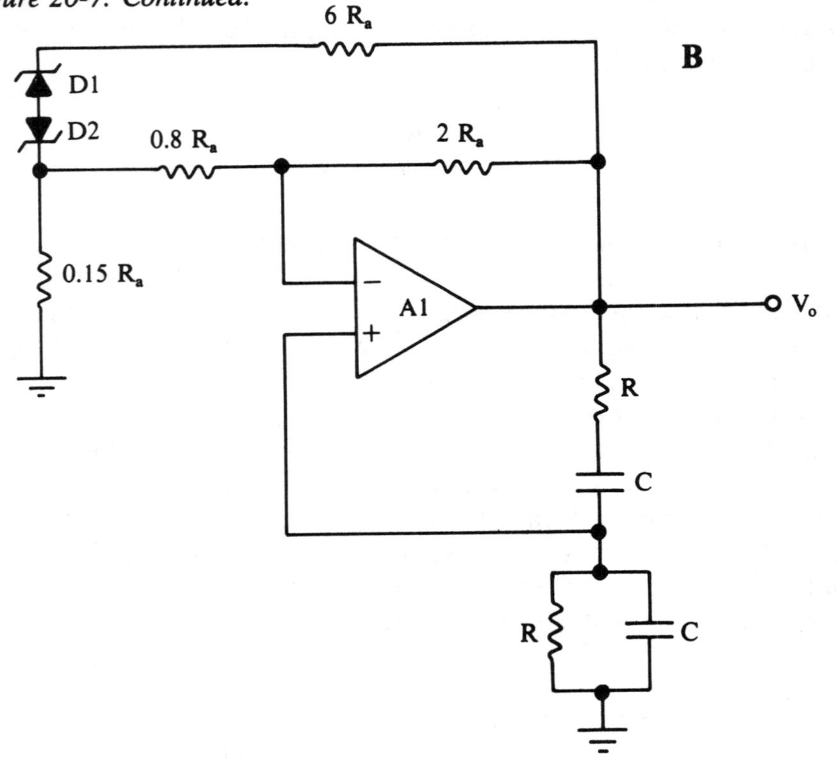

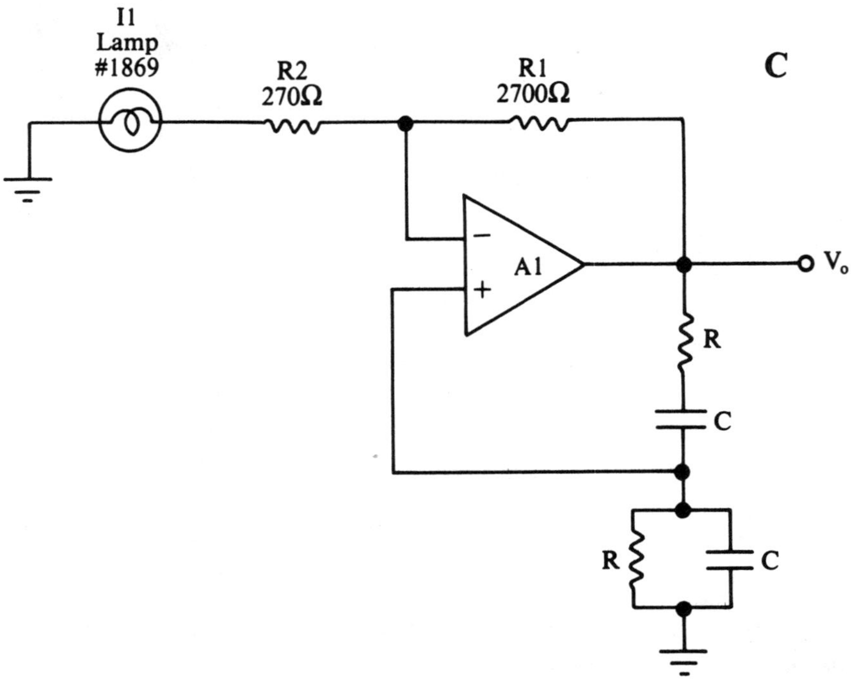

One final version of the gain-stabilized oscillator is shown in Fig. 20-7C. In this circuit, a small incandescent lamp is connected in series with resistor R2. When the amplitude of the output signal tries to increase above a certain level, the lamp will draw more current causing the gain to reduce. The lamp-stabilized circuit is probably the most popular form where stable outputs are required. A thermistor is sometimes substituted for the lamp.

Quadrature and biphasic oscillators

Signals that are *in quadrature* are of the same frequency, but are phase-shifted 90 degrees, with respect to each other. An example of quadrature signals are sine and cosine waves (Fig. 20-8A). One application for the quadrature oscillator is the demodulation of phase-sensitive detector signals in data-acquisition systems. The sine wave has an instantaneous voltage: $v = V\mathrm{Sin}(\omega_o t)$ and the cosine wave is defined by: $v = V\mathrm{Cos}(\omega_o t)$. Notice that the distinction between sine and cosine waves is meaningless, unless either both are present or if some other timing method is used to establish when "zero degrees" is supposed to occur. Thus, when sine and cosine waves are called for, it is in the context that both are present and a phase shift of 90 degrees is present between them.

The circuit for the quadrature oscillator is shown in Fig. 20-8B. It consists of two operational amplifiers, A1 and A2. Both amplifiers are connected as Miller integrators; A1 is a noninverting type and A2 is an inverting integrator. The output of A1 (V_{o1}) is assumed to be the sine-wave output. In order to make this circuit operate, a total of 360 degrees of phase shift is required between the output of A1, around the loop and back to the input of A1. Of the required 360-degree phase shift, 180 degrees are provided by the inversion that is inherent in the design of A2 (it is in the inverting configuration). Another 90 degrees is obtained because A2 is an integrator, so it inherently causes a 90-degree phase shift. An additional 90-degree phase shift is provided by RC network R_3C_3. If $R_1 = R_2 = R_3 = R$, and $C_1 = C_2 = C_3 = C$, the frequency of oscillation is:

$$F = \frac{1}{2\pi RC} \tag{20-23}$$

The cosine output (V_{o2}) is taken from the output of amplifier A2. The relative amplitudes are approximately equal, but the phase is shifted 90 degrees between the two stages.

A *biphasic oscillator* is a sine-wave oscillator that outputs two identical sine-wave signals that are 180 degrees out of phase with each other. The basic circuit is simple and is shown in block diagram form in Fig. 20-9. The biphasic oscillator consists of a sine-wave oscillator that is followed by a

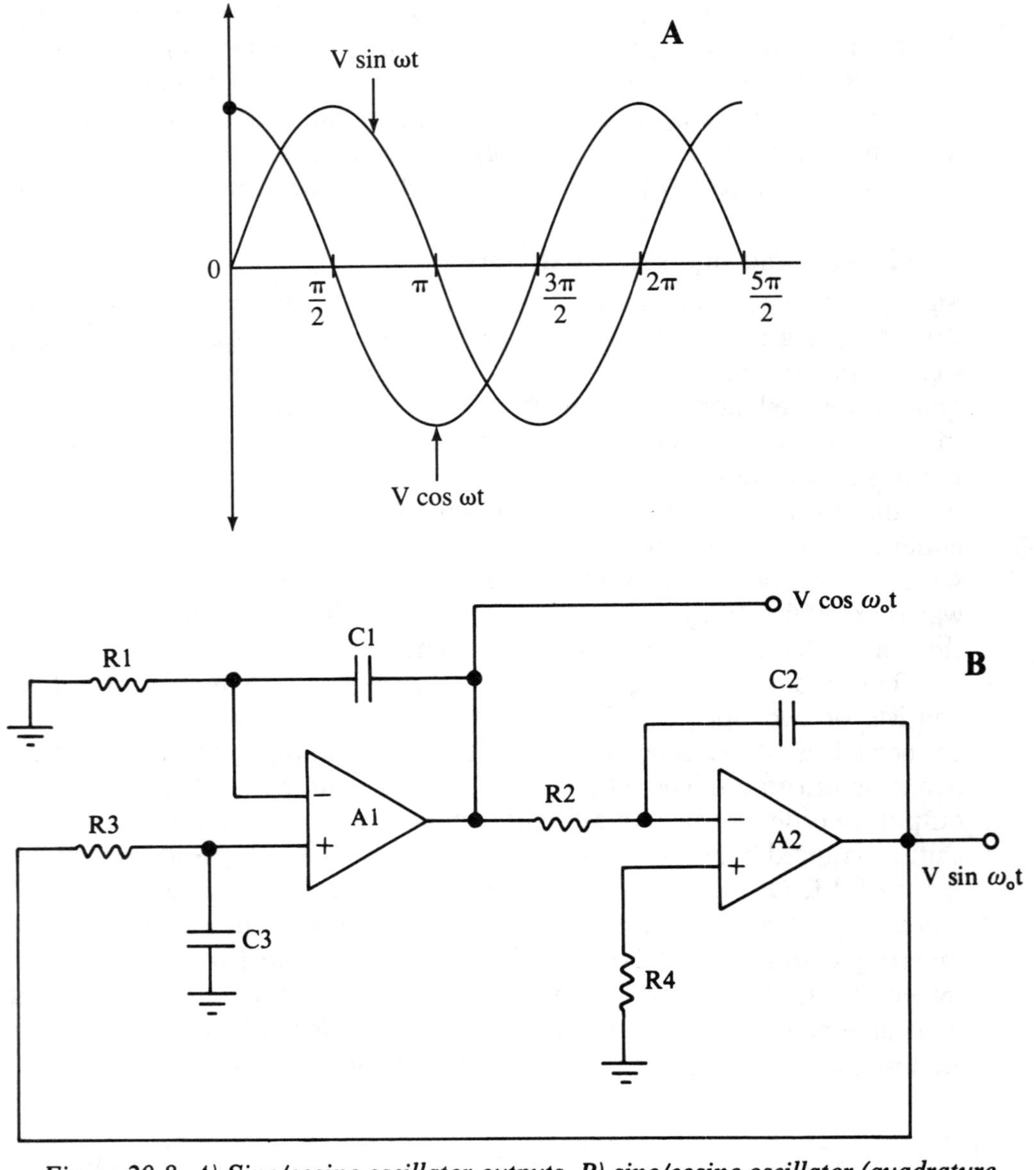

Figure 20-8. A) Sine/cosine oscillator outputs, B) sine/cosine oscillator (quadrature oscillator).

noninverting amplifier to produce the [VSin (ω_oT)] output, and an inverting amplifier that has a gain of -1 to produce the [VSin ($\omega_o T + \pi$)] output. Two amplifiers are needed to keep the phase error at a minimum. Otherwise, the propagation time of a single amplifier would cause a slight delay in the VSin ($\omega_o T + \pi$) signal. Biphasic oscillators are sometimes used in transducer-excitation applications in carrier amplifiers.

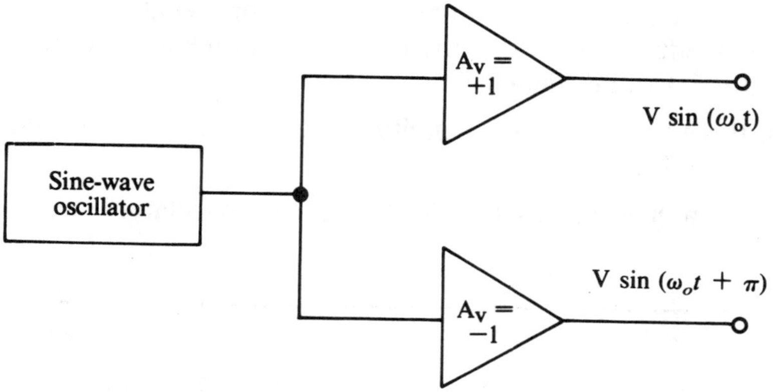

Figure 20-9. Biphasic oscillator.

Summary

The sine-wave oscillator generally consists of two parts: a feedback section, which selects the frequency of operation, and an amplifier section, which compensates for losses in the feedback section. For oscillation to occur, the product of the gains of these two sections must equal 1. In addition, the feedback must create a signal which will add to the original signal at a specific frequency, the frequency of oscillation. These two requirements are known as *Barkhausen's criteria.* Two common sine-wave oscillators are the phase-shift oscillator, which uses three resistors and three capacitors to select frequency, and the Wein-bridge oscillator, which uses a series RC network together with a parallel RC network to select frequency. Two other types of oscillators are quadrature oscillators, which generate two signals 90 degrees out of phase with one another, and biphasic oscillators, which create two signals 180 degrees out of phase with one another.

Questions

1. What are Barkhausen's criteria for oscillation?
2. For a particular oscillator, the feedback section has an attenuation of ⅓. What gain is needed for oscillation to occur?
3. Draw a diagram for a phase-shift oscillator. What is the attenuation of the feedback section?

4. Draw a diagram for a Wein-bridge oscillator. What is the attenuation of the feedback section? Why is the ratio of feedback to input resistor 2:1 rather than 3:1?
5. List three ways in which amplitude stability can be achieved in sine-wave oscillators.
6. What is a quadrature oscillator? a biphasic oscillator?

Problems

1. For the circuit in Fig. 20-3, $R_1 = R_2 = R_3 = 10$ kohms and $C_1 = C_2 = C_3 = 2.2\ \mu F$. **a.** Determine the necessary value of R_4 for oscillation to occur. **b.** Calculate the frequency of oscillation for the circuit.
2. Repeat Problem 1 for $R_1 = R_2 = R_3 = 50$ kohms and $C_1 = C_2 = C_3 = 0.01\ \mu F$.
3. For the circuit in Fig. 20-5, $R_1 = R_2 = 10$ kohms, $R_3 = 20$ kohms, $R_4 = 10$ kohms, and $C_1 = C_2 = 0.022\ \mu F$. Calculate the frequency of oscillation for the circuit.
4. Repeat Problem 3 for $R_1 = R_2 = 50$ kohms, $R_3 = 80$ kohms, $R_4 = 40$ kohms, and $C_1 = C_2 = 1\ \mu F$.

21 MONOSTABLE MULTIVIBRATORS

AFTER COMPLETING THIS chapter, you will be able to:

- □ Describe the characteristics of the monostable multivibrator
- □ Explain the different stages of the monostable multivibrator waveform

The monostable-multivibrator (MMV) has two permissible output states (high and low), but only one of them is stable. The MMV produces one output pulse in response to an input trigger signal (Fig. 21-1). In this case, a negative-going trigger pulse (V_t) results in a positive-going output pulse (V_o). The output pulse (V_o) has a duration, T, in which the output is in the *quasi-stable state*. The MMV is also known under several alternate names: *one shot*, *pulse generator*, and *pulse stretcher*. The latter name is derived from the fact that the output duration T is longer than the trigger pulse ($T > T_t$), as in Fig. 21-1.

Monostable multivibrators have a wide variety of applications in electronic circuits. Besides the pulse stretcher, the MMV also locks out unwanted pulses. Figure 21-2 shows that the output responds to only the first trigger pulse. The next two pulses occur during the active time, T, and so they are ignored. Such an MMV is said to be *nonretriggerable*. A common application of this feature is in switch contact "debouncing." All mechanical switch contacts bounce a few times on closure, which creates a short run of exponentially decaying pulses. If an MMV is triggered by the first pulse from the switch, and if the MMV remains quasi-active long enough for the bouncing pulses to die out, the MMV output signal becomes the debounced switch closure. The main requirement is that the MMV duration must be longer than the switch-contact bounce pulse train (5 mS is generally considered adequate for most switch types).

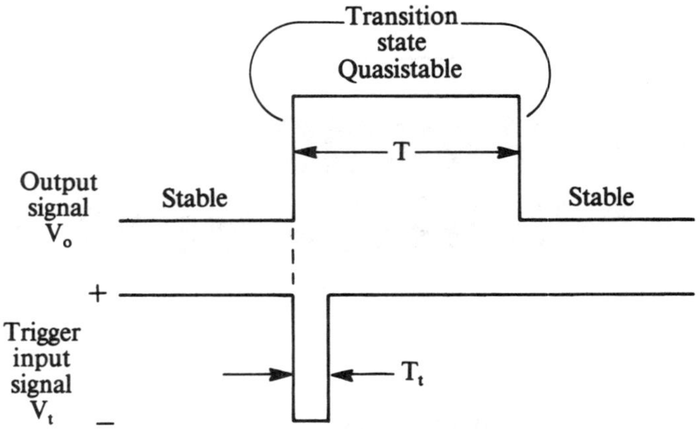

Figure 21-1. Monostable multivibrator (one shot) triggering operation.

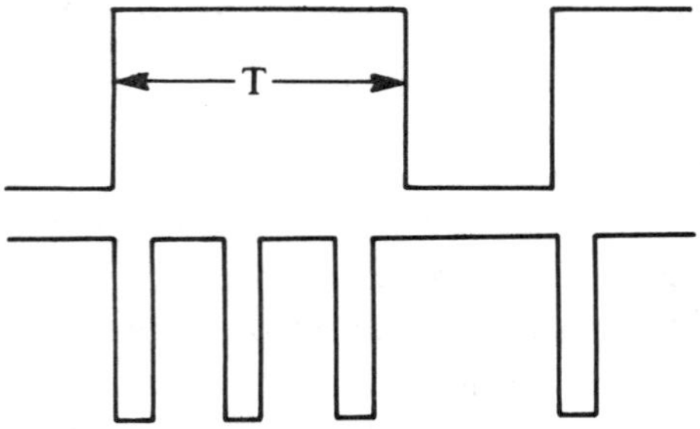

Figure 21-2. Effect of additional trigger pulses on a nonretriggerable circuit.

The range of possible MMV applications is too broad for a detailed description here, so the general categories are: pulse generation, pulse stretching, contact debouncing, pulse signal clean-up, switching, and synchronization of circuit functions (especially digital).

Figure 21-3A shows the circuit for a nonretriggerable monostable multivibrator that is based on an op amp; the timing diagram is shown in Fig. 21-3B. This circuit is based on the voltage-comparator circuit. When no feedback exists, the effective voltage gain of an op amp is its open-loop gain (A_{vol}). When both −In and +In are at the same potential, the differential input voltage (V_{id}) is zero, so the output is also zero. However, if $V_{(-In)}$ does not equal $V_{(+In)}$, the high gain of the amplifier forces the output to either its positive or negative saturation values. If $V_{(-In)} > V_{(+In)}$, the op amp sees a positive differential-input signal, so the output saturates at $-V_{sat}$. However, if $V_{(-In)} < V_{(+In)}$, the amplifier sees a negative differential-input signal and the output saturates to $+V_{sat}$. The operation of the MMV depends on the relationship of $V_{(-In)}$ and $V_{(+In)}$.

Four states of the monostable multivibrator that must be considered: stable state, transition state, quasi-stable state, and refractory state.

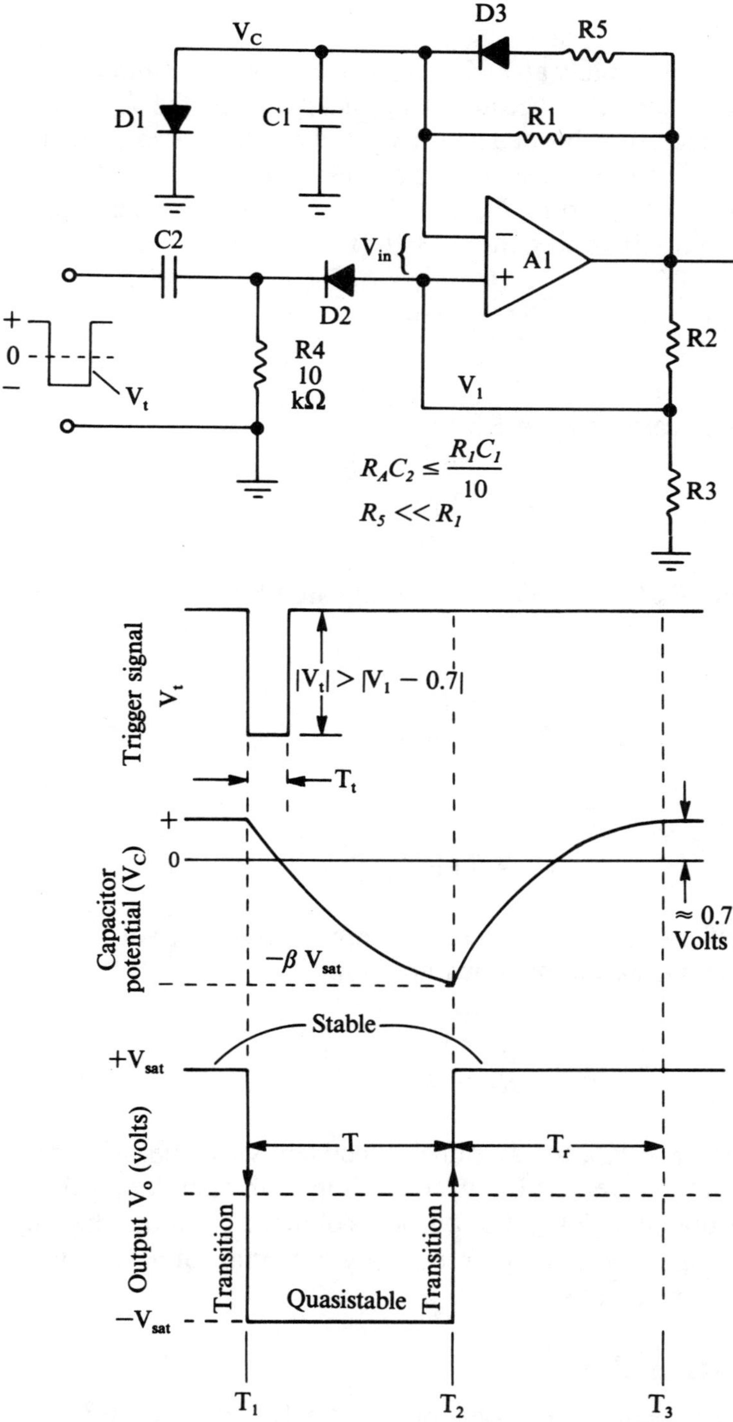

Figure 21-3. A) op-amp monostable, B) timing diagram.

Stable state

Output voltage V_o is initially at $+V_{sat}$. Capacitor C1 will attempt to charge in the positive-going direction because $+V_{sat}$ is applied to the R_1C_1 network. But, because diode D1 is shunted across C_1, the voltage across C_1 is clamped to $+V_{D1}$. For a silicon diode, such as the 1N914 or 1N4198, $+V_{D1}$ is about +0.7 Vdc. Thus, the inverting input (−In) is held to +0.7 Vdc during the stable state. The noninverting input (+In) is biased to a level V_1, which is:

$$V_1 = \frac{R_3\,(+V_{sat})}{R_2 + R_3} \tag{21-1}$$

or, in the special case of $R_2 = R_3$:

$$V_1 = \frac{+V_{sat}}{2} \tag{21-2}$$

The factor $(R_3/R_2 + R_3)$ is often designated by the Greek letter beta (β), so:

$$\beta = \frac{R_3}{R_2 + R_3} \tag{21-3}$$

therefore:

$$V_1 = \beta\,(+V_{sat}) \tag{21-4}$$

The amplifier (A1) sees a differential input voltage (V_{id}) of $V_1 - V_{D1}$, or $(V_1 - 0.7)$ volts. Using the previous notation:

$$V_{id} = \frac{R_3\,(+V_{sat})}{R_2 + R_3} - 0.7 \tag{21-5}$$

As long as $V_1 > V_{D1}$, the amplifier effectively sees a negative dc differential voltage at the inverting input, so (with its high open-loop gain, A_{vol}) it will remain saturated at $+V_{sat}$. For purposes of this discussion, the amplifier is a 741 that is operated at dc power-supply potentials of ±12 Vdc, so V_{sat} typically will be ±10 volts.

Transition state

The input trigger signal (V_t) is applied to the MMV of Fig. 21-3A through RC network R_4C_2. The general design rule for this network is that its time-constant should be not more than 10 percent of the time-constant of the timing network:

$$R_4C_2 < \frac{R_1C_1}{10} \tag{21-6}$$

At time T_1 (see Fig. 21-3B), trigger signal V_t makes an abrupt high-to-low transition to a peak value that is less than $(V_1 - 0.7)$ volts. Under this condition, the polarity of V_{id} is now reversed and the inverting input now sees a positive voltage: $(V_1 + V_t - 0.7)$ is less than V_{D1}. Output voltage V_o now snaps rapidly to $-V_{sat}$. The fall time of the output signal depends on the slew rate and the open-loop gain of op amp A1.

Quasi-stable state

The output signal from the MMV is the quasi-stable state between T_1 and T_2 in Fig. 21-3B. It is called *quasi-stable* because it does not change over $T = T_2 - T_1$, but when T expires, the MMV "times out" and V_o reverts to the stable state $(+V_{(sat)})$.

During the quasi-stable time, D1 is reverse-biased, and capacitor C1 discharges from +0.7 Vdc to zero and then recharges toward $-V_{sat}$. When $-V_o$ reaches $-V_1$, however, the value of V_{id} crosses zero and that change forces V_o to snap back to $+V_{sat}$.

From Eq. (21-6), it is possible to derive the timing equation for the MMV. The timing capacitor must charge from an initial value (V_{C1}) to a final value (V_{C2}) in time T. The question is: "What value of R_1C_1 will cause the required transitions?" Consider the case $R_2 = R_3$ $(V_1 = 0.5V_{sat})$:

$$R_1C_1 = \frac{-T}{Ln\left(\dfrac{V_{sat} - V_{C2}}{V_{sat} - V_{C1}}\right)} \tag{21-7}$$

$$R_1C_1 = \frac{-T}{Ln\left(\dfrac{V_{sat} - ((0.5)\,(V_{sat} + 0.7))}{V_{sat} - 0.7}\right)} \tag{21-8}$$

and, for the case where $V_{sat} = 10$ Vdc:

$$R_1C_1 = \frac{-T}{Ln\left(\dfrac{10\text{ Vdc} - ((0.5)\,(10\text{ Vdc} + 0.7))}{10\text{ Vdc} - 0.7}\right)} \tag{21-9}$$

$$R_1C_1 = \frac{-T}{Ln\left(\dfrac{10 \text{ Vdc} - 5.35 \text{ volts}}{10 \text{ Vdc} - 0.7 \text{ volts}}\right)} \tag{21-10}$$

$$R_1C_1 = \frac{-T}{Ln\left(\dfrac{4.65}{9.3}\right)} \tag{21-11}$$

$$R_1C_1 = \frac{-T}{Ln\ (0.5)} \tag{21-12}$$

$$R_1C_1 = \frac{-T}{-0.69} \tag{21-13}$$

Thus,

$$T = 0.69\ R_1C_1 \tag{21-14}$$

Equation (21-14) represents the special case in which $\beta = 1/2$ (i.e., $R_2 = R_3$). Although $R_2 = R_3$ is usually true for this class of circuit, R_2 and R_3 are not always equal. A more generalized expression is:

$$RC = \frac{T}{Ln\left(\dfrac{1 + 0.7\ V/V_{sat}}{1 - \beta}\right)} \tag{21-15}$$

in which:

$$\beta = \frac{R_3}{R_2 + R_3} \tag{21-16}$$

When the quasi-stable state times out, the circuit status returns to the stable state (where it remains dormant until it is triggered again).

Experiment 21-1

A monostable multivibrator is constructed so that $R_2 = 10$ kohms and $R_3 = 4.7$ kohms. If the active device (A1) is a 741 op amp and the dc power supplies are ± 12 Vdc, $\pm V_{sat} = \pm 10$ volts. Find the RC time constant to produce a 5-second output pulse. Also, propose values for sample components.

1. First, calculate β:

$$\beta = \frac{R_3}{R_2 + R_3}$$

$$\beta = \frac{4.7 \text{ kohms}}{(10 \text{ kohms} + 4.7 \text{ kohms})}$$

$$\beta = \frac{4.7}{14.7} = 0.32$$

2. Calculate R_1C_1:

$$RC = \frac{T}{Ln\left(\dfrac{1 + 0.7 \text{ volts}/V_{sat}}{1 - \beta}\right)}$$

$$R_1C_1 = \frac{5 \text{ seconds}}{Ln\left(\dfrac{1 + (0.7 \text{ volts}/10 \text{ volts})}{1 - 0.32}\right)}$$

$$R_1C_1 = \frac{5 \text{ seconds}}{Ln\left(\dfrac{1.07}{0.68}\right)}$$

$$R_1C_1 = \frac{5 \text{ seconds}}{Ln\ (1.57)}$$

$$R_1C_1 = \frac{5 \text{ seconds}}{0.453} = 11.04 \text{ seconds}$$

3. The time constant of R_1C_1 is 11.04 seconds. Select a standard-value capacitor and find the resistor values that will give the correct value of R_1C_1.

Candidates:

$C_1 = 10\ \mu\text{F}$, $R_1 = 1.1$ megohms (1 megohms)
$C_1 = 6.8\ \mu\text{F}$, $R_1 = 1.62$ megohms (1.5 megohms)
$C_1 = 4.7\ \mu\text{F}$, $R_1 = 2.35$ megohms (2.2 megohms)

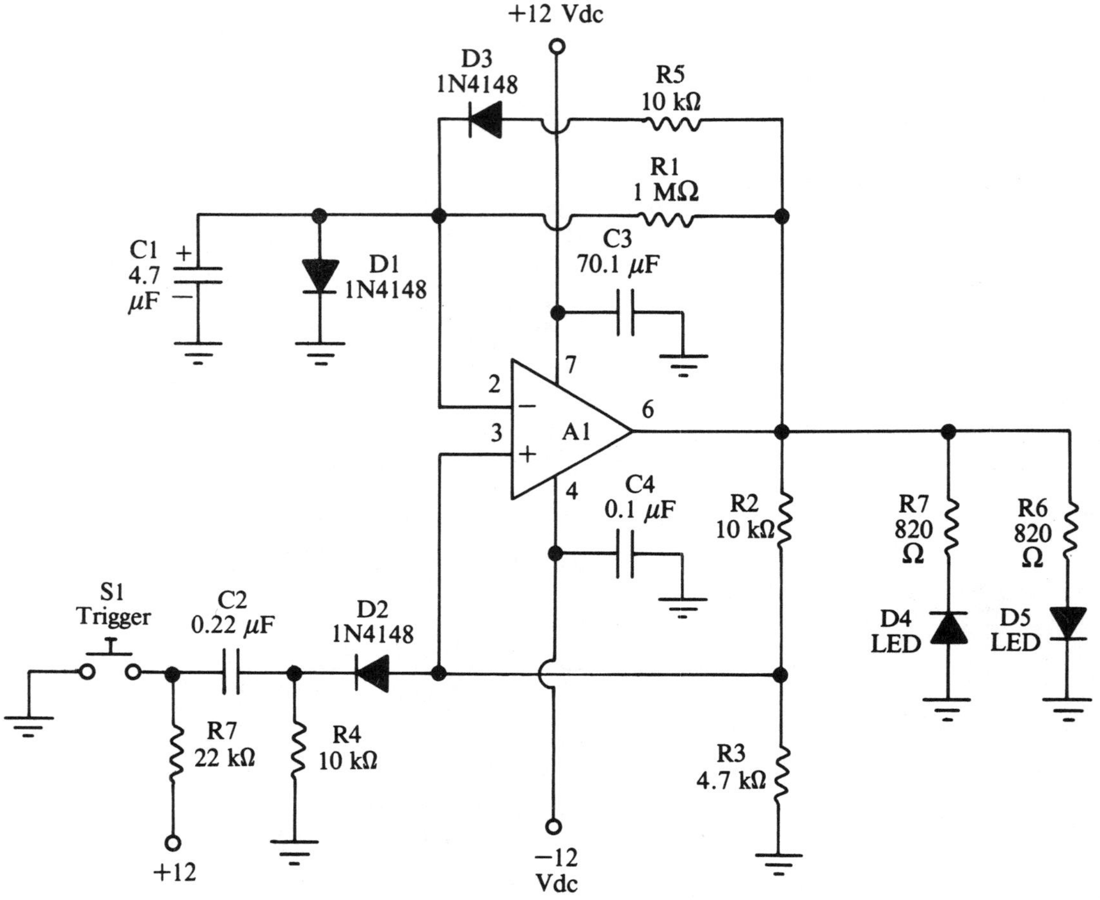

Experiment 21-1. Circuit for Experiment 21-1.

If the accuracy of the pulse is not critical, select one of the standard-value capacitors and the nearest standard-value resistor (shown in parenthesis).

Experiment 21-2

This experiment explores the monostable multivibrator (one-shot) circuit. You will not need an oscilloscope for this experiment, but one can be used to monitor either V_c or V_o.

1. Construct the circuit of Experiment 21-2 with the values shown for R_1C_1 (alternate values from Experiment 21-1 can be substituted, if you wish). The output indicator consists of two LEDs and their

associated current-limiting resistors (R6 and R7). A 741 op amp can be used for A1. Switch S1 and a pull-up resistor act as a source for trigger signals.

2. Turn the power on. Only one LED should light; this is the stable state.
3. Press S1 momentarily, then release. The output should snap to the other output state, which causes the lighted LED to extinguish and the unlit LED to light. This is the quasi-stable state.
4. The second LED should remain lit for about 5 seconds. Time the actual quasi-stable output duration with a sweep second hand.
5. After the one shot "times out" (i.e., returns to the stable state), trigger the circuit again (press S1). Release S1, then immediately press it again. If you time the circuit, you will find that the output duration does not increase following the second closing of S1.

Refractory period

At time t_2, output signal voltage V_o switches from $-V_{sat}$ to $+V_{sat}$. Although the output has timed out, the MMV is not yet ready to accept another trigger pulse. The *refractory state* between t_2 and t_3 is characterized by having the output in the stable state, while the input is unable to accept a new trigger input stimulus. The refractory period must await the discharge of C1 under the influence of the output voltage, to satisfy $V_1 < (V_1 - 0.7)$ volts.

The monostable-multivibrator circuit will be visited again—in the form of the 555 IC timer in chapters 22 and 23.

Summary

The monostable multivibrator (MMV), or one shot, has one stable state and one unstable, or quasi-stable, state. It is normally in the stable state and requires an external trigger to move it into the quasi-stable state. It stays in this state for a period of time determined by external components and then returns to its stable state. The four different parts of the waveform are the stable state, the transition state (or rise time), the quasi-stable state, and the refractory period, during which it has returned to the stable state but is not yet ready to receive a new trigger.

Questions

1. What is a monostable multivibrator? What are other names for this circuit?
2. What are the four parts of the timing cycle for a monostable multivibrator? What determines the length of each?

Problems

1. For the circuit in Fig. 21-3B, $R_2 = R_3 = 20$ kohms. If $R_1 = 100$ kohms and $C_1 = 0.1\ \mu F$, determine the length of the quasi-stable period of the multivibrator.
2. Repeat Problem 1 for $R_1 = 50$ kohms and $C_1 = 0.022\ \mu F$.
3. A pulse train consisting of 0.1-ms pulses 2 ms apart is used to trigger a monostable multivibrator whose quasi-stable state lasts 5 ms. What is the frequency of the output signal from the MMV?
4. Repeat Problem 3 for 2-ms pulses 50 ms apart triggering a monostable multivibrator whose quasi-stable state lasts 120 ms.

22 THE 555 IC TIMER

AFTER COMPLETING THIS chapter, you will be able to:

- ☐ Describe the internal circuits of the 555 timer
- ☐ Discuss the monostable operation of the 555 timer
- ☐ Discuss the astable operation of the 555 timer

The IC timer represents a class of chips that are extraordinarily well behaved and very easy to use. These timers are based on the properties of the series RC timing network (see chapter 16) and the voltage comparator. A combination of voltage-comparator circuits and digital circuits are used inside the 555 chips. Although several devices are on the market, the most common and best known is the 555 device. The 555 is now made by a number of different semiconductor manufacturers, but it was originated by Signetics, Inc., in 1970. Today, the 555 remains one of the most widespread, bestselling ICs on the market, rivaling even some general-purpose op amps in numbers sold.

The original Signetics products included the SE-555, which operated at a temperature range of −55 to +125°C, and the NE-555, which operated at 0 to +70°C. Several different designations are now commonly used for 555s that are made by other companies, including 555 and LM-555. A dual 555-class timer is also marketed as a 556. A low-power CMOS version of the 555 is also marketed as the LMC-555.

The 555 is a multi-purpose chip that will operate at dc power-supply potentials from +5 to +18 Vdc. The temperature stability of these devices is on the order of 50 PPM/°C (i.e., 0.005%/°C). The output of the 555 can either sink or source up to 200 mA of current. It is compatible with TTL devices (when the 555 is operated from +5-Vdc power supply), CMOS devices, op amps, other linear IC devices, transistors, and most classes of solid-state devices. The 555 will also operate with most passive electronic components. Several factors contribute to the popularity of the 555 device. Besides the versatile nature of the device, its operation is straightforward and circuit designs are generally simple. Like the general-purpose op amp, the 555 usually works in a predictable manner, according to the standard published equations.

The 555 operates in two different modes: monostable (one shot) and astable (free running). Figure 22-1A shows the astable-mode output from pin 3 of the 555. The waveform is a series of square waves that can be varied in duty cycle from 50 to 99.9 percent, and in frequency from less than 0.1 Hz to more than 100 kHz. Monostable operation (Fig. 22-1B) requires that a trigger pulse is applied to pin 2 of the 555. The trigger must drop from a level $>2(V+)/3$ down to $<(V+)/3$. Output pulse durations from microseconds up to hours are possible. The principal constraint on longer operation is the leakage resistance of the capacitor that is used in the external timing circuit.

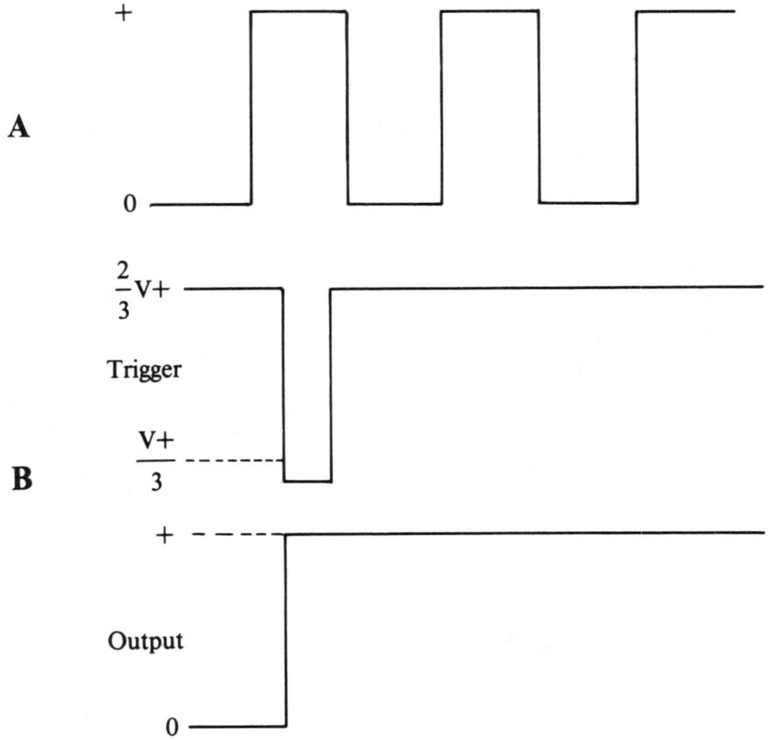

Figure 22-1. A) Square-wave output from 555, B) monostable output from 555.

Pinouts and internal circuits of the 555 IC timer

The package for the 555 device is shown in Fig. 22-2. Most 555s are sold in the 8-pin miniDIP package as shown, although some are found in the 8-pin

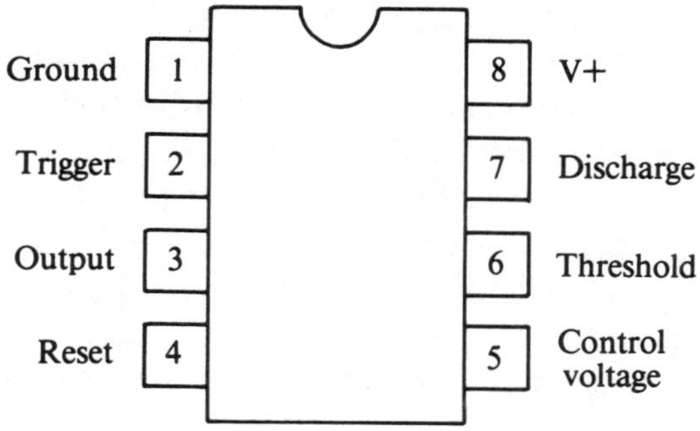

Figure 22-2. 555 in miniDIP package.

metal-can IC package. The latter are mostly the military-specification temperature-range SE-555 series. The pinouts are the same on both miniDIP and metal-can versions. The internal circuitry is shown in block form in Fig. 22-3. The following stages are found: two voltage comparators

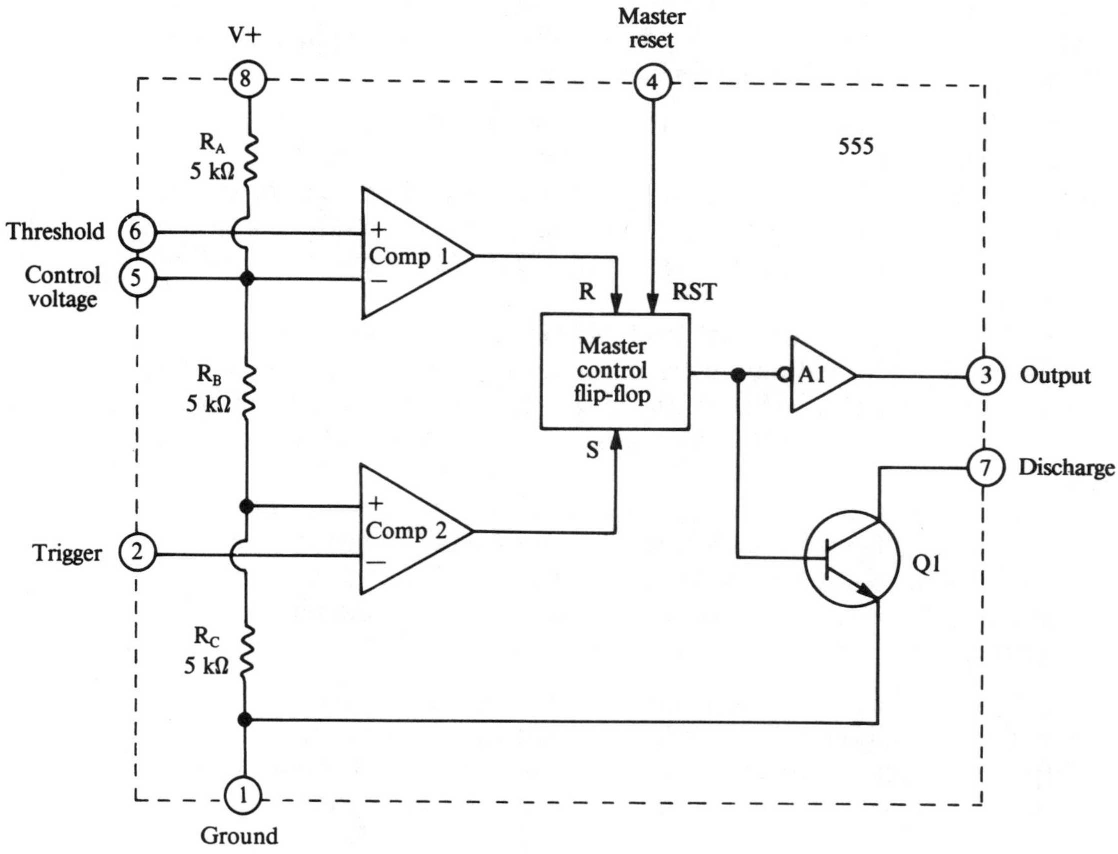

Figure 22-3. Internal circuit of 555.

(COMP1 and COMP2), a reset/set (RS) control flip-flop (which can be reset from outside the chip through pin 4), an inverting output amplifier (A1), and a discharge transistor (Q1). The bias levels of the two comparators are determined by a resistor voltage divider (R_a, R_b, and R_c) between $V+$ and ground. The inverting input of COMP1 is set to $2(V+)/3$ and the noninverting input of COMP2 is set to $(V+)/3$. Figures 22-2 and 22-3 show the pinouts of the 555. They should be memorized. In the following descriptions, the term *high* implies a level $> 2(V+)/3$, and *low* implies a grounded condition ($V = 0$), unless otherwise specified. These pins serve the following functions:

Ground (pin 1) This pin is the common reference point for all signals and voltages in the 555 circuit, both internal and external to the chip.

Trigger (pin 2) The trigger pin is normally held at a potential $>2(V+)/3$. In this state, the 555 output (pin 3) is low. If the trigger pin is brought low to a potential $<(V+)/3$, the output (pin 3) abruptly switches to the high state. The output remains high as long as pin 2 is low, but the output does not necessarily revert back to low immediately after pin 2 is brought high again (see operation of the following threshold input).

Output (pin 3) The output pin of the 555 is capable of either sinking or sourcing current up to 200 mA. This operation is in contrast to other IC devices, in which the outputs of various devices will either sink or source current, but not both. Whether the 555 output operates as a sink or a source depends on the external-load configuration. Figure 22-4 shows both types of operation.

In Fig. 22-4A, the external load (R_L) is connected between the 555 output and $V+$. Current only flows in the load when pin 3 is low. In that condition, the external load is grounded through pin 1 and a small internal source resistance, R_{S1}. In this configuration, the 555 output is a current sink.

The operation that is depicted in Fig. 22-4B is for the case where the load is connected between pin 3 of the 555 and ground. When the output is low, the load current is zero. When the output is high, however, the load is connected to $V+$ through a small internal resistance R_{S2} and pin 8. In this configuration, the output serves as a current source.

Reset (pin 4) The reset pin is connected to a preset input of the 555 internal control flip-flop. When a low is applied to pin 4 the output of the 555 (pin 3) switches immediately to a low state. In normal operation, it is common to connect pin 4 to $V+$ in order to prevent false resets from noise impulses.

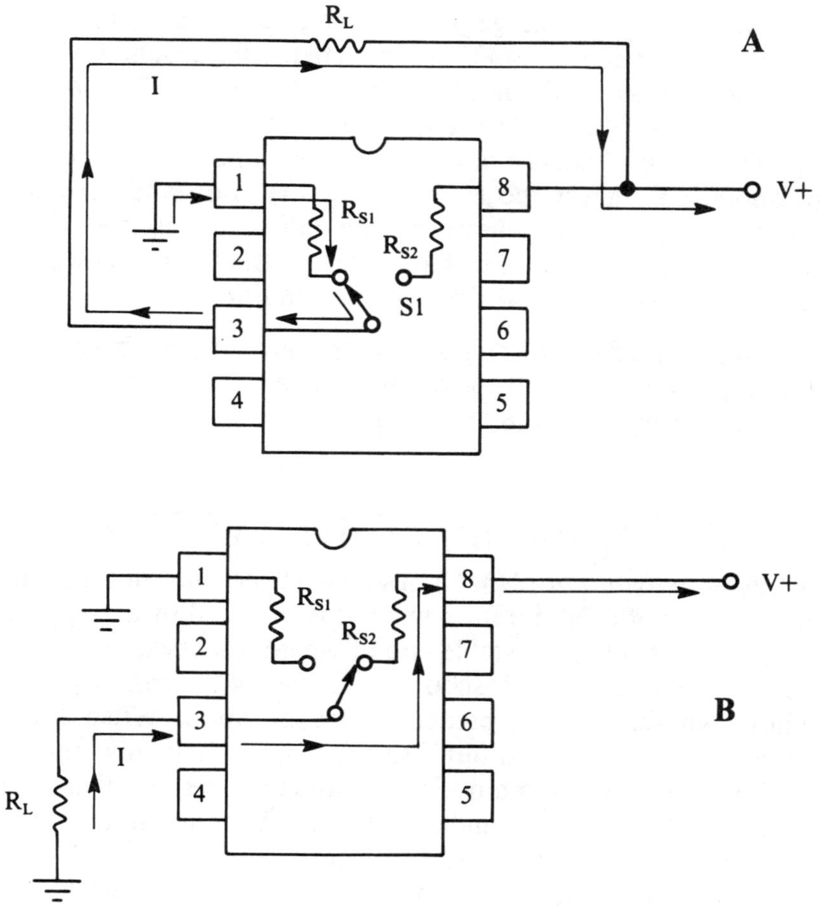

Figure 22-4. A) Current sink operation, B) current source operation.

Control voltage (pin 5) This pin normally rests at a potential of $2(V+)/3$ as a result of an internal resistive voltage divider (see R_a through R_c in Fig. 22-3). Apply an external voltage to this pin or connect a resistor to ground to change the duty cycle of the output signal. If not used, then pin 5 should be decoupled to ground through a 0.01-to-0.1-μF capacitor.

Threshold (pin 6) This pin is connected to the noninverting input (+In) of comparator COMP1, and it is used to monitor the voltage across the capacitor in the external RC timing network. If pin 6 is at a potential of $<2(V+)/3$, the output of the control flip-flop is low and the output (pin 3) is high. Alternatively, when the voltage on pin 6 is $\geq 2(V+)/3$, the output of COMP1 is high and chip output (pin 3) is low.

Discharge (pin 7) The discharge pin is connected to the collector of npn transistor Q1, and the emitter of Q1 is connected to the ground (pin 1). The base of Q1 is connected to the not-Q output of the control flip-flop. When the 555 output is high, the not-Q output of the control flip-flop is low, so Q1 is turned off. The CE resistance of Q1 is very high under this condition, so does not appreciably affect the external circuitry. But when the control flip-flop not-Q output is high, however, the 555 output is low and Q1 is biased on hard. The CE path is in saturation, so the CE resistance is very low. Pin 7 is effectively grounded under this condition.

V+ power supply (pin 8) The dc power supply is connected between pin 1 (ground) and pin 8 (positive). In good practice, a 0.1-to-10-μF decoupling capacitor is used between pin 8 and ground.

Monostable operation of the 555 IC timer

A *monostable multivibrator* (*MMV*), also called the one shot, produces a single output pulse of fixed duration when it is triggered by an input pulse. Op-amp versions of the monostable circuit where discussed in chapter 21. The output of the one shot will snap high following the trigger pulse, and will remain high for a fixed, predetermined duration. When this time expires, the one shot is "timed-out," so the output snaps low again. The output of the one shot will remain low indefinitely, unless another trigger is applied to the circuit. The 555 can be operated as a monostable multivibrator by suitably connecting the external circuit.

Figure 22-5 shows the operation of the 555 as a monostable multivibrator. In order to make the operation of the circuit easier to understand, Fig. 22-5A shows the internal circuitry as well as the external circuitry. Figure 22-5B shows the timing diagram for this circuit (Fig. 22-5C shows the same circuit in the more conventional schematic diagram format).

The two internal voltage comparators are biased to certain potential levels by a series voltage dividers that consist of internal resistors $R_a R_b$, and R_c. The inverting input of voltage comparator COMP1 is biased to $2(V+)/3$, while the noninverting input of COMP2 is biased to $(V+)/3$. These levels govern the operation of the 555 device in whichever mode is selected. An external timing network ($R_1 C_1$) is connected between V+ and the noninverting input of COMP1 via pin 6. Also connected to pin 6 is 555 pin 7, which has the effect of connecting the transistor across capacitor C1. If the transistor is turned on, the capacitor looks into a very low resistance short circuit through the CE path of the transistor.

When power is initially applied to the 555, the voltage at the inverting input of COMP1 will go immediately to $2(V+)/3$ and the noninverting

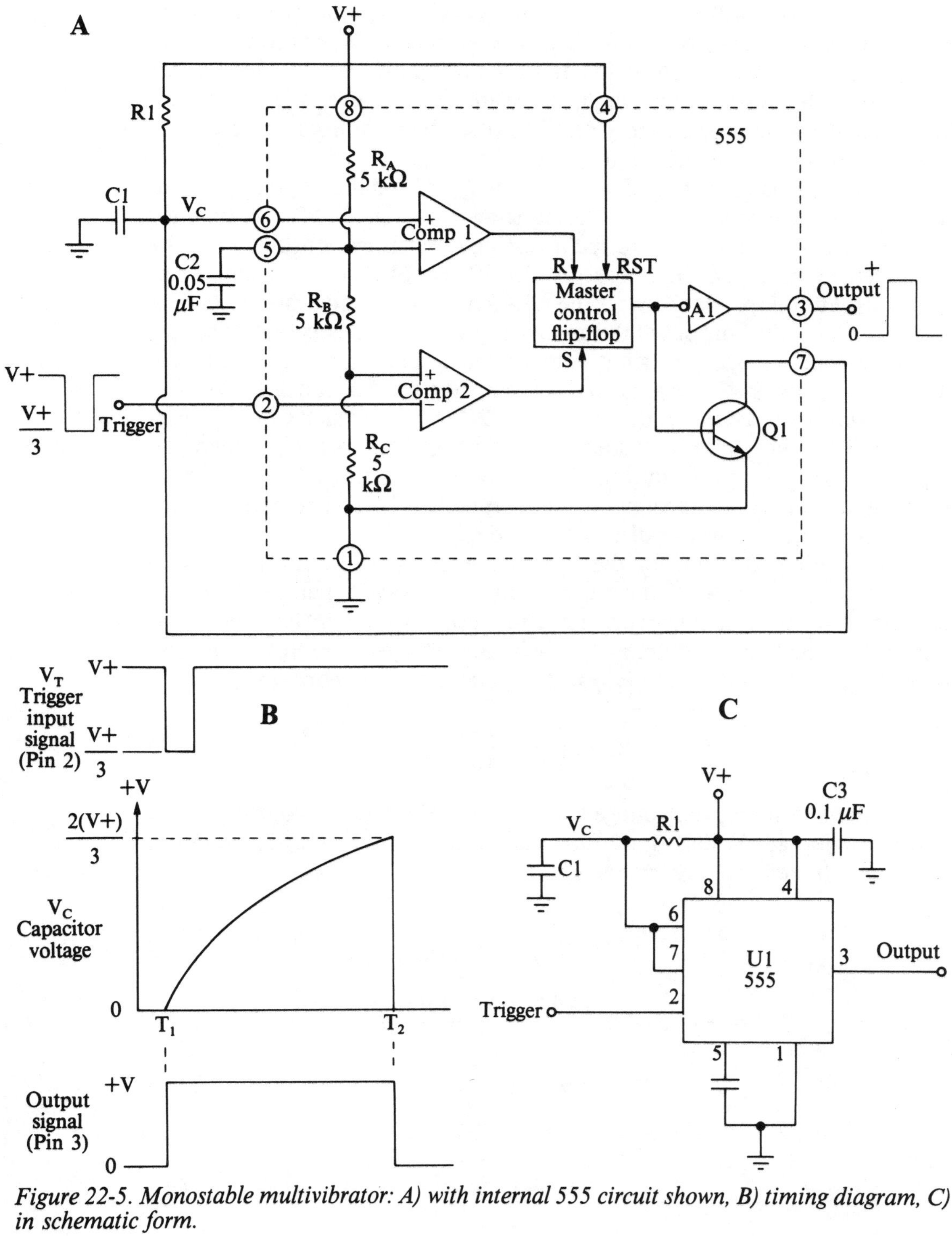

Figure 22-5. Monostable multivibrator: A) with internal 555 circuit shown, B) timing diagram, C) in schematic form.

input of COMP2 will go to $(V+)/3$. The control flip-flop is in the reset condition, so the not-Q output is high. Because this flip-flop is connected to output pin 3 through an inverting amplifier (A1), the output is low at this point. Also, because not-Q is high, transistor Q1 is biased into saturation, which creates a short circuit to ground across external timing capacitor C1. The capacitor remains discharged in this condition ($V_c = 0$).

If a trigger pulse is applied to pin 2 of the 555, and if that pulse drops to a voltage that is less than $(V+)/3$, as shown in Fig. 22-5B, comparator COMP2 sees a situation where the inverting input is less positive than the noninverting input, so the output of COMP2 snaps high. This action sets the control flip-flop, which forces the not-Q output low, and the 555 output high. The low at the output of the control flip-flop also means that transistor Q1 is now unbiased, so the short across the external capacitor is removed. The voltage across C1 begins to rise (see Figs. 22-5B and 22-5D). The voltage will continue to rise until it reaches $2(V+)/3$, at which time comparator COMP1 will snap high and cause the flip-flop to reset. When the flip-flop resets, its not-Q output drops low again, terminates the output pulse, and returns the capacitor voltage to zero. The 555 remains in this state until another trigger pulse is received.

The timing equation for the 555 can be derived in exactly the same manner as the equations that are used with the operational amplifier MMV circuits. The basic RC network equation was covered in chapter 16; it relates the time that is required for a capacitor voltage to rise from a starting point (V_{C1}) to an end point (V_{C2}) with a given RC time constant:

$$T = -RC\ Ln\left(\frac{V - V_{C2}}{V - V_{C1}}\right) \tag{22-1}$$

In the 555 timer, the voltage source is V+, the starting voltage is zero, and the trip-point voltage for comparator COMP1 is $2\,(V+)/3$. Equation (22-1) can therefore be rewritten as:

$$T = -R_1C_1\ Ln\left(\frac{V - V_{C2}}{V - V_{C1}}\right) \tag{22-2}$$

$$T = -R_1C_1\ Ln\left(\frac{(V+) - 2\,(V+)/3}{V+}\right) \tag{22-3}$$

$$T = -R_1C_1\ Ln(1 - 0.667) \tag{22-4}$$

$$T = -R_1C_1\ Ln\ (0.333) \tag{22-5}$$

$$T = 1.1R_1C_1 \tag{22-6}$$

Input triggering methods for the 555-MMV circuit

The 555-MMV circuit is triggered by bringing pin 2 from a positive voltage down to a level that is less than $(V+)/3$. The circuit can be triggered if a pulse is applied from an external signal source, or through other means. Figure 22-6 shows the circuit for a simple pushbutton switch trigger circuit. A pull-up resistor (R2) is connected between pin 2 and V+. If normally-open (n.o.) pushbutton switch S1 is open, the trigger input is held at a potential very close to V+. When S1 is closed, however, pin 2 is brought low to ground potential. Because pin 2 is now at a potential less than $(V+)/3$, the 555 MMV will trigger. This circuit can be used for contact debouncing.

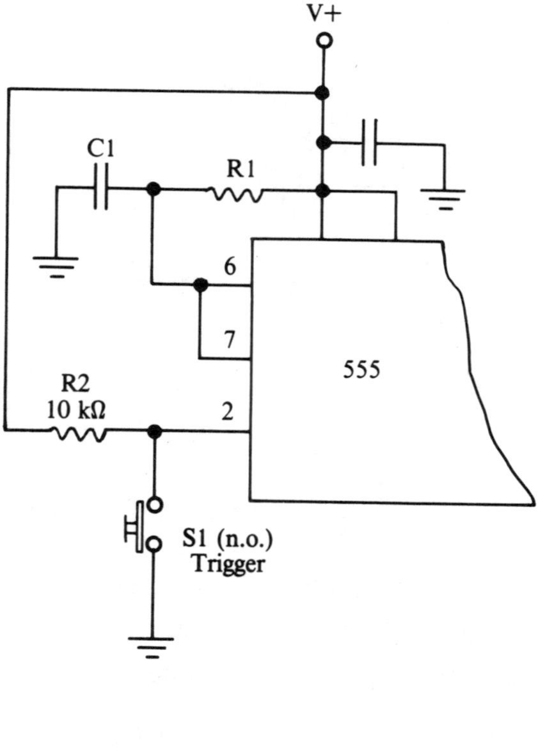

Figure 22-6. Pushbutton triggering of 555.

Experiment 22-1

This experiment examines the operation of the monostable-multivibrator (one shot) configuration of the 555 IC timer. The time duration of the output signal is set long enough that you can explore the operation without an oscilloscope. Those readers who use a 'scope can reduce the values to $C_1 = 0.05\ \mu F$, $R_{1A} = 33$ kohms, and $R_{1B} = 20$ kohms.

The output indicator is an LED in series with a current-limiting resistor (R3, 820 ohms). The 555 device can serve as both a *current source* and a *current sink*. In this circuit, the current-source capability is used, so the LED and the resistor are connected from pin 3 to ground.

1. Build the circuit of Experiment 22-1. Switch S1 is a normally open (n.o.) pushbutton type that closes when pressed.

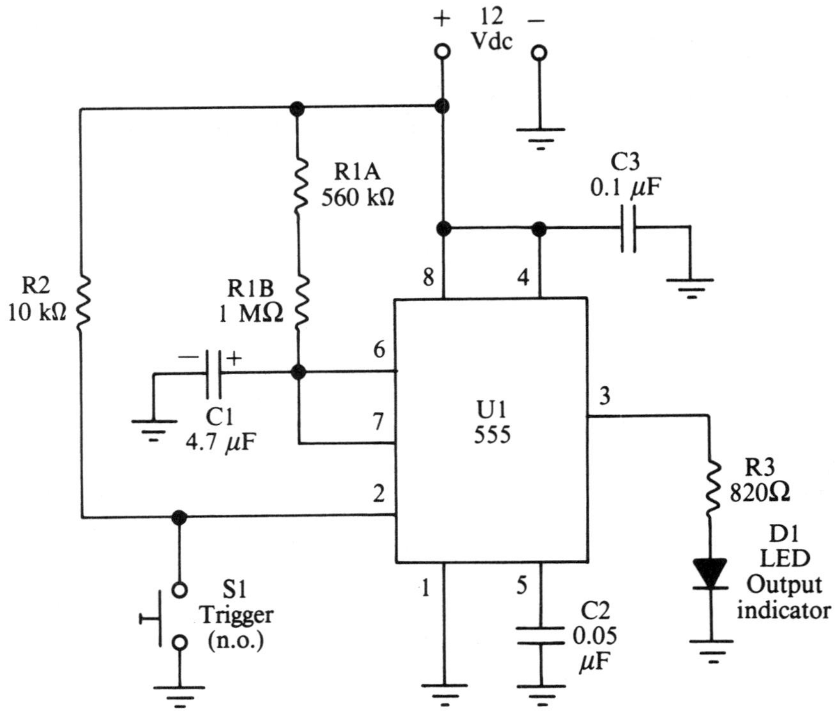

Experiment 22-1. Circuit for Experiment 22-1.

2. Turn on the circuit; the LED should be off.
3. Set R1B to the minimum value (R_{1B} = 0 ohms).
4. Calculate the output time duration with the equation:

$$T = 1.1\ (R_{1A} + R_{1B})\ C_1$$

5. Connect a high-impedance voltmeter across C1.
6. Press S1, while watching the voltmeter and the output LED. The output duration can be measured with a watch or clock that is equipped with a sweep second hand.

The LED should light for a period of time that is approximately equal to the duration calculated on page 372 (the difference is caused by measurement errors and tolerances of the resistor and capacitor values).

7. Repeat the process using various settings of R1B from minimum to maximum value.

A circuit for inverting the trigger pulse applied to the 555 is shown in Fig. 22-7. In this circuit, an npn bipolar transistor is used in the common-emitter mode to invert the pulse. Again, a pull-up resistor is used to keep pin 2 at V+ when the transistor is turned off. But when the positive-polarity trigger pulse is received at the base of transistor Q1, the transistor saturates, which forces the collector (and pin 2 of the 555) to near ground potential.

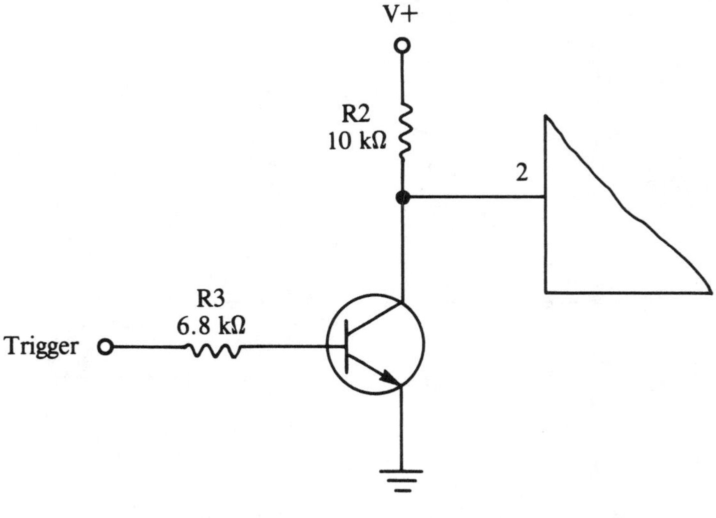

Figure 22-7. Inverted-pulse triggering.

Figure 22-8 shows two ac-coupled versions of the trigger circuit. In these circuits, a pull-up resistor keeps pin 2 normally at *V*+. But when a pulse is applied to the input end of capacitor C4, a differentiated (i.e., spiked) version of the pulse is created at the trigger input of the 555. Diode D1 clips the positive-going spike to 0.6 or 0.7 volts below *V*+ and passes only the negative-going pulse to the 555. If the negative-going spike can counteract the positive bias that is sufficiently provided by R2 to force the voltage lower than (*V*+)/3, the 555 will trigger. A pushbutton switch version of this same circuit is shown in Fig. 22-8B.

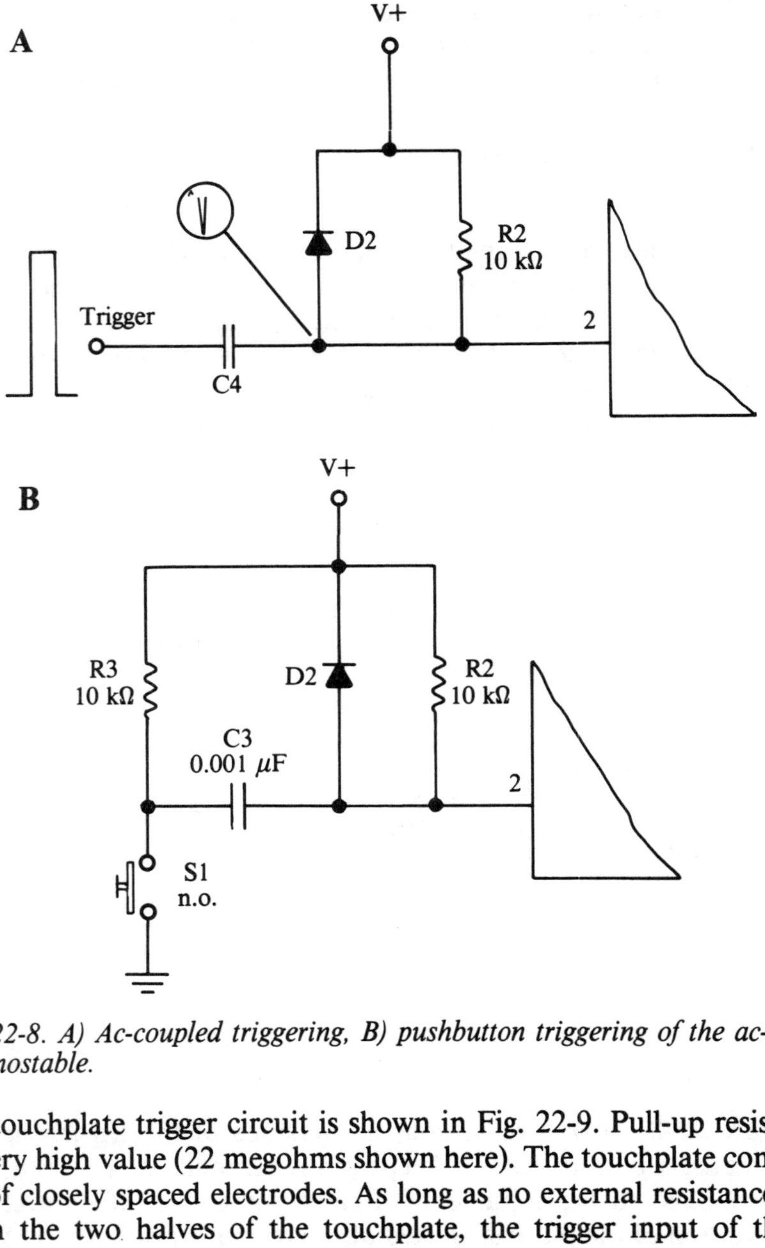

Figure 22-8. A) Ac-coupled triggering, B) pushbutton triggering of the ac-coupled 555 monostable.

A touchplate trigger circuit is shown in Fig. 22-9. Pull-up resistor R2 has a very high value (22 megohms shown here). The touchplate consists of a pair of closely spaced electrodes. As long as no external resistance exists between the two halves of the touchplate, the trigger input of the 555 remains at *V+*. But when a resistance is connected across the touchplate, voltage *V1* drops to a very low value. If the average finger resistance is about 20 kohms, the voltage drops to:

$$V_1 = \frac{(V+)\ (20\ \text{kohms})}{(R_2 + 20\ \text{kohms})} \qquad (22\text{-}7)$$

Figure 22-9. Touch-plate triggering.

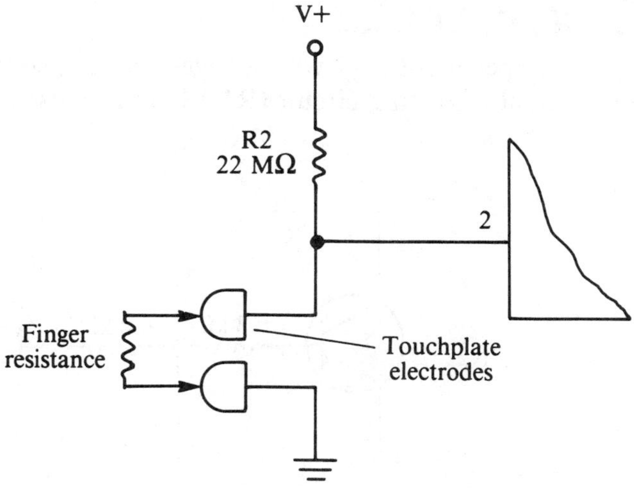

When R_2 = 22 megohms, the voltage drops to 0.0009 (V+), which is certainly less than (V+)/3.

The same concept is used in the liquid level detector shown in Fig. 22-10. Once again, a 22-megohm pull-up resistor is used to keep pin 2 at V+ under normal operation. When the liquid level rises sufficiently to short the electrodes, however, the voltage on pin 2 (V_1) drops to a very low level, which forces the 555 to trigger.

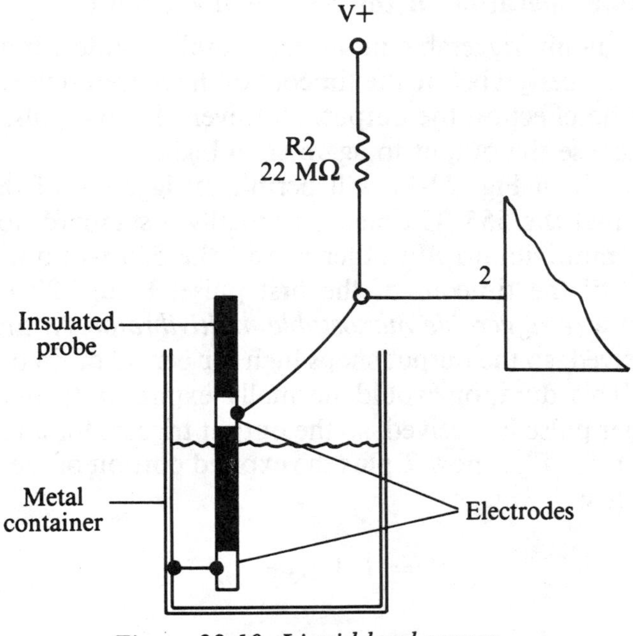

Figure 22-10. Liquid-level sensor.

Experiment 22-2

Repeat Experiment 22-1 with a trigger circuit, such as Fig. 22-9, instead of the manual triggering circuit (R1/S1) that is used in Experiment 22-1.

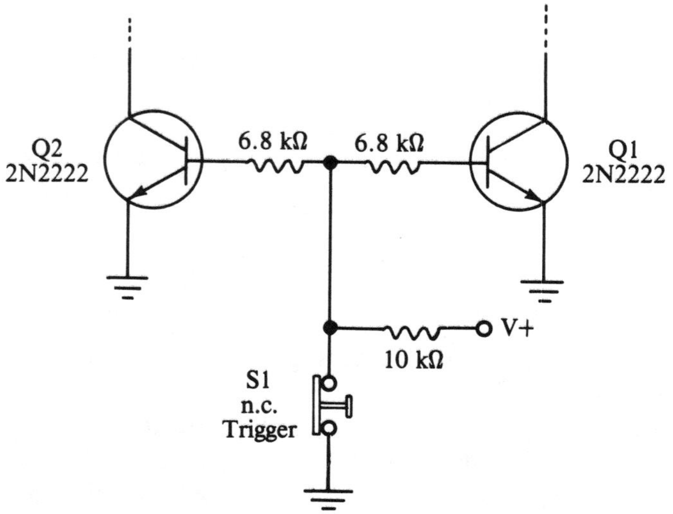

Experiment 22-2. Circuit for Experiment 22-2.

Retriggerable operation of the 555 MMV circuit

The 555 is a nonretriggerable monostable multivibrator. If additional trigger pulses are received before the timeout of the output pulse, the additional pulses have no effect on the output. However, the first pulse after timeout occurs will cause the output to again snap high.

The circuit in Fig. 22-11 will permit retriggering of the 555 device. Remember that the 555 IC timer is normally a standard nonretriggerable monostable multivibrator. In other words, the 555 will not retrigger once triggered, until the timeout of the first pulse. Figure 22-11A shows the operation of a *retriggerable monostable multivibrator.* At time t_1, a trigger pulse is received, so the output snaps high for period of time T, determined by 1.1RC. This duration would normally expire at t_3, but at time t_2, a second trigger pulse is received, so the output triggers for another period T. The total period, T', is now T plus the expired portion of the original pulse. In other words:

$$T' = T + (t_2 - t_1) \qquad (22\text{-}8)$$

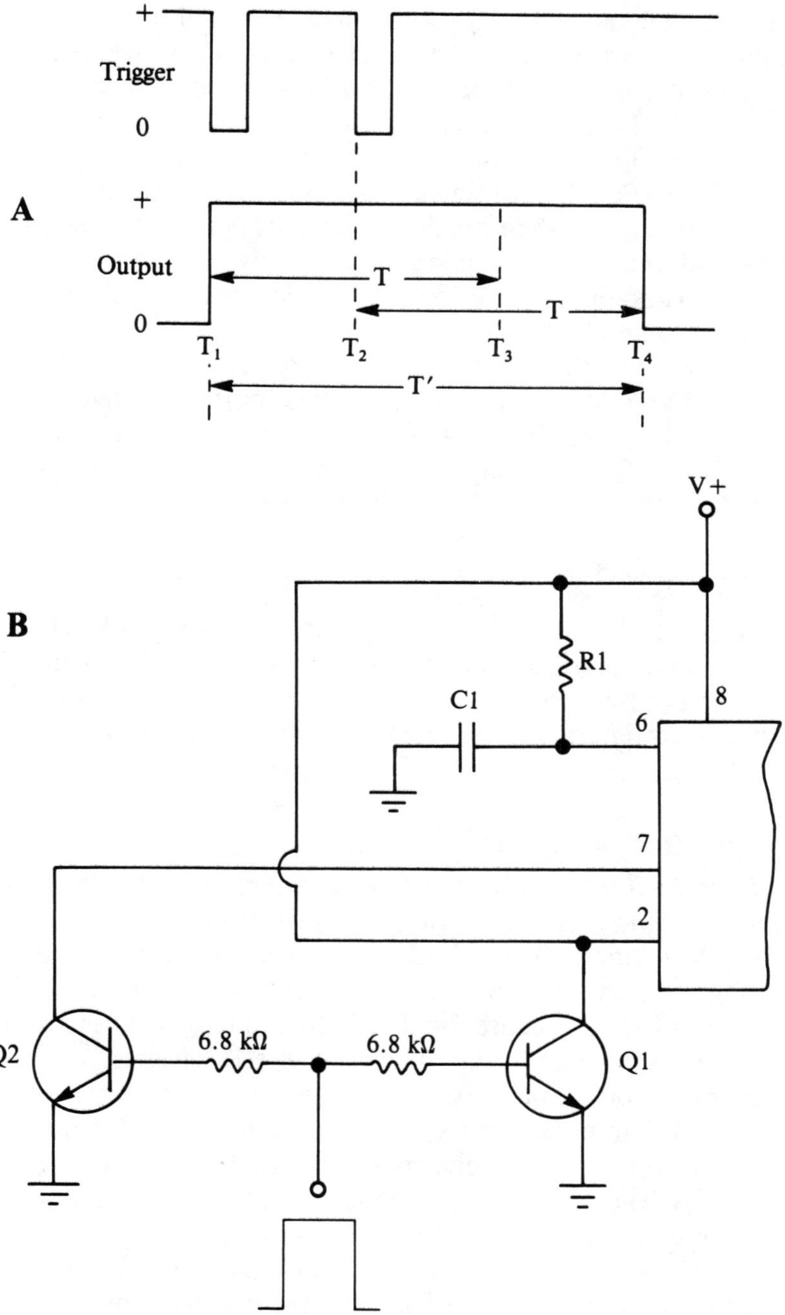

Figure 22-11. A) Retriggerable operation of the 555, B) retriggerable monostable multivibrator.

Figure 22-11B shows a trigger-pin circuit that will make a 555 into a retriggerable monostable multivibrator. An external npn transistor (Q2 in Fig. 22-11B) is connected with the CE path across timing capacitor C1. In this sense, it mimmicks the internal discharge transistor seen earlier. A second transistor, Q1, is connected to the trigger input of the 555 in a manner similar to Fig. 22-7 (discussed earlier). The bases of the transistors form the trigger input. When a positive pulse is applied to the combined trigger line, both transistors become saturated. Any charge in C1 is immediately discharged and pin 2 of the 555 is triggered when the collector of Q1 is dropped to less than $(V+)/3$.

As long as no further trigger pulses are received, this circuit behaves like any other 555-MMV circuit. But if a trigger pulse is received before the timeout (defined by the previous equation), the transistors are forward-biased once again. Q1 retriggers the 555, while Q2 dumps the charge that built up in the capacitor. Thus, the 555 retriggers.

Experiment 22-3

Do Experiment 22-1, but use a trigger circuit, such as Fig. 22-11B, in place of the original manual trigger circuit (R2/S1). Remember that you will need a positive-going pulse from a square-wave generator or from a circuit (such as the original trigger circuit) that uses a normally closed switch for S1 (see Experiment 22-2).

Astable operation of the 555 IC timer

An *astable multivibrator* (*AMV*) is a free-running circuit that produces a square-wave output. The 555 can be connected to produce a variable duty-cycle AMV circuit (Fig. 22-12). The internal stages of the 555 are shown in Fig. 22-12A and the schematic version is shown in Fig. 22-12B. This circuit is an AMV because the threshold and trigger pins (6 and 2) are connected together, which forces the circuit to self-retrigger.

Under initial conditions at turn-on, the voltage across timing capacitor C1 is zero, while the biases on COMP1 and COMP2 are (as usual) set to $2(V+)/3$ and $(V+)/3$, respectively, by the internal resistor voltage divider (R_a, R_b, and R_c). The output of the 555 is high under this condition, so C1 begins to charge through the combined resistance ($R_1 + R_2$). On discharge, however, transistor Q1 shorts the junction of R_1 and R_2 to ground, so the capacitor discharges only through R2. The resulting waveform is shown in Fig. 22-12C. The time that the output is high is t_1 and the low time is t_2. The period (T) of the output square wave is the sum of these two durations: $T = (t_1 + t_2)$. As with all similar RC-timed circuits, the equation that sets

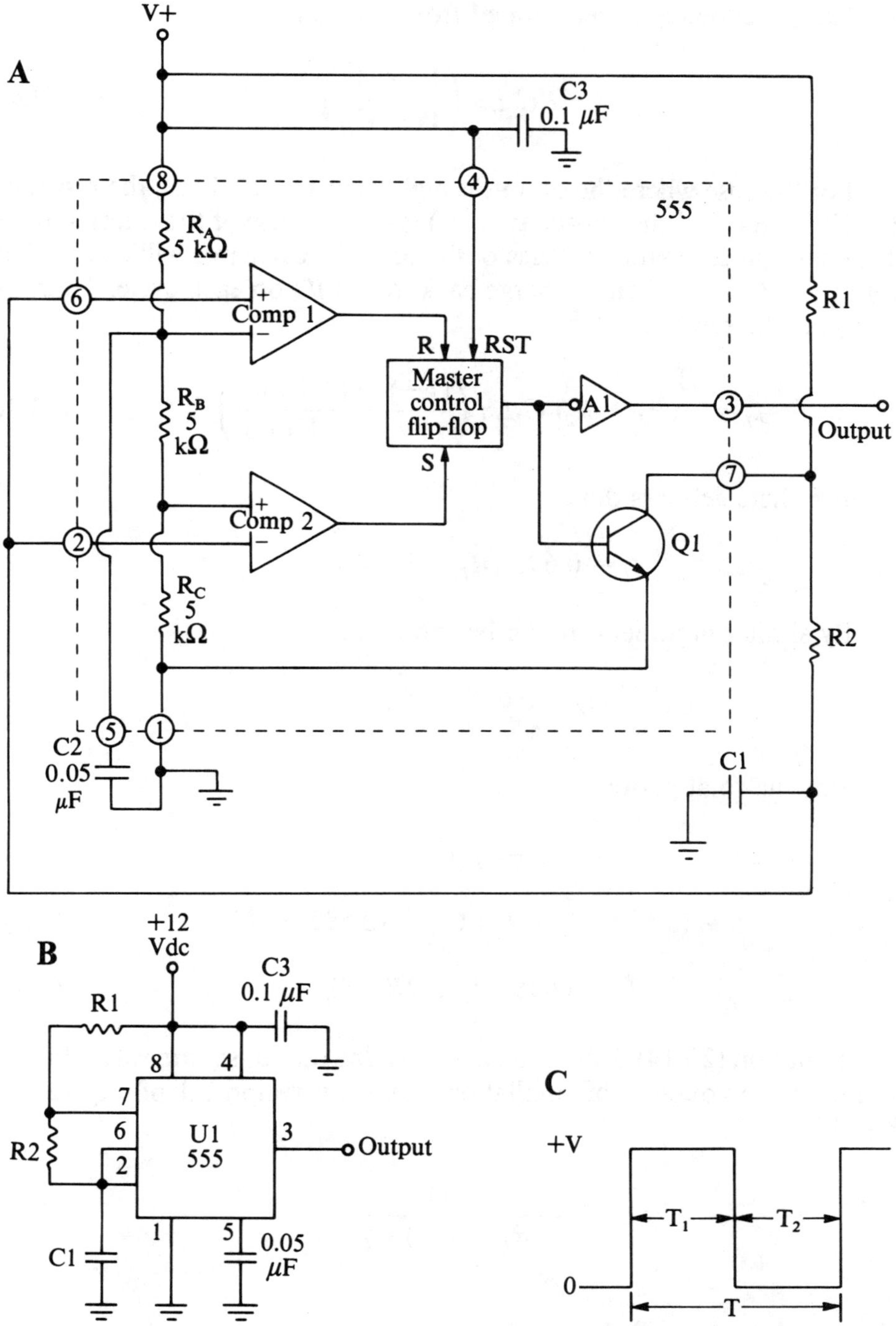

Figure 22-12. A) Astable 555 multivibrator with internal circuit shown, B) schematic version, C) pulse relationships.

oscillating frequency is determined from the equation:

$$T = -RC\ Ln\left(\frac{V - V_{C2}}{V - V_{C1}}\right) \tag{22-9}$$

For the case where the output is high (t_2 in Fig. 22-12C), the resistance R is $(R_2 + R_2)$ and the capacitance (C) is C_1. Because of the internal biases of the voltage comparator stages of the 555, the capacitor will charge from $(V+)/3$ to $2(V+)/3$, then discharge back to $(V+)/3$ on each cycle. Thus, Eq. (22-9) can be rewritten:

$$t_1 = -(R_1 + R_2)\ C_1 Ln\left(\frac{(V+) - (2(V+)/3}{(V+) - ((V+)/3}\right) \tag{22-10}$$

or, once the algebra is done:

$$t_1 = 0.695\ (R_1 + R_2)\ C_1 \tag{22-11}$$

By similar argument, it can be shown that:

$$t_2 = 0.695\ R_2\ C_1 \tag{22-12}$$

For the total period T:

$$T = t_1 + t_2 \tag{22-13}$$

$$T = (0.695\ (R_1 + R_2)\ C_1) + (0.695\ R_2\ C_1) \tag{22-14}$$

$$T = 0.695\ (R_1 + 2R_2)\ C_1 \tag{22-15}$$

Equation (22-14) defines the period of the output square wave. In order to find the frequency of oscillation, take the reciprocal of Eq. (22-15), $F = 1/T$:

$$F = \frac{1.44}{(R_1 + 2R_2)\ C_1} \tag{22-16}$$

Experiment 22-4

This experiment demonstrates the astable configuration of the 555 IC timer. The frequency is set low enough to permit the use of an LED output indicator circuit (D1/R3). If you wish to use an oscilloscope, replace the output indicator with a 10-kohm load resistor to ground.

1. Build the circuit of Experiment 22-4. Make sure that the polarity of C1 is observed or the circuit will not work properly.

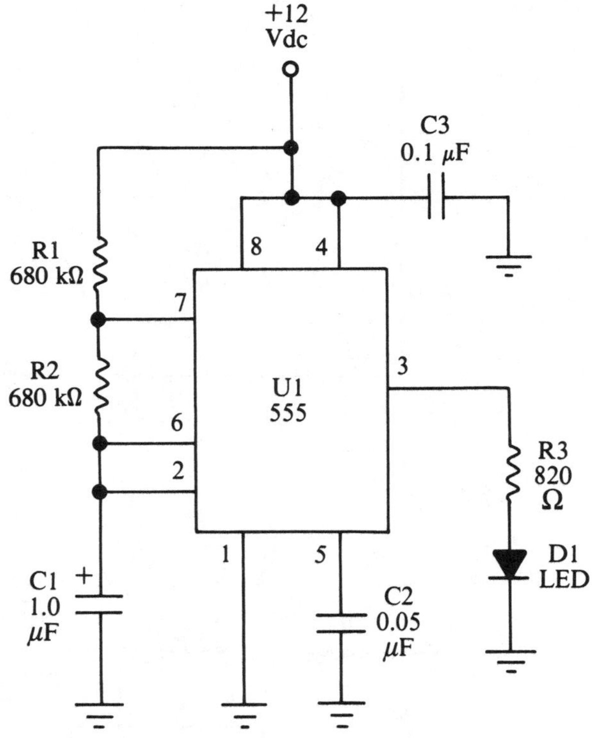

Experiment 22-4. Circuit for Experiment 22-4.

2. Turn on the dc power to the circuit and observe the LED. It should be blinking at a frequency determined by:

$$F = \frac{1.44}{(3 \times 680{,}000 \text{ ohms})\,(1 \times 10^{-6} \text{ farads})}$$

3. Set the scale of a dc voltmeter to read 10 volts full-scale.
4. Connect the voltmeter in parallel with capacitor C1. Use an FET-input analog voltmeter or a digital voltmeter. An ordinary analog VOM might overload the circuit and produce an unusable result.
5. Observe the capacitor charging timing, in conjunction with the blinking of the LED.

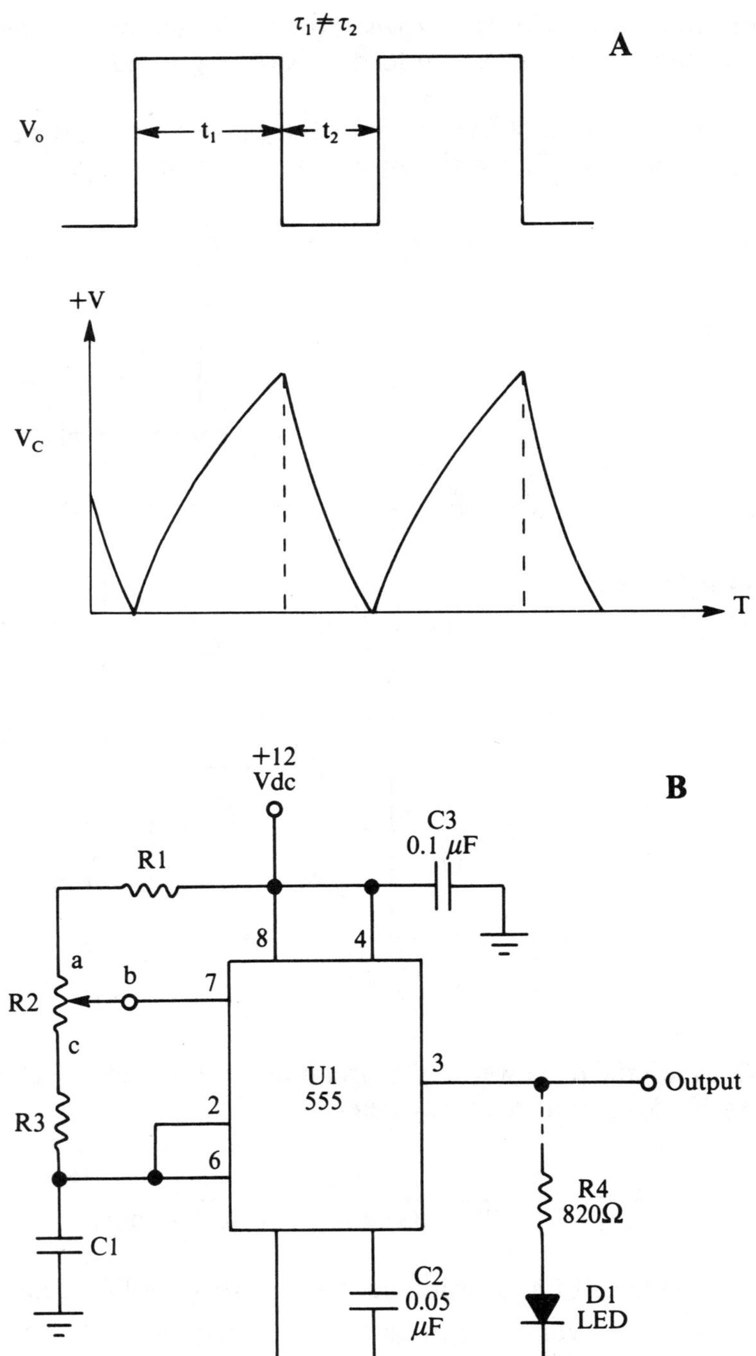

Figure 22-13. A) Timing relationships, B) variable duty-cycle circuit.

Duty cycle of 555 astable multivibrator

Time segments t_1 and t_2 are not equal in most cases, so the charge and discharge times for capacitor C1 are also not equal (see Fig. 22-13A). The duty cycle of the output signal is the ratio of the high period to the total period (t_1/T). Expressed as a percent:

$$\%DC = \frac{R_1 + R_2}{R_1 + 2\ R_2} \qquad (22\text{-}17)$$

Various methods are used to vary the duty cycle. First, a voltage can be applied to pin 5 (control voltage). Second, a resistance can be connected from pin 5 to ground. Both of these tactics alter the internal bias voltages that are applied to the comparator.

Alternatively, you can also divide the external resistances R_1 and R_2 into three values. Figure 22-13B shows a variable duty factor 555 AMV that uses a potentiometer (R2) to vary the ratio of the charge and discharge resistances.

The oscillation frequency of this circuit is found from the following equation. It is the same as the previous equations, except that it accounts for the resistance of the potentiometer.

$$F = \frac{1.44}{[(R_1 + R_{2ab} + 2\ (R_{2cb} + R_3)\ C_1} \qquad (22\text{-}18)$$

Where:

R_{2ab} is the resistance between terminals *a* and *b* on potentiometer R2
R_{2bc} is the resistance between terminals *b* and *c* on potentiometer R2

Experiment 22-5

Perform Experiment 22-5 with the modified circuit shown in Fig. 22-13B. Use the following values for R_1, R_2, and R_3:

$R_1 = R_3 = 470$ kohm
$R_2 = 500$ kohms (multiturn is preferred)

A 555 sawtooth generator circuit

A sawtooth waveform (Fig. 22-14A) starts at a given potential (usually zero) at time t_1 and rises linearly to some value V at t_2, then drops abruptly back to the initial condition. The circuit for a 555-based sawtooth generator (Fig. 22-14B) is simple, and is based on the 555 timer IC. The basic circuit is the

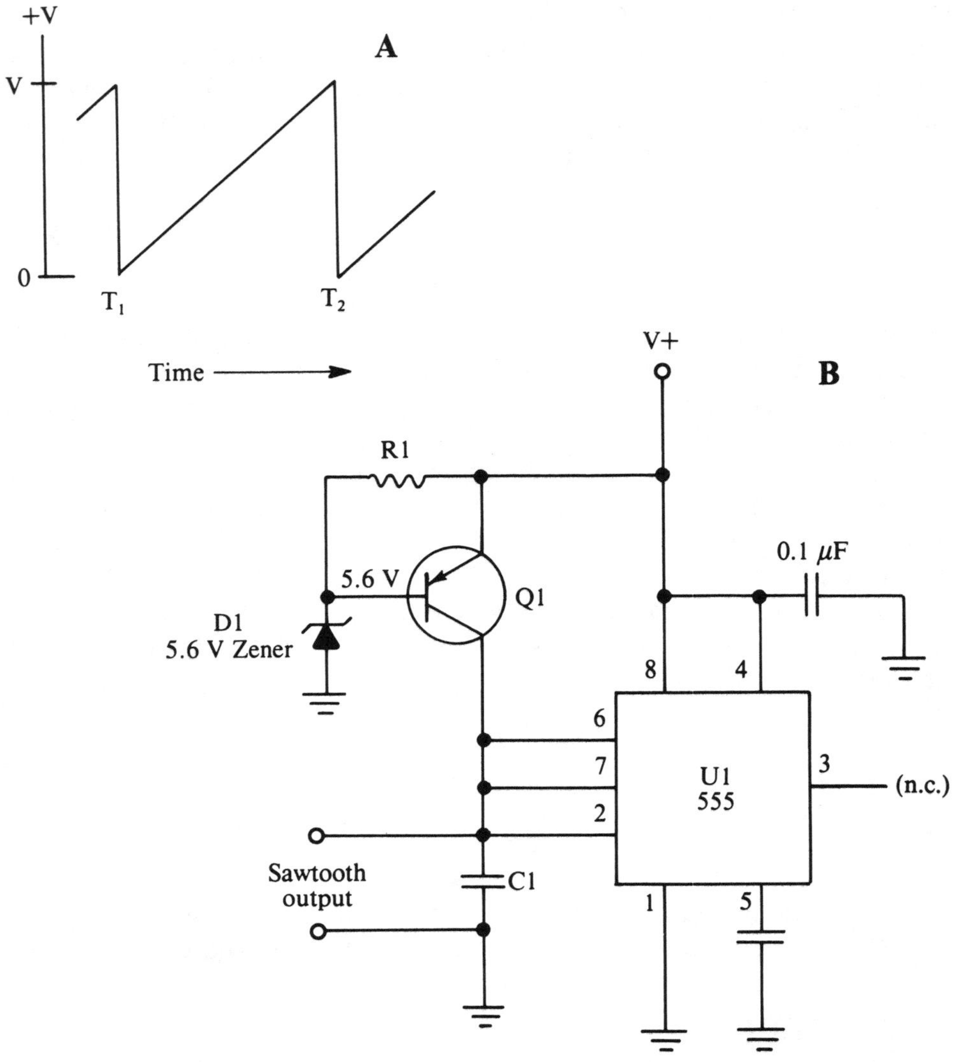

Figure 22-14. A) Sawtooth waveform, B) 555 sawtooth circuit.

monostable multivibrator configuration of the 555, in which one of the timing resistors is replaced with a transistor that is operated as a current source (Q1). Almost any audio small-signal pnp silicon replacement transistor can be used, but the 2N3906 was used here. The zener diode is a 5.6-Vdc unit. Note that the output is taken from pins 6 and 7, rather than the regular chip output (pin 3), which is not used.

The circuit as shown is a one shot multivibrator. The 555 is triggered when pin 2 is brought to a potential less than ⅔ of the supply potential.

When a pulse is applied to pin 2 through differentiating network R_1C_1, the device will trigger because the negative-going slope meets the triggering criteria. To make an astable sawtooth multivibrator drive the input of this circuit, either a square wave or a pulse train must produce at least one pulse for each required sawtooth. The circuit in Fig. 22-14B, a nonretriggerable monostable multivibrator, will ignore subsequent trigger pulses during the one shot's "refractory" period.

Summary

The 555 IC is an extremely flexible timer, normally packaged as an 8-pin miniDIP. It consists of a resistor network, two op amp comparators, and an RS flip-flop whose output is connected to both a transistor and an inverter. Depending on the external connections made to the 555, it can act as either a monostable or an astable multivibrator. The monostable multivibrator requires an external trigger for operation; a touch switch is a common source of this trigger. The astable multivibrator is self-triggering.

Questions

1. What is a 555 timer?
2. What are the two main modes of operation for the 555 timer?
3. Identify the eight pins of the 555 timer and state the function of each.
4. Draw a diagram of the internal circuitry for the 555 timer.
5. Draw a 555 timer connected as a monostable multivibrator. Use a block diagram for the timer.
6. Draw a 555 timer connected as an astable multivibrator. Use a block diagram for the timer.
7. Show how a positive-going pulse can be used to trigger the 555 timer.
8. Show how the duty cycle for an astable multivibrator can be varied.

Problems

1. For the 555 timer in Fig. 22-3, V+ is 12 volts, and a control voltage of 6 volts is placed at pin 5. If this timer is used in the astable multivibrator mode, what is the range of voltages for **a.** the output? **b.** the capacitor?

2. For the 555 timer in Fig. 22-3, V+ = 15 volts, and a control voltage of 5 volts is placed at pin 5. If this timer is used in the astable multivibrator mode, what is the range of voltages for **a.** the output? **b.** the capacitor?

3. For the circuit in Fig. 22-5C, V+ = 15 volts, R_1 = 10 kohms and C_1 = 0.01 μF. Calculate **a.** the maximum capacitor voltage **b.** the length of the output pulse

4. Repeat Problem 3 for V+ = 5 volts, R_1 = 33 kohms and C_1 = 2.2 μF.

5. For the circuit in Fig. 22-12B, V+ = 15 volts, R_1 = 22 kohms, R_2 = 180 kohms, and C_1 = 2 μF. Calculate **a.** the maximum and minimum capacitor voltage **b.** time high **c.** time low **d.** frequency of oscillation

6. Repeat Problem 5 for V+ = 5 volts, R_1 = 100 kohms, R_2 = 20 kohms, and C_1 = 0.5 μF.

7. For the circuit in Fig. 22-15, +V = 15 volts, R_1 = 20 kohms, R_2 = 75 kohms, and C = 0.5 μF. Assume that D_1 is an ideal diode and calculate **a.** time high **b.** time low **c.** frequency of operation

8. For the circuit in Fig. 22-15, +V = 5 volts, R_1 = 180 kohms, R_2 = 51 kohms, and C = 0.22 μF. Assume that D_1 is an ideal diode and calculate **a.** time high **b.** time low **c.** frequency of operation

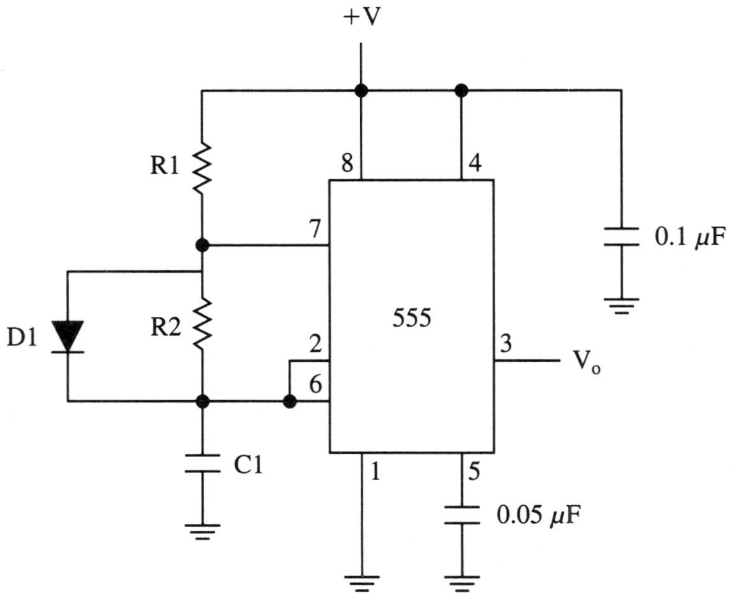

Figure 22-15.

23 555 APPLICATIONS

AFTER COMPLETING THIS chapter, you will be able to:

- □ Explain pulse detection and counting
- □ Discuss 555 interfacing

The 555 timer is one of the most widely used chips on the market. The 555 is popular because it's very cheap, it's easily available, and it's very well-behaved in circuits (the timing equations actually work!). A person cannot be involved in the electronics hobby for long before learning that the 555 can be used as a square-wave generator (astable multivibrator) and as a monostable one-shot multivibrator. This chapter looks at some of the unusual applications that you can do with the 555 timer. If you have not done so, please read chapter 16 (on RC networks) and chapter 22 (on the operation of the 555). This chapter does not contain any experiments, but the circuits can be either built as projects or incorporated into larger projects as you please.

Missing pulse detector

The missing-pulse detector remains dormant as long as a series of trigger pulses are received, but it will produce an output pulse when an expected pulse is missing. These circuits are used in a variety of applications including alarms. For example, in a bottling plant, cans are packaged into six-packs. As each can passes a photocell, a pulse is generated to the input of a missing-pulse detector. If a pulse is not received, however, the machine knows that the count is one can short, so it issues an alarm or a corrective action. Similarly, in a wildlife photography system, an infrared LED is modulated or chopped with a pulse waveform. As long as the pulse is received at the sensor, the circuit is dormant. But if an animal passes through the IR beam, even briefly, a missing-pulse detector will sense its presence and issue an output that fires a camera and an electrical shutter control.

Figure 23-1 shows the circuit for a missing-pulse detector that is based on the 555. This circuit is the standard 555 MMV, except that a discharge transistor is shunted across capacitor C1. When a pulse is applied to the input, it will trigger the 555, turn on Q1, and cause the capacitor to discharge. After the first input pulse, the output of the 555 snaps high and remains high until a missing pulse is detected.

Circuit action can be seen in Fig. 23-1B. At times t_1 and t_2, input pulses are received. As long as $(t_2 - t_1)$ is less than the time required for C1 to charge up to $2(V+)/3$, the 555 will never timeout. But if a pulse is missing, as at t_3, the capacitor voltage continues to rise to the critical $2(V+)/3$, threshold value. When V_c reaches this point, the 555 will timeout and force its output low. The output remains low until a subsequent input pulse is received (t_4), at which time Q1 turns on again and forces the capacitor to discharge. The cycle can then continue as before.

Pulse positioner

A *pulse positioner* is a circuit that allows the timing of a pulse to be adjusted to coincide with some external event. For example, in some instrumentation circuits, a short pulse must be positioned to a certain point on a sine wave (e.g., the peak). The pulse positioner could be triggered from the zero-crossing of the sine wave. Then, it could be adjusted to place the output pulse where it is needed.

Figure 23-2A shows the concept of pulse positioning with two one shots, OS1 and OS2 (Fig. 23-2B). The repositioned pulse is not actually the original pulse, but it is a recreated pulse with similar characteristics. The input pulse is used to trigger OS1. The duration of this one shot is fixed to the delay that is required of the repositioned pulse. If the delay must be variable, resistor R1 is made variable. When OS1 times-out, it will trigger OS2. The output pulse of OS2 is set to the parameters of the original input phase. An inverter circuit is used to make the output of OS2 have the same polarity as the trigger pulse at the input of OS1. To an outside observer, the pulse appears to have been repositioned, but it was merely recreated at the time T (the delay period in Fig. 23-2B).

Tachometry

The word *tachometry* is used to designate the measurement of a repetition rate. In the automotive tachometer, for example, the instrument counts the pulses that are produced by the ignition coil to measure the engine speed in RPM. In medical instruments, it is often necessary to measure factors (such as heart or respiration rate electronically) with tachometry circuits. A heart-rate meter (cardiotachometer) measures the heart rate in beats per minute

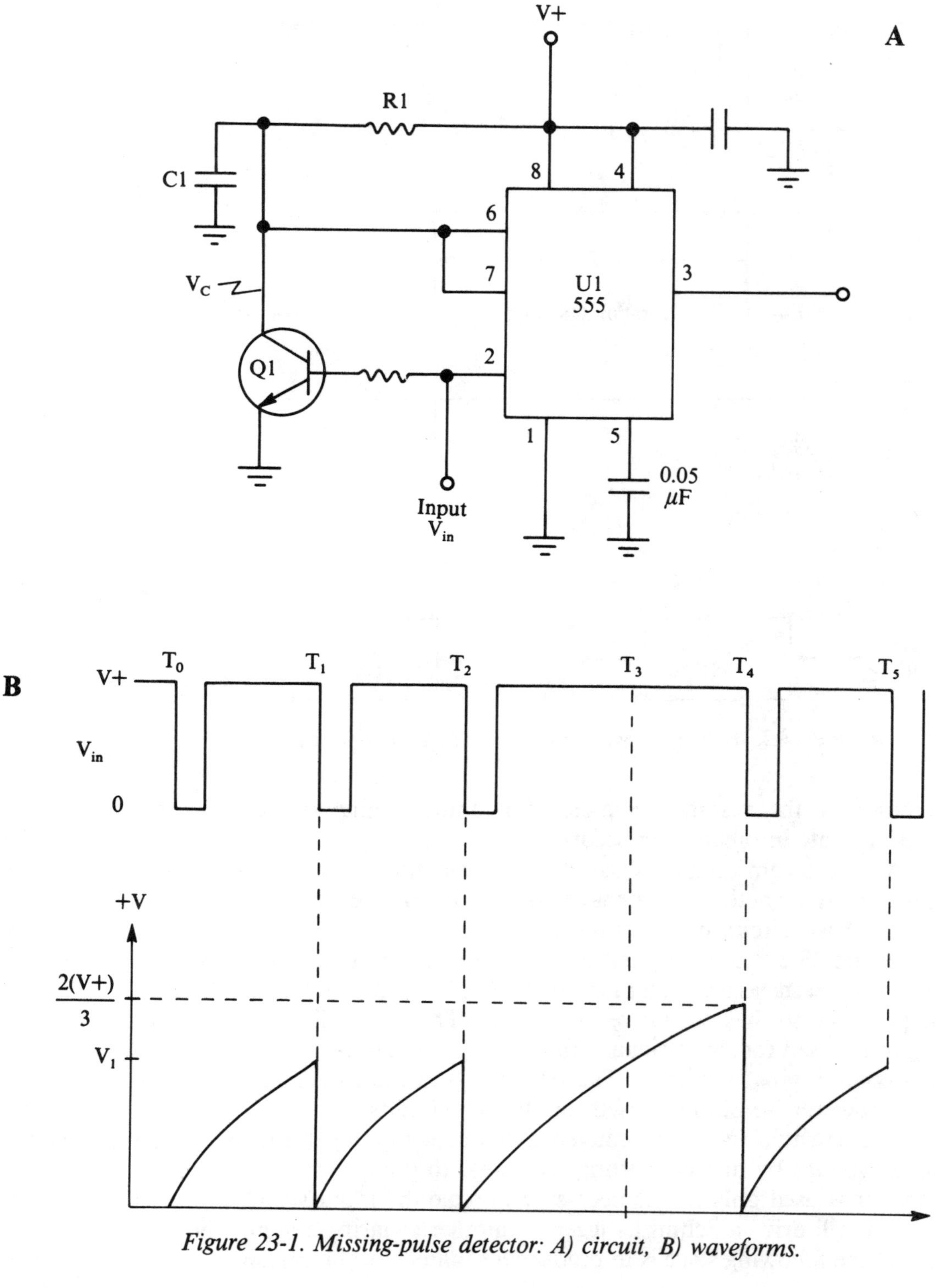

Figure 23-1. Missing-pulse detector: A) circuit, B) waveforms.

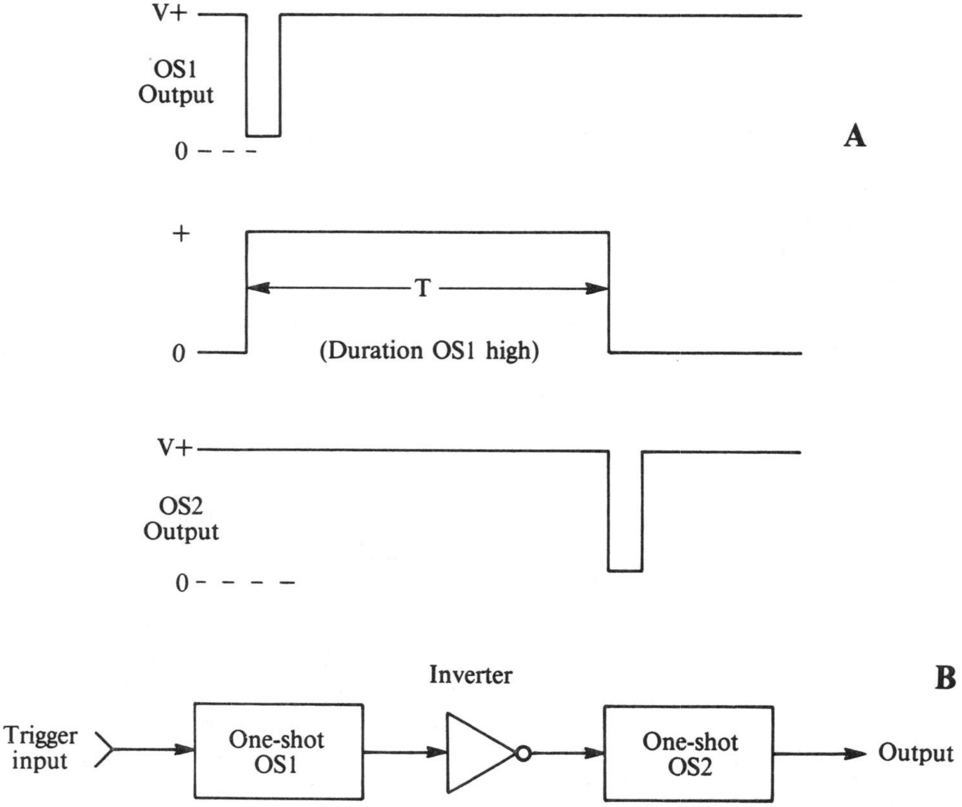

Figure 23-2. A) Pulse stretching, B) how it's done (block diagram).

(BPM) and the respiration meter (pneumotachometer) measures the breathing rate in breaths per second.

A certain commonality exists among nondigital tachometer circuits. It doesn't matter whether the rate is audio, sub-audio, or even above the audio rate, the basic circuit design is the same.

Figure 23-3 shows the basic tachometer circuit in block-diagram form. Not all of the stages are present in all circuits, but some of them are basic to the problem, so they are universally found. The ac amplifier and Schmitt trigger are used for input signal conditioning, so they are used only where such conditioning is needed. The one shot and the Miller integrator are basic, however, so both are used for all such circuits.

The essential idea is to convert a frequency or repetition rate to an analog voltage by first converting the signal to pulse form. The ac input amplifier is used only if it is necessary to scale the input signal to a level where it will drive a Schmitt trigger or another squaring circuit. The purpose of the following stage is to produce a square-wave output signal at the

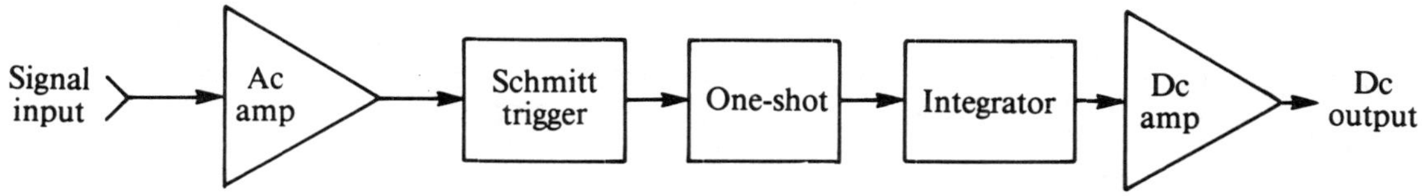

Figure 23-3. Tachometer circuit in block form.

same frequency as the input signal. The purpose of the stages in Fig. 23-3 is to produce a dc voltage output that is proportional to the input frequency or pulse-repetition rate. The integrator is designed to produce an output voltage that is the time average of the input signal. That is, the integrator output is proportional to the area under the input signal. The tachometer designer creates a situation in which the only variable is the frequency or repetition rate of the input signal. Variation obscures the results.

The output pulse of a one shot has a constant amplitude and constant duration. The area under the pulse is the product of the amplitude and duration, so from pulse to pulse, the area does not change. If the one shot is constantly retriggered by the input signal, the total area under the resultant pulse train is a function of only the number of pulses. Therefore, the time average of the integrator output will be a dc voltage that is proportional to the input frequency.

Figure 23-4 shows a practical application of the tachometer principle. This circuit was used to demodulate the AF modulated signal from an instrumentation telemetry set. A similar circuit (not based on the 555) was once popular as a "coilless" FM detector in communications and broadcast receivers. These *pulse-counting detectors* operate at 10.7 MHz (a commonly used FM IF frequency in receivers).

The circuit shown in Fig. 23-4 was used to demodulate a human electrocardiograph (ECG) signal that was transmitted over telephone lines. The ECG is an analog voltage waveform that was used to frequency modulate an audio voltage-controlled oscillator (VCO) at the transmit end. Normally, the ECG has too low a Fourier frequency content (0.05 to 100 Hz) to pass over the restricted passband of the telephone lines (300 to 3000 Hz). But when used to frequency modulate a 1500-Hz carrier, however, the signal could pass easily over telephone circuits.

The demodulator circuit is shown in Fig. 23-4A. The input waveshaping function is performed by an LM-311 voltage comparator. The job of the LM-311 is to square the 200 mV peak-to-peak sine-wave input signal so that it is capable of triggering the 555 (U2). In this mode, the LM-311 operates basically as a zero-crossing detector circuit. The output of the 555 is a pulse train that has constant amplitude and duration. These pulses vary only in repetition rate, which is the same as the frequency of the input

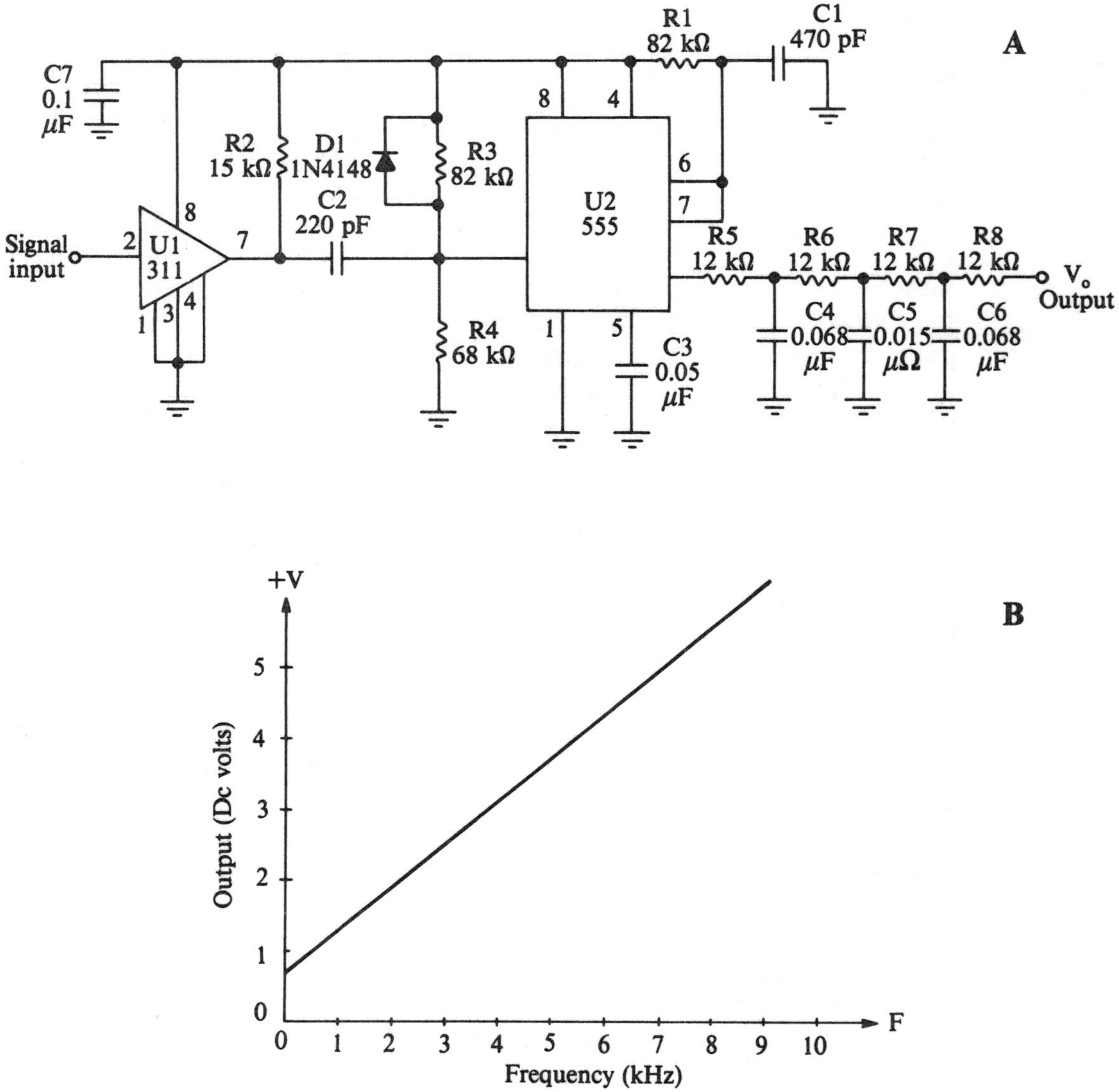

Figure 23-4. A) 555 tachometer, B) output voltage vs. input-frequency characteristic.

signal. The 555 output pulses are integrated in a passive RC integrator ($R_5 - R_7/C_4 - C_6$). The output of the integrator is a dc voltage that is a linear function of the input frequency (see Fig. 23-4B). This dc voltage can be scaled, if necessary, to any desired level.

A related circuit is shown in Fig. 23-5. This 555-based tachometer is used to measure AF over three ranges: dc to 50 Hz, dc to 500 Hz, and dc to

5000 Hz. The circuit uses the same form of input-signal conditioning as the previous circuit, and it uses a 555 as the one shot. The integration function is the combination of RC network R4/C4 and the mechanical inertia of the meter (M1) movement.

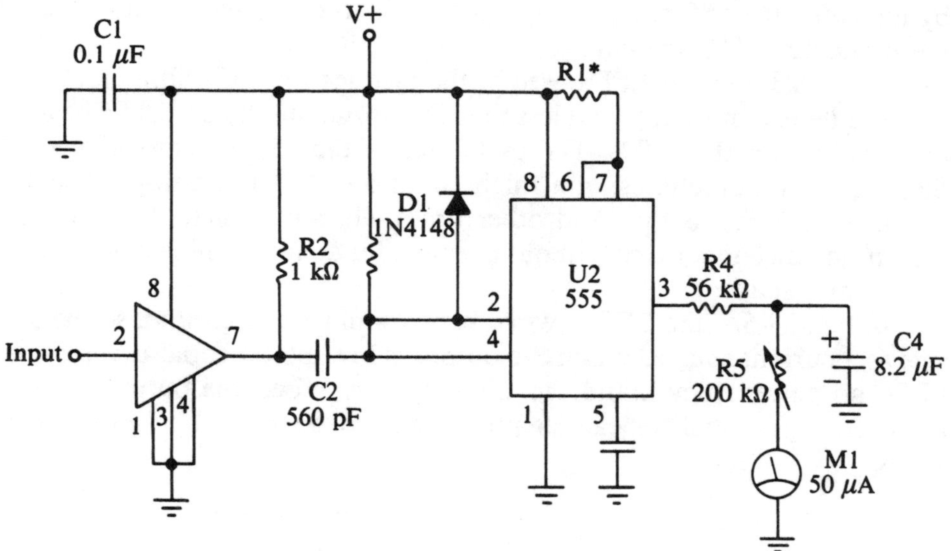

Figure 23-5. AF meter.

The calibration ranges are:

Range (Hz)	R1 Value
50	1 megohm
500	100 kohm
5,000	10 kohms

Interfacing the 555 to outside circuits

A lot has been written over the years about the 555, so this chapter takes a different approach. I assume that you either know the basics of 555 (or can find out), so this section concentrates on the interface between the 555 device and the "outside world."

Output circuits for the 555 MMV circuit

The 555 output terminal (pin 3) will serve as either a *current source* or a *current sink*, depending on how you wire the circuit. This ability is contrary to common practice with other ICs, so it extends the usefulness of the 555. The output will sink or source up to 200 mA of current. It can be made TTL-compatible by setting *V*+ = 5 Vdc. CMOS compatibility is achieved by matching the 555 power-supply potentials to the levels that are used in the particular CMOS circuit.

Figure 23-6 shows LEDs used as the load for the 555. Although LEDs are used here, almost any load less than 200 mA could theoretically be used to substitute for the LEDs. The usefulness of the 555 is demonstrated by these circuits. Sometimes, you might want an LED indication when the output of the 555 is low, and other times when it is high. The 555 can accommodate either event without the need for an intervening open-collector inverter stage.

In Fig. 23-6A, the LED is wired between pin 3 and ground, so the 555 acts as a current source. When the output is low, no potential is across the LED, no current flows, and the LED is dark. When the output is high, however, a potential appears at pin 3, current flows in the LED, and it glows.

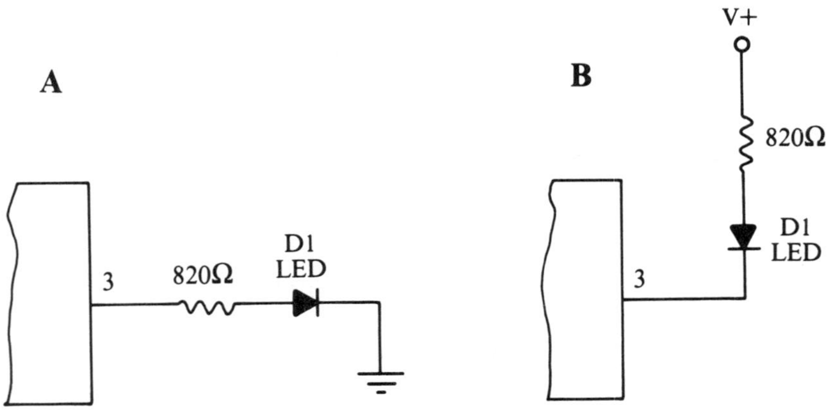

Figure 23-6. LED interfacing: A) current source, B) current sinking.

Figure 23-6B shows the opposite connection. Here, the LED is wired between the output terminal and V+, so the 555 output acts as a current sink. When the output is low, the cathode end of the LED is essentially grounded through a small resistance, so the LED is on. When the output is high, on the other hand, the potential at both ends of the LED is close to *V*+, so no differential voltage exists and no current flows.

In both cases of Fig. 23-6, the resistor in series with the LED limits the current flowing in the LED. On most unmarked LEDs, the maximum safe current is 15 mA (0.015 A), so set resistor R2 to (V+)/0.015 or greater.

When the current in the load exceeds the 200-mA capacity of the 555 output terminal, you can add an external transistor switch (Fig. 23-7) to handle the higher current load (represented by R_L). The 555 output is used to turn the transistor on and off. In Fig. 23-7A, the transistor is an npn type. Thus, when the output of the 555 is low, there is no bias voltage applied to the transistor base-emitter (BE) junction, and that keeps the transistor turned off. When the 555 output is high, the transistor is biased on hard, and it turns on. The "cold" end of the load, connected to the Q1 collector is thereby grounded and current flows. The value of the base resistor on Q1 depends on the load current and the transistor beta, and it can be found experimentally.

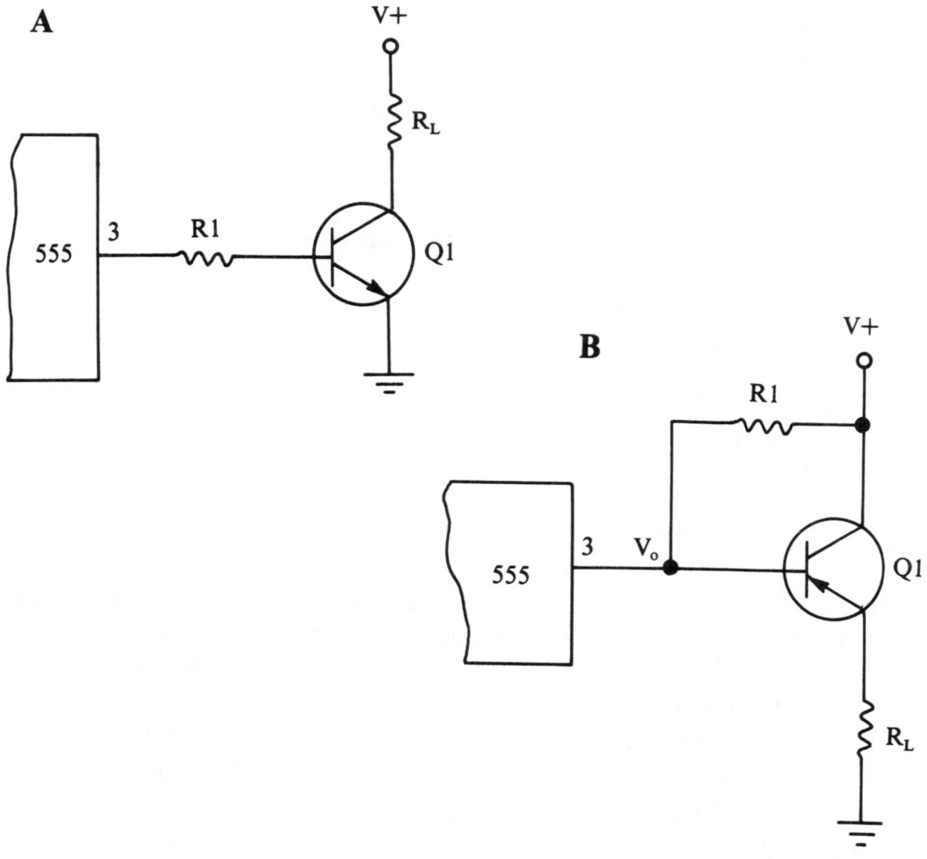

Figure 23-7. A) Using a transistor to handle larger current loads, B) pnp-transistor version.

A pnp transistor is similarly used in Fig. 23-7B. In the pnp transistor, the base must be less positive than the emitter, so this circuit turns on when the output of the 555 is low. When the output is high, the emitter and base are close to the same potential, so no action occurs.

Even higher currents can be accommodated by using an electromechanical relay as a load for the 555 output (Fig. 23-8). In addition, the relay makes it possible to use the 555 in a low-voltage dc circuit with other electronics to control a high-voltage load circuit.

Select a relay with a coil dc voltage rating of 18 volts or less (5, 6, and 12 Vdc are common), and match that voltage to the *V*+ used on the 555. For example, if you are using the 555 at +12 volts, select a 12-Vdc relay. Also, make sure that the coil current is less than 200 mA. If you don't see the coil current rating, calculate it from the coil resistance (the most commonly listed relay specification): $I_{coil} = (V+)/R_{coil}$.

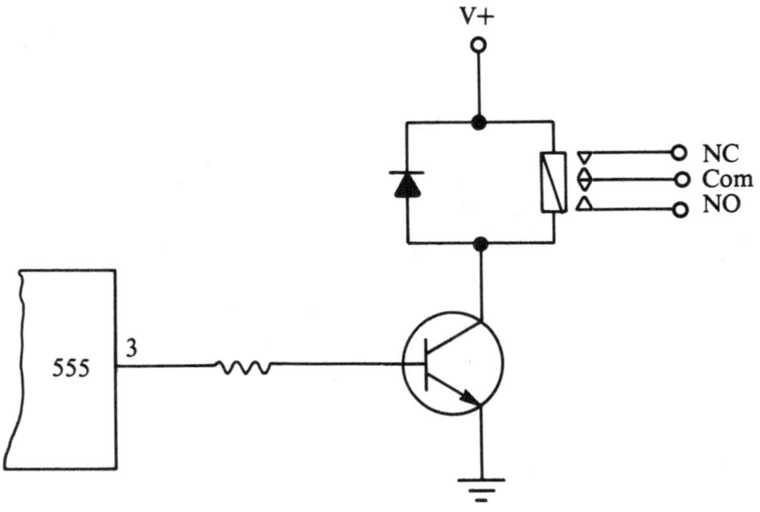

Figure 23-8. Relay interfacing to the 555.

In Fig. 23-8 the relay coil is shunted with a 1N4007 rectifier diode. These diodes are used for spike suppression when the relay coil is de-energized. The *counterelectromotive force* (*CEMF*) created at that point can be high voltage (which can ruin the 555 and other electronics) and will have a reverse polarity, with respect to *V*+. Thus, the diodes are normally reverse-biased, except when a large CEMF "inductive kick" spike from the relay is received. Note: These diodes are *not* optional.

Figure 23-9 shows a method for solving a problem that is sometimes seen in 555 relay drivers and certain other 555 circuits, in which either digital pulses or noise spikes are seen. The spikes can get inside the 555 on

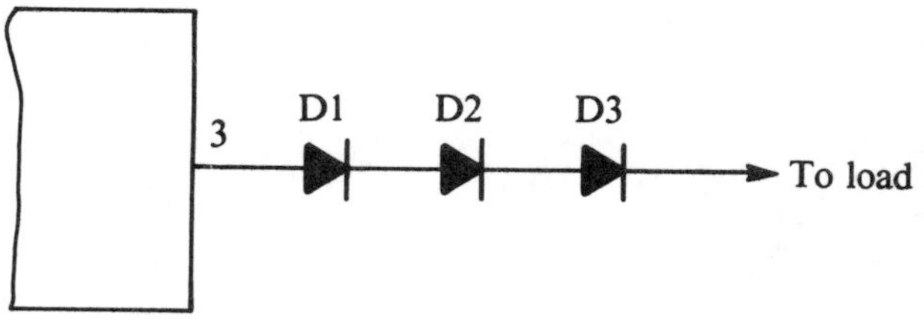

Figure 23-9. Protecting the 555 output.

the output terminal and force the internal digital electronics of the chip to reset. These diodes are a crude means to isolate the 555 output. If you've experienced seemingly flaky operation of a 555 (555s don't usually show flaky operations!), you might consider if the problem is external pulses coupled through pin 3. If that is the case, this circuit (Fig. 23-9) might be the solution. It's worked a few times for me and it might also work for you.

Summary

The 555 timer has many applications, either in monostable or astable mode. In the missing pulse detector, the timer is triggered only when one in a series of pulses is missing. The tachometer converts the pulse repetition rate to a dc voltage, whose value is proportional to the rate. The 555 can be used as a current source or a current sink. In either case it is capable of handling a maximum of 200 mA of current. When larger currents are required, relays may be used together with a transistor driver.

Questions

1. Describe the action of a 555 timer as a missing pulse detector.
2. What is tachometry?
3. Draw a block diagram of a tachometer circuit.
4. For the circuit in Fig. 23-4A, the input signal has a frequency of 4 kHz. What is the output voltage?
5. What is the difference between current sourcing and current sinking?
6. What is the maximum current capability for the 555 timer? Show how the 555 can be used as a timer for a load which requires more than this amount.

ANSWERS TO ODD-NUMBERED PROBLEMS

Chapter 2

1. 11 μC
3. 6.64 pF
5. **a.** 24,000 ± 5% ohms
   b. 22,800–25,200 ohms

Chapter 5

1. 162.6 V
3. 11.5 V
5. 17 V
7. 0.22
9. 5×10^{-6}

Chapter 8

1. −4 V
3. −1.2 V
5. 5 kohms
7. 0.005 μF

Chapter 9

1. 9 V
3. **a.** 2.75 V
   b. 18.5 Hz
5. 490 kohms, 0.04 μF, 0.004 μF
7. 66.7

Chapter 10

1. 46 dB
3. −7.5 V
5. **a.** 1.25 V
   b. 2.5 V

Chapter 11

1. 210, 30
3. 0.047 V
5. 15.9 pF

Chapter 12

1.

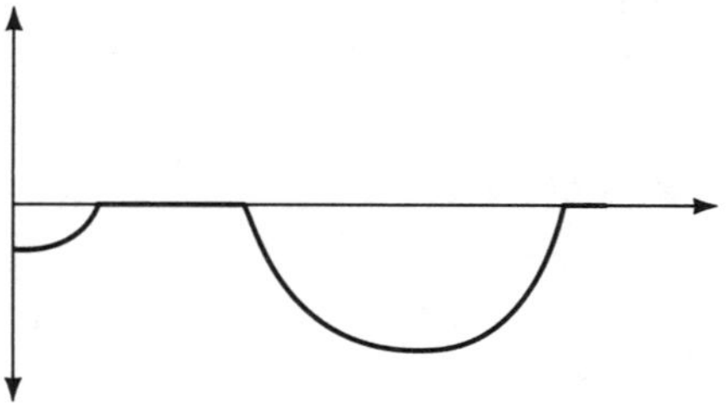

3.

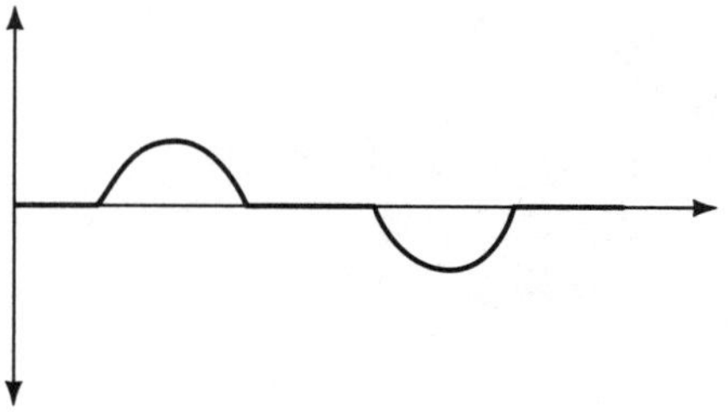

Chapter 13

1. 250 mS
3. 4.32 mA

Chapter 14

1. **a.** −10
   b. 24 μA
3. 7.2 Hz

Chapter 15

1. 0.44 V–4.7 V
3. 17
5. 3.18 Hz
7. a. 0 V
   b. −0.27 V

Chapter 16

1. 10.76 V
3. 0.75 V
5. 0.49 ms
7. 4.97 seconds

Chapter 17

1. a. 30
   b. inverting
3. 564 pF

Chapter 18

1. a. 11 V
   b. ± 9.17 V
   c. 208 Hz
3. See Fig. 18-9A

Chapter 19

1. 17.6 ms, 56.8 Hz
3. 0.11 ms
5. 1.53 mA

Chapter 20

1. a. 290 kohms
   b. 2.96 Hz
3. 723 Hz

Chapter 21

1. 6.9 ms
3. 167 Hz

Chapter 22

1. a. 0, +12
   b. 3 to 6 V
3. a. 10V
   b. 0.11 ms
5. a. 10 V, 5 V
   b. 280 ms
   c. 249 ms
   d. 1.89 Hz
7. a. 6.93 ms
   b. 26.0 ms
   c. 30.4 Hz

INDEX